TRANSFORMATIONS AND GEOMETRIES

The Appleton-Century Mathematics Series

Raymond W. Brink and John M. H. Olmsted, Editors

A First Year of College Mathematics, 2nd ed., by Raymond W. Brink
Algebra—College Course, 2nd ed., by Raymond W. Brink
Analytic Geometry, rev. ed., by Raymond W. Brink
College Algebra, 2nd ed., by Raymond W. Brink
Plane Trigonometry, 3rd ed., by Raymond W. Brink
Spherical Trigonometry by Raymond W. Brink
An Introduction to Matrices, Vectors, and Linear Programming by Hugh G. Campbell
Matrices with Applications by Hugh G. Campbell
Modern Basic Mathematics by Hobart C. Carter
Theory of Ordinary Differential Equations by Randal H. Cole
Elementary Concepts of Modern Mathematics by Flora Dinkines
Parts also available individually under the following titles:
Part I, *Elementary Theory of Sets*
Part II, *Introduction to Mathematical Logic*
Part III, *Abstract Mathematical Systems*
Introduction to the Laplace Transform by Dio L. Holl, Clair G. Maple, and Bernard Vinograde
Introductory Analysis by V. O. McBrien
College Geometry by Leslie H. Miller
A Second Course in Calculus by John M. H. Olmsted
Advanced Calculus by John M. H. Olmsted
Basic Concepts of Calculus by John M. H. Olmsted
Calculus with Analytic Geometry (2 volumes) by John M. H. Olmsted
Intermediate Analysis by John M. H. Olmsted
Prelude to Calculus and Linear Algebra by John M. H. Olmsted
Real Variables by John M. H. Olmsted
Solid Analytic Geometry by John M. H. Olmsted
The Real Number System by John M. H. Olmsted
Analytic Geometry by Edwin J. Purcell
Calculus with Analytic Geometry by Edwin J. Purcell
Analytic Geometry and Calculus by Lloyd L. Smail
Calculus by Lloyd L. Smail
The Mathematics of Finance by Franklin C. Smith

DAVID GANS
New York University

TRANSFORMATIONS AND GEOMETRIES

APPLETON-CENTURY-CROFTS
New York/EDUCATIONAL DIVISION
MEREDITH CORPORATION

To My Wife

750-2

Library of Congress Card Number: 68-8960

PRINTED IN THE UNITED STATES OF AMERICA

390-34620-9

Preface

The writer has long believed that the study of geometric transformations and the ideas related to them deserves an earlier place in the curriculum. The concept of a geometric transformation is extremely simple, very much in the spirit of contemporary mathematics, and, when studied in combination with two other simple and important ideas, group and invariant, is probably without equal as a means of giving the student a breadth of view in geometry.

At present, judging from available textbooks on the subject, the systematic study of geometric transformations and related ideas is not usually begun until the last undergraduate or first graduate year, and is therefore conducted at the pace and with the degree of sophistication expected in those years. This is unfortunate educationally, for there are important aspects of the subject that are elementary and best handled at a lower level; either they are omitted from upper level courses because of their elementary character, or are treated hastily and incompletely for the same reason. A good student, for example, can take a course based on almost any one of the above-mentioned texts and still not be able to write down the equations of the Euclidean motions, the simplest of geometric transformations. And he will also be ignorant of other elementary and basic facts about the subject. This is not necessarily a fault of the course or the text, for each has the right to determine its subject matter and may be of high quality within that framework. The fact remains, nevertheless, that the student is the loser in not having had the opportunity to learn about transformations earlier in his education.

The present text was written to help remedy this situation. It develops the subject at a less advanced level and in greater detail than is customary, and is suitable for students of quite modest backgrounds. The main prerequisite for the book is a good knowledge of analytic geometry, together with the algebra and trigonometry that go hand in hand with that subject. Some acquaintance with the limit concept, such as is often acquired in a course of analytic geometry, may also be helpful to the student, but it is not essential. Calculus is not used in the book.

The specific objectives of the book are (1) to introduce the student to transformational thinking in geometry, (2) to show how new types or systems of geometry can be obtained by the use of transformations, and (3) to study one of these systems, projective geometry, in detail.

The discussion begins with a presentation of the concepts of transformation, group, and invariant in very simple situations within the Euclidean plane (Chapter I). In the chapters which follow, it is shown that the properties of geometric figures in this plane are quite varied and can be classified by means of groups of transformations and their invariants. One is thus led to see that different types of geometry can be distinguished within the Euclidean plane, such as metric, affine, projective, and topological geometries, each being associated with a certain kind of transformation and concerned with geometric properties of a certain type. These transformations and geometries of the Euclidean plane are studied in Chapters II through VII. In connection with affine geometry, the further idea that it can be developed independently of Euclidean geometry, on the basis of its own axioms, is also considered, and a set of such axioms is given. After the unsatisfactory character of projective geometry of the Euclidean plane is noted, the concept of the projective plane is developed and projective geometry is studied within the new plane. This forms the subject of Chapters VIII through X.

Both algebraic and nonalgebraic methods are used throughout the book. Although our concern is mainly with plane geometries, achieving our objectives sometimes requires that we consider simple three-dimensional situations.

An aspect of the book worth noting is that topological ideas are discussed before the projective plane is presented. This is not a customary procedure, but it serves a useful purpose. Ordinarily, the development of the projective plane from the Euclidean plane by the introduction of ideal points is done abstractly and the student who wishes to make it concrete for himself has the hopeless task of trying to visualize the addition of a common ideal point to two Euclidean lines which are everywhere equidistant! But if he had prior knowledge of certain simple models of the two planes, something easily acquired from a little study of topological transformations, he could visualize the introduction of ideal points and see the planes in their correct relation. Such visualization, moreover, can be helpful at various places in the further study of projective geometry.

Each chapter of the book is logically dependent on all those which precede it, so that a course based on the book should ordinarily consist of a block of consecutive chapters, including Chapter I. The book can also be used for "outside" reading on special topics, for example, Euclidean plane motions, affine geometry, or topological transformations.

Since the chapter numbers always appear at the tops of pages, they are not used in the numbering of sections, theorems, and definitions, as is so common today. A chapter number is mentioned in a discussion only when reference is made to a different chapter. Thus, "see §1" means "see §1 of the current chapter" and "see V, 1.2" means "see statement 1.2 in Chapter V." This is the second italicized statement in §1 of Chapter V. If it is a definition, it will be so labeled; if unlabeled, it will be a theorem.

The writer takes this opportunity to express his appreciation to the Ford Foundation for the faculty fellowship which helped to get the manuscript started, to New York University for the leave granted him in order to complete it, to Dr. Raymond W. Brink for many helpful criticisms of the manuscript, to his students for what he learned from their probing questions, to his typist (who wishes to remain anonymous) for the great skill and patience shown in handling an often illegible manuscript, and to the staff of Appleton–Century–Crofts for the efficient and friendly way in which they converted it into a book.

D. G.

Contents

Chapter I
GENERAL INTRODUCTION

SECTION

1.	A bit of history	1
2.	Definition of a geometric transformation	2
3.	The inverse of a transformation	5
4.	The compounding of transformations	6
5.	Groups of transformations	7
6.	Geometric invariants	9
7.	Transformations of the plane into itself	11
8.	Linear transformations	13

Chapter II
MOTIONS OF THE EUCLIDEAN PLANE

1.	Introduction	16
2.	Some general properties of motions	16
3.	Motions and congruence	19
4.	Translations	22
5.	Inverses and resultants	28
6.	Equations of translations in vector form	31
7.	Rotations	33
8.	Equations of rotations about the origin	37
9.	Equations of rotations in matrix form	40
10.	Equations of the general rotation	46
11.	Resultants of translations and rotations	49
12.	The group of displacements	51
13.	Reflections	54
14.	Equations of reflections	56

SECTION

15. Other opposite motions 59
16. Equations of the group of motions 64
17. Further remarks on congruence 66

Chapter III
TRANSFORMATIONS OF SIMILARITY

1. Introduction 71
2. General properties of similarity transformations 71
3. A key similarity transformation 73
4. Equations of the similarity group 75
5. Extension of the notion of similar figures 78
6. Metric geometry 80

Chapter IV
AFFINE TRANSFORMATIONS

1. Introduction 83
2. The affine group 84
3. A key affine transformation 86
4. A key affine transformation (continuation) 91
5. Resolution of the general affine transformation 95
6. Affine properties 97
7. Affine geometry 101
8. Collinearity and concurrence 103
9. Affine equivalence 106
10. Affine properties and 1-1 transformations 110
11. Axioms for affine geometry 113
12. Distance in affine geometry 116

Chapter V
PROJECTIONS, A TRANSITION

1. Introduction 123
2. Parallel projection of a line 123
3. Parallel projection of a plane 125
4. Parallel projections and affine transformations 128
5. Central projections 130

SECTION

6. Central projection of a line on an intersecting line 132
7. Central projection of a plane on an intersecting plane 137
8. Projective properties 144
9. The values of cross-ratios 144
10. Harmonic division 148
11. Cross-ratio of concurrent lines 149
12. Harmonic division of concurrent lines 155
13. Cross-ratio of parallel lines 156
14. Cross-ratio and the conic sections 158
15. Applications of projections 160

Chapter VI
PROJECTIVE TRANSFORMATIONS

1. Introduction 165
2. Definition of a projective transformation 165
3. Some implications of the definition 168
4. Equations of projective transformations 170
5. The projective group 176
6. Projective transformations and projections 178
7. Conic sections 180
8. Projective equivalence 181
9. Projective geometry of the Euclidean plane 182

Chapter VII
TOPOLOGICAL TRANSFORMATIONS

1. Introduction 189
2. Topological transformations of the plane 190
3. Example of a nonaffine topological transformation 193
4. Topological properties of curves 196
5. More general topological transformations 199
6. Homeomorphs of lines and circles 201
7. Topological transformations and order 207
8. Homeomorphs of the plane 209
9. Models of the plane 212
10. More on the circular model of the plane 216
11. Surfaces not homeomorphic to the plane 220
12. The projective plane 225
13. A bounded model of Euclidean space 228

Chapter VIII

THE PROJECTIVE PLANE

SECTION

1.	Introduction	230
2.	Ideal points	230
3.	Extended planes	232
4.	Model of an extended plane	234
5.	The ideal plane; projective planes	235
6.	Projective space	237
7.	Projections viewed more broadly	238
8.	Collinearity, concurrence, duality	241
9.	Cross-ratio and ideal elements	243
10.	Order on a projective line	248
11.	Figures in the projective plane	251
12.	Harmonic properties of complete figures	254
13.	The construction of harmonic conjugates	256
14.	The Theorem of Desargues	258
15.	Other perspective figures	261
16.	The Theorem of Pappus	262
17.	Connections with Euclidean geometry	266
18.	Projective conics	270
19.	Transformations of a projective plane into itself	275
20.	Other methods of developing projective geometry	279

Chapter IX

ANALYTIC PROJECTIVE GEOMETRY

1.	Introduction	280
2.	Homogeneous coordinates of points	280
3.	Equations of projective lines	286
4.	Linear combination of points	291
5.	Linear combination and cross-ratio	294
6.	Equations of a projectivity in a projective plane	296
7.	Geometries of the projective plane	303
8.	Equations of projective conics	308
9.	Equations of tangents to projective conics	315
10.	Projective curves of higher degree	319
11.	Homogeneous coordinates of lines	321
12.	Equations of points	322
13.	Linear combination of lines	325
14.	Projective transformations in line coordinates	326

SECTION

15. Line curves 328
16. Nonhomogeneous line coordinates 329
17. Line curves of the second degree 331
18. Line conics as projective figures 337
19. Correlations 339
20. The summation notation 343
21. Vector and matrix notations 347
22. Collineations as topological transformations 353

Chapter X
PROJECTIVE DESCRIPTIONS OF CONICS

1. Introduction 356
2. A projective view of point conics 356
3. Projective correspondences 358
4. A second projective view of point conics 362
5. The Theorem of Pascal 365
6. Line conics and cross-ratio 370
7. The Theorem of Brianchon 373

APPENDIX: DETERMINANTS 376
ANSWERS TO SELECTED EXERCISES 386
BIBLIOGRAPHY 395
INDEX 397

CHAPTER I

General Introduction to Transformations

1. A Bit of History. The geometry with which the reader is already well acquainted is known as Euclidean geometry, after Euclid, who was the first to organize into a logical system the major facts and methods of geometry known to the Western world. He did this in his famous *Elements*, written about 300 B.C.

Until the seventeenth century geometry was studied and developed mainly in the nonalgebraic, so-called synthetic manner of Euclid and was concerned largely with the *metric* properties of figures, that is, properties involving the measurement of lengths, angles, areas, and volumes. From then on at least two new developments within the framework of Euclidean geometry can be noted:

(1) the use of algebraic methods in connection with coordinate systems;

(2) an increasing interest in the *nonmetric* properties of figures.

The first of these led to the now familiar analytic geometry, whose importance for mathematics and its applications to the physical and social sciences can hardly be exaggerated. The second, less familiar than the first, was also very important for the future of mathematics. Not only did it lead in time to a better understanding of Euclidean geometry and greatly improved presentations of that subject, but it culminated in the creation of new and useful systems of geometry concerned primarily with nonmetric properties. Notable among these systems are **projective geometry, affine geometry,** and **topological geometry** (or **topology**).

Nonmetric properties are easily illustrated, for they are quite varied and occur frequently throughout Euclidean geometry. Some are embodied in prominent statements, such as *two points determine a line**, and *two lines meet once or not at all.* Most of them, however, are less well-recognized and may even be taken for granted because they seem so obvious. That *each line in a plane separates the plane into two parts*, and that *a line which subdivides an angle of a triangle meets the opposite side* are examples of such nonmetric properties. As these few illustrations might suggest, nonmetric properties are often very basic.

The ability of mathematicians to classify nonmetric properties and to distinguish them in a precise way from metric properties was greatly improved as

*Unless the contrary is indicated, *line* will always mean *straight line* in this text.

a result of a plan set forth in 1872 by the noted German mathematician Felix Klein. Fundamental in the plan is the concept of a *group of geometric transformations*, which we explain later in the chapter. At this point we wish merely to note that among the various types of nonmetric properties that can be distinguished by means of this plan are the *projective*, the *affine*, and the *topological properties*, and it is these properties which are studied in the new systems of geometry mentioned earlier. The degree of a curve, for example, is a projective property of the curve, whereas the property of a curve being open or closed is topological. The length of a curve is, of course, one of its metric properties. The meeting or not meeting of a pair of lines is an affine property of the lines. Also affine is the property of one point being between two others. The precise definitions of these four types of properties as they would be given on the basis of the Kleinian plan are discussed in later chapters.

The growing interest in nonmetric properties at the start of the seventeenth century, to which we have referred, was directed mainly at the projective properties. At first this study was regarded simply as a new development within Euclidean geometry, and hence more or less dependent on metric ideas. By the latter half of the nineteenth century, however, so much progress had been made that this dependence was no longer necessary, and the subject could stand as a separate system called projective geometry. In a somewhat similar way the study of affine and topological properties was at first a part of Euclidean geometry, and only later developed into the separate systems known as affine geometry and topology. The latter is still a subject of intensive investigation.

Although this book is concerned only with the foregoing ideas and their elaboration, it is worthwhile to round out this brief historical survey by mentioning that other systems of geometry besides those already referred to also evolved during the nineteenth century, notably those associated with the names, Lobachevsky and Riemann. The latter systems also germinated within the Euclidean framework, but their concern is still primarily with metric properties, which they handle in an unconventional way. Although transformations and the Kleinian idea played no part in the discovery and early development of the geometries of Lobachevsky and Riemann, subsequent study by means of transformations revealed an intimate connection between these systems and projective geometry.

2. Definition of a Geometric Transformation. The term *transformation* occurs in a variety of mathematical situations and often means simply a change made in an equation or expression to facilitate some process, such as finding a root, drawing a graph, or computing an integral. Another meaning, the one referred to in the preceding section, is that of a functional relation between geometric objects. The statement "y is a function of x" is usually interpreted to mean that there are two sets of numbers such that to each value x in one set there corresponds a definite value y in the other set. But it is also consistent with the broad use of the term *function* now current to regard the statement "y is a function of x" as meaning that there are two sets of geometric objects (points, lines, circles, etc.) such that to each object x in one set there corresponds a definite object y in the other. In other words, the variables in a functional relation may be any kind of mathematical objects. When the objects are geometrical, however, it is customary to call the functional relation a **transformation,** or a **mapping.** For example, let P be a variable point which may take all positions on a circle, and let P' be the point midway between P and the center of the circle. P' is then a function of P since to each position of P there corresponds a definite position of P', and this functional relation is called a transformation or a mapping.

The terminology and notation for transformations differ in various ways from those of other functional relations. The geometric objects that play the role of independent variables are called the **originals** or **models,** and those constituting the dependent variables are called the **correspondents** or **images.** In the above example the various positions of point P are the originals, while those of P' are the images. The usual functional notation is retained for transformations except that the familiar f is often replaced by some more suggestive letter, such as T. In our example, then, we can write $P' = T(P)$. This can be read in the usual functional way; or else we can say that *T transforms P into P'*, or that *P' is the image of P under T.** Functional symbolism is also used in a slightly extended way. If we denote the set of all the originals in the example by C and the set of all images by C', then we may write $C' = T(C)$ and say† that *T transforms C into C'*, or that *T maps C onto (or on) C'*. There are other transformations of C into C' besides the one just mentioned. For example, noting that C' is a circle, we can take the point P'' on it which is diametrically opposite P' to be the image of P (Fig. I, 1). Denoting this transformation by S to distinguish it from the first, we can then write $P'' = S(P)$.

Transformations can also be designated without using the functional notation. Thus, the transformation T of the preceding example could simply be referred to as *the transformation $P \rightarrow P'$*, where the arrow is read "into" or "goes into."

The geometric objects used in a transformation may be of any sort. A set of points can, for example, be mapped on a set of lines, as when the image of a variable point on a circle is taken to be the tangent to the circle at that point. Or, lines can be mapped on lines as, for example, when the correspondent of a variable tangent to a circle is the line perpendicular to the tangent at the latter's point of contact with the circle. Similarly, lines can be mapped on parabolas,

---

*Alternative statements with a more dynamic flavor such as *T sends P into P'* or *P goes into P'* are also often used and serve to make the terminology more varied. One must be careful, however, not to be led by them into the error of supposing that physical motion has a place in the mathematical concept of a transformation, or that the latter is anything more than a correspondence.

†Our use of "into" and "onto" is to be noted.

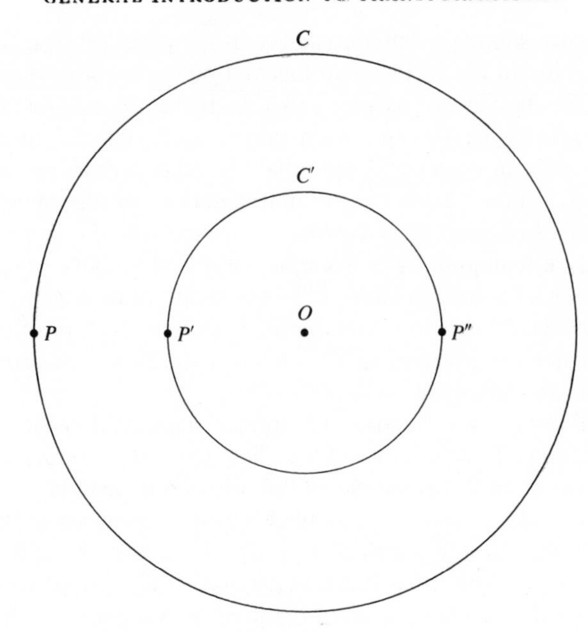

Fig. I, 1

parabolas on sine curves, and so on. We shall be concerned mainly with **point transformations,** which are transformations in which the originals and images are points.

In each example mentioned thus far the images and the originals are distinct sets of objects. This is not necessary. In particular, the two sets may coincide. When this occurs we have a **transformation of a set into itself** or a **mapping of a set on itself.** This does not mean that each object necessarily goes into itself, although this can happen. A mapping of a circle on itself* is achieved, for example, by defining the image of each point on it to be the diametrically opposite point. This mapping is called a *reflection* of the circle in its center. A different mapping of the circle on itself results from taking the image of each point to be the point which is a quartercircumference away in the counterclockwise direction. This mapping is called a *rotation through* 90°, or $\frac{1}{2}\pi$ *radians*, without any implication, of course, that there is any physical movement.

In neither of these two examples does any point coincide with its image. A mapping of the circle onto itself in which this does occur is obtained by reflecting the circle in a diameter; here each endpoint of the diameter goes into itself. A geometric object which is its own image in a mapping is called a **fixed element** of the mapping. Thus, in the example just given, there are two *fixed*

*When we speak of *mapping one geometric figure on itself or on another geometric figure*, we mean that we are mapping the set of points forming the first figure on itself or on the set of points forming the second figure.

points. The mapping of a set on itself in which each member of the set is its own image is called the **identical mapping** (of the set), or simply the **identity** (with respect to the set). We always denote this mapping by the letter I. Thus, for the set of points on a circle, I means the mapping which leaves each point fixed, and for the set of tangents to a circle it means the mapping in which each tangent is its own image.

EXERCISES

1. Describe two different transformations of the points of a line g into the points of a parallel line g'.

2. Describe two mappings, different from I, of the set of tangents to a circle onto itself.

3. Describe two different mappings of the set of points of a parabola onto itself in which the vertex is a fixed point.

3. The Inverse of a Transformation. The type of transformation defined in §2 is called a **single-valued transformation** since each object has a single image. Multiple-valued transformations also exist but do not concern us. Our interest, in fact, lies only with the simplest of single-valued transformations, namely, those in which no two originals have the same image. Thus, each original has just one image, and each image has but one original. Such transformations are said to be **one-to-one.** All the examples of §2 are of this type. An example of a single-valued transformation not of this type is the mapping of a circle on one of its diameters in which the image of each point is the foot of the perpendicular from the point to the diameter.

Reversing the roles of original and image in a one-to-one transformation T clearly yields another one-to-one transformation. The latter is called the **inverse** of T and is denoted by T^{-1}, a symbol which is analogous to the inverse sine symbol $\sin^{-1}$. Thus, if $T(X) = X'$, then $T^{-1}(X') = X$. If T, for example, is the transformation $P \to P'$ of §2 in which the circle C is mapped on the circle C', then T^{-1} is the transformation $P' \to P$, which maps C' on C. To give another example, if T is the rotation of a circle through 90°, then T^{-1} is the rotation of the circle through $-90°$.

If T is a one-to-one mapping of a set A on a set B, it clearly achieves the same *pairing* of the members of A and B as does T^{-1}. Such a simple pairing of the elements in one set with those in another, such that each element appears in only one pair and there is no designation of originals and images, is called a **one-to-one correspondence.** From each such correspondence, then, we can always obtain *two* one-to-one mutually inverse mappings by choosing the elements of one set or the other to be the originals.

4. The Compounding of Transformations. Henceforth, all transformations dealt with will be one-to-one except when the contrary is stated. Also, we shall usually write one-to-one as 1-1.

Interchanging the images and originals in a transformation is called **inverting** the transformation, and it is one of the important ways in which new transformations are obtained from given ones. Another way, which we call **compounding,** consists of combining two transformations in a certain way so as to obtain a third.

To explain this, let us suppose that S maps a set A on a set B, and that T maps B on a set C. Then, through S, each member of A has a unique image in B; and, through T, each member of B has a unique image in C. The combined effect of S and T is, thus, to make a unique member of C correspond to each member of A, or, in other words, to determine a transformation of A into C. If we denote this transformation by U, then U is called the **resultant** of S and T in that order, and S and T are called the **components** of U. Instead of introducing a new letter U to represent the resultant transformation, we may simply denote it by ST. Although this symbol has two parts, it is to be thought of as representing a single entity—the resultant transformation. In this respect the symbol resembles many other familiar symbols, for example, $\frac{2}{3}$ and $\sin x$.

If $A = C$ in the above discussion, that is, if A and C are the same set, then $S(A) = B$ and $T(B) = A$, with the result that ST is a mapping of A on itself. In particular, if T is the inverse of S, then ST is the identical mapping of A, that is, the mapping of A on itself in which each member is its own image. In this case we can write $ST = I$. This equation can also be written $SS^{-1} = I$ since $T = S^{-1}$.

If S, T are transformations of a set A into itself, then usually $ST \neq TS$, that is, the transformation which results from compounding S, T in one order is generally different from the one which results from compounding them in the opposite order. In other words, *the compounding of transformations is not generally commutative*. For example, let S be the reflection of a circle with center O in a diameter MN, and let T be the rotation of the circle through 90°. Then ST and TS represent different mappings of the circle on itself. To verify that they are different, one need only note that if M' is the image of M under ST, then the image of M under TS is the point opposite M'.

On the other hand, *the compounding of transformations always obeys the associative law*. That is, the equation $(RS)T = R(ST)$ always holds, where R, S, T are mappings of a set A on itself. This equation states that the resultant of RS and T in that order and the resultant of R and ST in that order are the same transformation. To prove this, let A_1 be any member of A, and suppose that

$R(A_1) = A_2$, $S(A_2) = A_3$, $T(A_3) = A_4$. Since RS sends A_1 into A_3, and T sends A_3 into A_4, $(RS)T$ will send A_1 into A_4. Similarly, since R sends A_1 into A_2, and ST sends A_2 into A_4, $R(ST)$ will send A_1 into A_4. Hence $(RS)T$ and $R(ST)$ represent the same mapping of A on itself. The associative law can be extended to apply to any finite number of transformations. One may, therefore, insert parentheses at will when compounding transformations, provided that the order of the transformations is left unchanged. Thus, if Q is a fourth mapping of A on itself, then $(QR)(ST) = [Q(RS)]T$.

There is nothing to prevent any of the above transformations from being compounded with itself. The resultant TT of compounding T with itself is written T^2, TTT is written T^3, and so forth. Thus, if we consider mappings of a circle on itself, and T is a rotation through 90°, then T^2 is a rotation through 180°, T^3 a rotation through 270°, and T^4 a rotation through 360°. The latter leaves each point of the circle fixed and hence is I. It could just as well be called a rotation through 0°.* Similarly, T^5 may be called either a rotation through 450° or a rotation through 90°.

EXERCISES

1. If S, T are reflections of a circle in a pair of perpendicular diameters, respectively, verify that ST is the reflection of the circle in its center.

2. If S, T are rotations of a circle through 30° and 40°, respectively, describe the mappings ST; TS; S^{-1}; S^{13}.

3. If R is the reflection of a circle in its center, what are R^2; R^3; R^{-1}?

4. Prove that $(QR)(ST) = [Q(RS)]T$, where Q, R, S, T are mappings of a set on itself.

5. If $RS = T$, prove that $S = R^{-1}T$, where R, S, T are mappings of a set on itself.

5. Groups of Transformations. Although the word *set* has been used thus far to mean an aggregate or collection of geometric objects such as points and lines, one may use it more broadly, of course, to mean an aggregate of any sort. In particular, then, it can refer to an aggregate of transformations. The mappings of some given set on itself, such as were considered in §4, are an important example of such an aggregate. Within a set of mappings one can often distinguish subsets of mappings. Among the many possible mappings of a circle on itself, for example, are the reflections in its diameters and the rotations, just to mention two subsets. As further examples of sets of mappings we may men-

*The student who is puzzled by this statement is perhaps forgetting that a transformation is simply a correspondence, and not a physical process.

tion the mappings of one circle on another, or of one plane on another, or of a plane on itself. This last set of mappings concerns us a great deal throughout the book.

Often a set of transformations of a geometric figure into itself has the following two properties, which make the set self-contained with respect to inverting and compounding, and it is then called a **group of transformations:**

(1) the inverse of each member of the set is also a member of the set;

(2) the resultant of any two members (distinct or not) of the set is also a member of the set.

From (1) it follows, of course, that if S belongs to the set, so does S^{-1}, and from (2) we see that if S and T are transformations of the set, so are S^2, S^3, etc., as well as ST. Likewise, the identity transformation I is seen to be a member of every group of transformations since I is the resultant of S and S^{-1}.* We shall use the word *group* to mean *group of transformations.*

An obvious example of a group is the set of *all* transformations of any geometric figure into itself. If the figure is a line or circle, for example, this group is very large, that is, it contains a great many transformations. An example of a very small group is the set consisting of the rotations of a circle through 0 and π radians, respectively; the inverse of each of these rotations is itself, the resultant of each with itself is a rotation through 0, and the resultant of the two is a rotation through π. The rotations of a circle through 0, $\frac{1}{2}\pi$, π, $\frac{3}{2}\pi$ also form a group, the second and last being inverses of each other, and the resultant of the last two, for example, being the second. Since the rotations through 0, π are a subset of the rotations through 0, $\frac{1}{2}\pi$, π, $\frac{3}{2}\pi$, the group in our first example is called a *subgroup* of the group in our second example. The latter, in turn, is a subgroup of other groups. One of the latter is the group consisting of the eight rotations through the angles $n\pi/4$, where $n = 0, 1, 2, \ldots, 7$. The set consisting of the rotations through 0 and $\frac{1}{2}\pi$ is not a group, for it does not contain the inverse of the last, nor the resultant of the last with itself.

It is clear that the inverting and compounding of transformations belonging to a group can never lead to transformations which are outside of the group. For this reason a group of transformations is said to be **closed** with regard to these operations.

EXERCISES

1. Which of the following sets are groups?

(a) the rotations of a circle through $\frac{1}{2}\pi$, π, $\frac{3}{2}\pi$, 2π;

(b) the rotations of a circle through 0, π;

(c) I and the reflections of a circle in two perpendicular diameters, respectively;

*Since the compounding of the members of a group of transformations obeys the associative law, students who are familiar with the *general* notion of groups can see that a group of transformations is also a group in the more general sense.

(d) I and the reflections of a circle in all its diameters;
(e) I and the reflection of a circle in a diameter;
(f) I and the reflection of a circle in its center.

2. Show that no set of reflections of a circle in its diameters is a group.

3. A group of transformations is called *cyclic* if it contains a transformation T such that every transformation in the group is given by T^n, where n is a positive integer. Find a cyclic group of rotations of a circle containing just two members; just three.

4. A group of transformations is called *commutative* if $ST = TS$, where S and T are any transformations of the group. Show that every group of rotations of a circle is commutative.

6. Geometric Invariants. We are now in a position to clarify further some of the remarks made in §1, particularly those dealing with Klein's use of transformations in studying and classifying geometric properties.

A key idea in this approach is that usually in a mapping of a set A on a set B some geometric properties of the objects in A are also properties of their images in B, whereas other properties in A are not possessed in B. In brief, we say that each mapping **preserves** some properties, or leaves them **invariant**, and that it fails to preserve, or **destroys**, others. In the reflection of a circle in a diameter, for example, if P', Q' are the images of any two points P, Q, then distance $P'Q'$ = distance PQ, and the mapping is therefore said to *preserve distance* or to *leave distance invariant*. On the other hand, if the order or sense of points P, Q, R on the circle is clockwise, then the sense of their images P', Q', R' on the circle is counterclockwise, and the mapping therefore does not preserve sense. A rotation of the circle, however, *preserves sense as well as distance*. In the mappings of a circle on a smaller circle considered in §2, distance is not invariant. It is easily seen, though, that in these mappings equal distances go into equal distances. These mappings are therefore said to preserve *the equality of distances*.

In each of the above examples a circle was transformed into a circle, and thus the property of being a circle was preserved. That this property can be destroyed was seen in §3, where we mapped a circle onto one of its diameters. Since this mapping is not 1-1, let us now give one that is. Consider a square inscribed in a circle with center O (Fig. I, 2). As the image of an arbitrary point P of this circle let us take the point P' of the square in which the latter meets the radius OP. In this way we obtain a mapping $P \rightarrow P'$ of the circle on the square, and the mapping is clearly 1-1. Distance is not invariant here, nor is the equality of distance. Sense, however, is preserved, and so is another non-metric property of a circle, namely, the property of being a *simple closed figure*. This means that a square is like a circle in being a closed figure which does not intersect itself. The numerical symbol 8, for example, is also a closed figure, but is not a simple closed figure because it intersects itself.

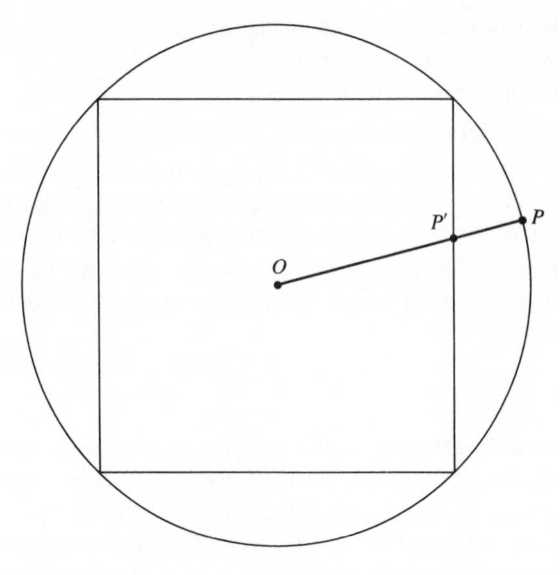

Fig. I, 2

It is not difficult to exhibit transformations possessing other *metric invariants* than those mentioned above, for example, the magnitude of an angle or the size of an area. Similarly, many additional *nonmetric invariants* can easily be illustrated. For our present purposes, however, it is sufficient to note that one of the interesting problems in the study of such invariants is to discover the transformations which possess metric invariants, and, for example, to identify the set of transformations leaving distance invariant, or the set leaving angle-size invariant. The remaining transformations would then be characterized by their nonmetric invariants, and the problem here would be to discover, in particular, which of these transformations leave the degree of a curve unchanged, which send a simple closed figure into a simple closed figure, and the like. In this way one is led to various sets of transformations, each set being concerned with nonmetric invariants of a particular type. Affine transformations, for example, are concerned with *affine invariants*, projective transformations with *projective invariants*, and so on. As was mentioned in §1, these terms and the entire procedure will be clarified in later chapters. In addition we shall see that the sets of transformations referred to above are usually groups.

EXERCISES

For each of the following transformations mention some properties which remain invariant, and some which do not remain invariant.

1. The reflection of a circle in its center.

2. The transformation of the parabola $y = x^2$ into the line $y = -1$ by *orthogonal projection*, in which the image of a point P of the parabola is the foot of the perpendicular from P to the line.

3. The transformation of the parabola $y = x^2$ into the parabola $y = -x^2$ by reflection in the x-axis.

7. Transformations of the Plane Into Itself. Although it will be useful from time to time throughout this book to bring simple spatial concepts into the discussion, our prime concern is with two-dimensional, or plane, geometry. Accordingly, after the present introductory chapter we shall study in detail various transformations of the plane into itself. By *the plane* is meant the Euclidean plane, which is studied in high school plane geometry and in plane analytic geometry. The following chapters may in fact be regarded as a continuation of these two subjects in a special direction.

We now give some examples of transformations of the plane into itself, none having been considered previously. Choose any point O of the plane and consider the transformation in which O goes into itself, and any other point A of the plane goes into the point A' midway between O and A. Each point of the plane then has a unique image, and clearly each point of the plane is the image of a unique point. Thus we have a 1-1 mapping of the plane onto itself. Another such mapping is obtained by again keeping O as a fixed point and taking A' to be one-quarter of the way from O to A. Or, with O still a fixed point, we could take A' so that A is midway between O and A'.

When coordinates are used, a transformation of the plane into itself is usually specified by means of equations which enable us to compute the coordinates of the image point when those of the original point are given. If a rectangular coordinate system is used, and the coordinates of a variable point and its image with reference to this system are (x, y) and (x', y'), respectively, then the equations

$$x' = 3x + 2, \qquad y' = 4y - 7, \tag{1}$$

for example, determine such a transformation. For, given a pair of values of x and y, a unique pair of corresponding values of x' and y' can be computed from these equations. Hence each point of the plane has a unique image. Thus, the image of $(0, 0)$ is $(2, -7)$. Conversely, each point of the plane is the image of some point; for, corresponding to each pair of values of x' and y', a unique pair of values of x and y is determined by the equations. To find the latter values conveniently we first solve the equations for x and y:

$$x = \tfrac{1}{3}x' - \tfrac{2}{3}, \qquad y = \tfrac{1}{4}y' + \tfrac{7}{4}. \tag{2}$$

Thus, if $x' = 5$ and $y' = 9$, substitution in equations (2) gives $x = 1$, $y = 4$. That is, $(5, 9)$ is the image of $(1, 4)$.

In analytic geometry, equations like (1) and (2) are sometimes used for changing from one coordinate system to another. Then, (x, y) are the coordinates of a point in one system and (x', y') are the coordinates of the *same* point in the other system. In our work, on the other hand, a *single coordinate system* is used, so that the x-axis and the x'-axis are the same, and likewise the y-axis and the y'-axis.

In the discussion preceding the above remark we showed that equations (1) determine a 1-1 transformation $(x, y) \to (x', y')$ of the plane into itself. These equations, which are solved for the coordinates of the image point, are called the **explicit** form of the transformation. Equations (2) represent the same transformation since there is no change in the meaning of (x, y) and (x', y'). We refer to equations (2), which are solved for the coordinates of the original point, as the **implicit** form of the transformation. We use these terms in the same way for all transformations specified by equations.

If we change our point of view in the above equations and regard (x, y) as the image of (x', y'), then these same equations represent the inverse transformation, $(x', y') \to (x, y)$. The latter then has (2) as its explicit form, and (1) as its implicit form.

It may be helpful in clarifying these ideas if we give two examples in which the plane is not mapped on itself in a 1-1 manner. Consider, first, the transformation represented by the equations*:

$$(3) \qquad x' = x^3 - x, \qquad y' = y^3 - y.$$

Each given point (x, y) clearly goes into a definite point (x', y'). Conversely, let any point (x', y') be specified. When the value of x' is substituted we obtain a third-degree equation in x with real coefficients. Since this equation has at least one real root, we see that there is at least one value of x corresponding to x'. Similarly, there is at least one value of y corresponding to y'. Thus, to each given point (x', y') there corresponds at least one point (x, y). The transformation therefore maps the plane onto itself. This is not done in a 1-1 way, however, for $(0, 0)$ is the image of the five points $(0, 0)$, $(1, \pm 1)$, $(-1, \pm 1)$.

Next, consider the transformation specified by

$$x' = 2^x, \qquad y' = 3^y.$$

Each point (x, y) clearly has a unique image (x', y'). The latter point is necessarily in the first quadrant since 2^x and 3^y are both positive. Hence the entire plane is mapped onto the first quadrant, excluding the axes. This transformation is 1-1, for if (x', y') is any given point inside the first quadrant, and we substitute these values of x' and y' in the above equations, we obtain a unique pair of values of x and y, namely,

$$x = \log_2 x', \qquad y = \log_3 y'.$$

---

*When we speak of *the transformation* represented by equations that are solved for a pair of coordinates, we always mean the transformation in which these coordinates denote the image point. In other words, *transformations will be understood to be in explicit form.*

As these examples show, there are transformations of the plane into itself which are not 1-1, and also transformations of the plane into a part of itself which are 1-1.

EXERCISES

1. For the transformation represented by equations (1):
(a) find the image of $(-1, 2)$ and the original of $(6, -8)$;
(b) find and describe the images of the x-axis and the y-axis;
(c) find the fixed points, if there are any;
(d) prove by examples that distance is not invariant, that is, $AB = A'B'$ is not always true, where A, B are points with images A', B'.

2. Which of the following transformations map the plane on itself, and which map the plane on a part of itself? In each case also determine whether the transformation is 1-1.

(a) $x' = x^3, \quad y' = -y^3$; (b) $x' = x^2, \quad y' = y^3$;
(c) $x' = e^x, \quad y' = e^{-y}$; (d) $x' = \cos y, \quad y' = \sin x$.

3. It was seen that some points of the plane have no originals under the transformation $x' = 2^x$, $y' = 3^y$. Give the equations of a transformation in which some points have no images.

4. Prove that the set of all 1-1 transformations of the plane into itself is a group.

8. Linear Transformations. The transformation

$$x' = 3x + 2, \qquad y' = 4y - 7, \tag{1}$$

used for illustrative purposes in the preceding section, is a special case of an important type known as a *linear transformation*. A **linear transformation** is a transformation whose equations have the form

$$\begin{cases} x' = ax + by + m \\ y' = cx + dy + n\,, \end{cases} \tag{2}$$

where the coefficients are any real numbers such that $ad - bc \neq 0$. The reason for this condition will be made clear presently. Equations (1) are the special case of (2) in which $a = 3, b = 0, m = 2, c = 0, d = 4, n = -7$, and $ad - bc = 12$. In the chapters which follow we shall be much concerned with certain subsets of linear transformations and their geometric invariants. It will be helpful, in avoiding a repetition of algebraic work in those chapters, if we now prove that a linear transformation maps the plane on itself in a 1-1 manner and that its inverse is also a linear transformation. In doing this we, of course, regard the coefficients in (2) as constants.

First, let us note that equations (2) determine a unique pair of values of

x', y' corresponding to each given pair of values of x, y. Hence, each point of the plane has a unique image. Conversely, regardless of the values substituted for x', y', it is always possible to solve the resulting equations for x and y and thus obtain the point (x, y) whose image is (x', y'). This is done by writing (2) as

$$\text{(3)} \qquad \begin{cases} ax + by = x' - m \\ cx + dy = y' - n, \end{cases}$$

and then solving for x and y by elementary algebra. The result is

$$\text{(4)} \qquad x = \frac{d(x' - m) - b(y' - n)}{ad - bc}, \qquad y = \frac{a(y' - n) - c(x' - m)}{ad - bc}.$$

Since the denominators are not zero by definition of a linear transformation, we see that unique values of x and y are determined. Thus we have proved that:

8.1 *A linear transformation maps the plane on itself in a* 1-1 *way.*

It is clear now that the condition $ad - bc \neq 0$ is included in the definition of a linear transformation in order to guarantee that each point of the plane have a unique original. Several important geometric properties of a linear transformation are related to the value of $ad - bc$, as we shall see. This expression is called the **determinant of the linear transformation** since it is the two-rowed determinant* formed from the coefficients of x and y in equations (2):

$$\begin{vmatrix} a & b \\ c & d \end{vmatrix} = ad - bc.$$

If T denotes the linear transformation $(x, y) \to (x', y')$ represented by equations (2), then its inverse T^{-1} is the transformation $(x', y') \to (x, y)$ represented by equations (4). To show that T^{-1} is a linear transformation, let us replace $ad - bc$ by Δ in (4) and then write the latter as

$$\text{(5)} \qquad \begin{cases} x = \dfrac{d}{\Delta}x' - \dfrac{b}{\Delta}y' + \dfrac{bn - dm}{\Delta} \\[2ex] y = -\dfrac{c}{\Delta}x' + \dfrac{a}{\Delta}y' + \dfrac{cm - an}{\Delta}. \end{cases}$$

These are equations of the same form as (2), with the roles of x, y and x', y' interchanged. Also, the determinant Δ' formed from the coefficients of x' and y' is

$$\Delta' = \begin{vmatrix} \dfrac{d}{\Delta} & -\dfrac{b}{\Delta} \\[2ex] -\dfrac{c}{\Delta} & \dfrac{a}{\Delta} \end{vmatrix} = \frac{d}{\Delta} \cdot \frac{a}{\Delta} - \left(-\frac{b}{\Delta}\right)\left(-\frac{c}{\Delta}\right) = \frac{ad - bc}{\Delta^2} = \frac{1}{\Delta},$$

and hence is not zero. Thus, T^{-1} is a linear transformation.

*All the facts about determinants, and their application to solving systems of linear equations, that are used in this book are stated in the Appendix.

8.2 *The inverse of a linear transformation T is also a linear transformation, and the determinants of the two transformations are reciprocals. If T is given explicitly by equations (2), then T^{-1} is given explicitly by equations (5). The latter equations also represent T implicitly.*

To illustrate, suppose that T is

$$\begin{cases} x' = 2x + 3y - 2 \\ y' = x \phantom{{}+3y} + 5y + 4. \end{cases}$$

Then $\Delta = 7$ and $\Delta' = \frac{1}{7}$. The explicit form of T^{-1} and the implicit form of T are both given by

$$\begin{cases} x = \dfrac{5}{7}x' - \dfrac{3}{7}y' + \dfrac{22}{7}, \\[2ex] y = -\dfrac{1}{7}x' + \dfrac{2}{7}y' - \dfrac{10}{7}. \end{cases}$$

We do not at this time complete the proof that the set of all linear transformations is a group.

EXERCISES

1. Solve $2x + 3y = 1, \quad 4x - 5y = 6$ for x and y by determinants.

2. Solve $2x + 3y = x', \quad 4x - 5y = y'$ for x and y in terms of x' and y' by determinants.

3. Find the image of (5, 4) and the original of (1, 6) in the transformation $(x, y) \to (x', y')$ of Ex. 2.

4. If T denotes the transformation $(x, y) \to (x', y')$ of Ex. 2, give the equations of T^{-1} in explicit form, and hence of T in implicit form.

5. Obtain equations (4) from equations (3), showing all details.

6. Give the equations of the transformation (2), explicitly and implicitly, when the origin is a fixed point. Use your results to write the transformation $x' = 4x - 7y$, $y' = 3x + 2y$ in implicit form.

7. If possible find the points whose images are (6, 1) and (7, 9) in the transformation $x' = 2x - 3y + 5, \quad y' = -6x + 9y + 4$. Does every point have an original? Does every point have an image? Do your results contradict Theorem 8.1?

CHAPTER II

Motions of the Euclidean Plane

1. **Introduction.** In this chapter we consider the 1-1 transformations of the plane into itself which leave distance invariant, that is, which send any two points into two others the same distance apart.*

Such transformations are called **motions,** a name suggested by a certain resemblance between them and physical movements. In virtue of their distance-preserving property, motions have the further property of sending each geometric figure into a congruent one, and for this reason are also called **congruent transformations,** or **isometries.** These three names are also used more generally to mean *any* distance-preserving transformation, not merely of the plane into itself. According to this broader usage, then, the rotations of a circle and the reflections of a circle in its diameters considered in Chapter I are examples of one-dimensional motions. The motions we are about to study are two-dimensional and are known precisely as **motions of the plane,** or **plane motions.** In our discussion "motion" always means "plane motion." If a motion M sends points A, B, C into points A', B', C', we write $M(A, B, C) = A', B', C'$ instead of $M(A) = A'$, $M(B) = B'$, $M(C) = C'$.

All the points, lines, and other figures dealt with in this chapter lie in a single plane. All numbers are understood to be real. If A, B are distinct points, the distance between them is denoted by AB, and their segment by $\overline{AB}$. By the **segment** $\overline{AB}$ we mean the set consisting of A, B, and all other points C such that $AC + CB = AB$. We call A and B the *endpoints* of $\overline{AB}$, and each such point C an *interior point* of $\overline{AB}$. Also, we say that C is **between** A and B, and denote this relation by (ACB) or (BCA). It is clear from plane geometry that when $AC + CB = AB$ holds, the points A, B, C lie on a line, or are *collinear*.

2. **Some General Properties of Motions.** Before discussing specific types of motions and thus showing how motions differ from one another, we

* *Distance* will always mean undirected, or nonnegative, distance, as given by the formula $\sqrt{(x_1 - x_2)^2 + (y_1 - y_2)^2}$ when rectangular coordinates are used.

consider what is characteristic of motions as a class. In doing this we take for granted temporarily that motions do exist. As our first theorem, we prove:

2.1 *The set of all motions is a group.*

Proof. We must show (1) that the inverse of any motion is a motion, and (2) that the resultant of any two motions (not necessarily different) is a motion. Let M be any motion, that is, a 1-1 distance-preserving mapping of the plane on itself. The inverse transformation M^{-1} then exists, is 1-1, and maps the plane on itself. Also, M^{-1} preserves distance, for if A', B' are any two points and $M^{-1}(A', B') = A, B$, then $AB = A'B'$ since $M(A, B) = A', B'$, and M is a motion. M^{-1} is therefore a motion. Next, let N be any motion, not necessarily different from M. Then MN maps the plane on itself in a 1-1 way since M and N do. If $N(A', B') = A'', B''$, then MN sends A, B into A'', B'', and $AB = A''B''$ since $AB = A'B'$ and $A'B' = A''B''$. Thus, MN is a motion. This completes the proof.

In the above discussion N may, in particular, be M^{-1}. In this case the motion MN is MM^{-1}, the identity.

Next, consider any two* points A, B and let $M(A, B) = A', B'$. If C is a point on line AB, distinct from A and B, then one of these three points, say B, is between the others. Hence

$$AB + BC = AC. \tag{1}$$

Let $M(C) = C'$. Then substitution of equal distances in equation (1) gives

$$A'B' + B'C' = A'C'. \tag{2}$$

If C' were not on line $A'B'$, that is, if A', B', C' were vertices of a triangle, equation (2) would be contradicted since the sum of the lengths of any two sides of a triangle exceeds the length of the third side. Hence C' is on line $A'B'$. The same result is obtained if C is between A and B, or if A is between B and C. Thus, M sends each point of line AB into a point of line $A'B'$. Since A, B, C are any collinear points, and their images A', B', C' are also collinear, we see that *M preserves collinearity.*

Now let D be any point not on line AB, and let $M(D) = D'$. If D' were on line $A'B'$, one of the points A', B', D', say B', would be between the others, so that $A'B' + B'D' = A'D'$. This implies that $AB + BD = AD$, which contradicts that A, B, D are noncollinear. Hence D' is not on line $A'B'$. Thus, M sends each point not on line AB into a point not on line $A'B'$, or *M preserves noncollinearity.* Consequently, each point on line $A'B'$ must be the image of a point on line AB. We have now proved:

2.2 *Motions preserve the collinearity and noncollinearity of points, with the result that the image of a line is always a line.*

The relation of three points, in which one is between the others, is called

---

* In this book "two points" always means "two distinct points," "three points" means "three distinct points," etc., and likewise for other geometric objects.

the **between-relation.** As already noted, (ABC) or (CBA) symbolizes that B is between A and C. In the beginning of the proof of 2.2 we assumed that (ABC). This permitted us to write equation (1). We then deduced equation (2), which, since A', B', C' are distinct, showed that $(A'B'C')$. In other words, *M preserves the between-relation.* This property of M was an essential ingredient in the proof of 2.2. On this property depend many other properties of M, as we now show.

Since the segment $\overline{AB}$ consists of A, B, and the points between A and B, and M preserves the between-relation, each point of $\overline{AB}$ goes into a point of $\overline{A'B'}$. Conversely, as one can easily show, using the invariance of betweenness, each point of $\overline{A'B'}$ is the image of a point of $\overline{AB}$. It then follows that $M(\overline{AB}) = \overline{A'B'}$. Thus, *the image of a segment is a segment*. Consequently, *the image of a triangle is always a triangle*, vertices going into vertices and sides into sides. For, let ABC be a triangle and let $M(A, B, C) = A', B', C'$. Then A', B', C', being noncollinear by 2.2, are the vertices of a triangle $A'B'C'$, whose sides $\overline{A'B'}$, $\overline{B'C'}$, $\overline{A'C'}$ are the images of the sides $\overline{AB}$, $\overline{BC}$, $\overline{AC}$ of triangle ABC. The latter triangle is therefore mapped onto triangle $A'B'C'$.

A point is said to be *inside*, or *interior to*, a triangle if it is between two points on different sides of the triangle. *The inside*, or *the interior*, of a triangle is the set of all points that are inside the triangle. We can now see that the motion M of the preceding paragraph maps the inside of triangle ABC onto the inside of triangle $A'B'C'$. For let D be a point inside triangle ABC (Fig. II, 1). Then D is between two points on different sides of the triangle, say A and E. Hence (ADE) and (BEC). If $M(D, E) = D', E'$, then, because of the invariance of betweenness, we have $(A'D'E')$ and $(B'E'C')$. That is, D' is interior to triangle $A'B'C'$.

Another consequence of the invariance of the between-relation is that *M sends angles into angles.* To show this we must first define a *ray*. If A, B are distinct points, the **ray** AB is the set of points consisting of $\overline{AB}$ and all points X such that (ABX). A is the *endpoint* of the ray. Now we return to Fig. II, 1. By definition, $\angle BAC$ is the figure consisting of the two rays AB and AC, and a similar statement holds for $\angle B'A'C'$. Let F be a point of ray AC not on $\overline{AC}$.

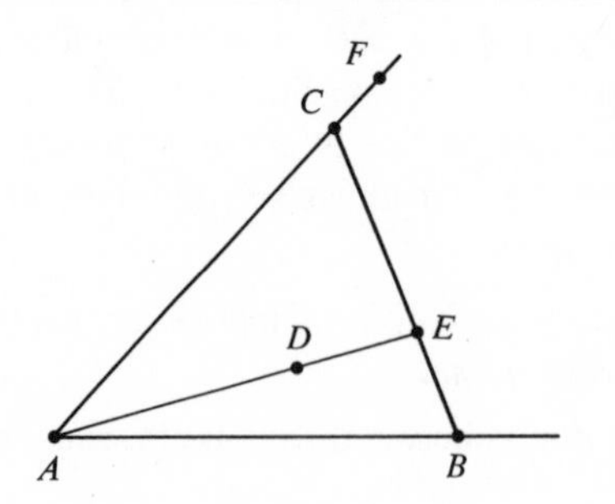

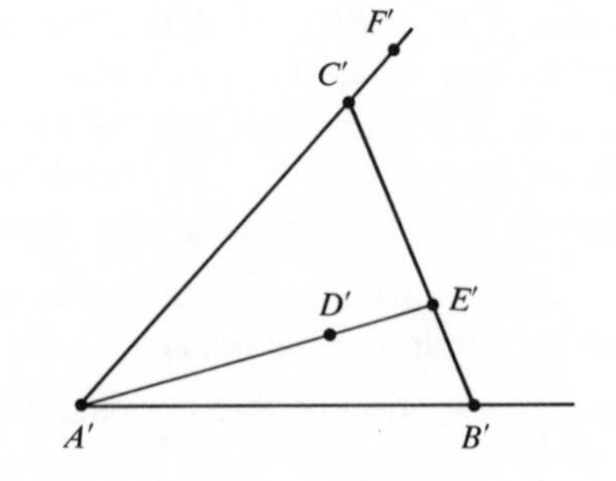

Fig. II, 1

Then $AC + CF = AF$. If $M(F) = F'$, then, since distance is preserved, $A'C' + C'F' = A'F'$. Hence $(A'C'F')$, so that F' is a point of ray $A'C'$ not on $\overline{A'C'}$. The same kind of argument shows, conversely, that each point of ray $A'C'$ not on $\overline{A'C'}$ is the image of a point of ray AC not on $\overline{AC}$. Since we saw earlier that $\overline{AC}$ goes into $\overline{A'C'}$, it follows that ray $A'C'$ is the image of ray AC. Similarly, ray $A'B'$ is the image of ray AB. Hence $\angle B'A'C'$ is the image of $\angle BAC$.

2.3 *Motions preserve the between-relation, with the result that the image of a segment, ray, angle, or triangle is a segment, ray, angle, or triangle, respectively.*

As already noted, the properties of motions stated in the earlier Theorem 2.2 are themselves consequences of the invariance of the between-relation.

EXERCISES

1. Prove that a motion sends intersecting lines into intersecting lines, and parallel lines into parallel lines.

2. Prove that a motion sends an n-sided polygon into an n-sided polygon.

3. Show that in mapping a line g on a line g' a motion sends each half-plane determined by g into a half-plane determined by g'. (Two points, neither of which is on a line g, are said to belong to the same or different *half-planes* determined by g according as g does not or does meet their segment.)

4. Prove that a motion maps the exterior of a triangle onto the exterior of the image triangle. (A point not belonging to a triangle or its interior is said to be *exterior to*, or *outside*, the triangle, and the set of all such points is called the *exterior*, or *outside*, of the triangle.)

5. Show that in mapping one angle on another a motion sends the interior of the first into the interior of the second. (A point is said to be *interior* to an angle if it is between two points on different sides of the angle. The set of all such interior points is called *the interior*, or *the inside*, of the angle.)

6. Prove that the set of all motions leaving a given figure fixed is a group.

3. Motions and Congruence. In §1 we stated that motions send each figure into a congruent figure. Since this important property of motions was not mentioned in §2, let us examine it now, considering first the figures of elementary geometry. In elementary geometry two figures are called *congruent* if they can be made to coincide.

As we saw, a motion sends a segment $\overline{AB}$ into a segment $\overline{A'B'}$, with A, B going into A', B'. Since, moreover, these segments have equal lengths, they can be made to coincide by the methods of elementary geometry, and so are congruent. Thus, *a motion always sends a segment into a congruent segment.*

As for triangles, we have already seen that a motion sends a triangle ABC into a triangle $A'B'C'$ in such a way that sides $\overline{AB}$, $\overline{AC}$, $\overline{BC}$ go into sides $\overline{A'B'}$, $\overline{A'C'}$, $\overline{B'C'}$, respectively. Since $AB = A'B'$, $AC = A'C'$, $BC = B'C'$, the sides of the two triangles are equal, respectively, and the triangles are congruent by a basic theorem of elementary geometry. Thus, *a motion always sends a triangle into a congruent triangle.*

Next, consider the effect of a motion M on a circle γ with center O and radius r. If $M(O) = O'$, let γ' denote the circle with center O' and radius r. Each point of γ, being at distance r from O, goes into a point at distance r from O', that is, into a point of γ'. Conversely, each point P' of γ' is the image of a point of γ. For if P' is the image of P, then $OP = O'P' = r$, which implies that P is on γ. Thus $M(\gamma) = \gamma'$. Now, γ and γ' have equal radii and are therefore congruent according to a theorem of elementary geometry. Therefore, *a motion always sends a circle into a congruent circle.*

Continuing in this way we could dispose of the other situations in elementary geometry where two figures are called congruent. At most this would involve the consideration of parallelograms, n-sided polygons, and occasional statements such as *all right angles are congruent* and *all straight angles are congruent.* In other words, elementary geometry has nothing to say about the congruence of ellipses, parabolas, cycloids, sine curves, etc., beyond the general statement that congruent figures are figures which can be made to coincide. Since this definition serves elementary geometry so well in proving the standard theorems on congruent triangles and circles, one might suppose that it could also be useful in proving analogous theorems for all figures. And indeed it could be, if one were willing to continue basing congruence on so intuitive a definition and giving proofs which involve moving figures around as if they were physical objects. This procedure, however, is not strictly mathematical, useful though it may be in elementary instruction, and we therefore do not use it. Instead we adopt the following alternative definition of congruence:

3.1 Definition. *Two figures are called congruent if, and only if, one can be mapped on the other by a motion.*

This statement replaces the intuitive idea of *physical* movement contained in the high school definition by the *mathematical* idea of a distance-preserving mapping. Although the motivation for introducing the new definition is the need to handle the figures of higher geometry, the definition also applies to the figures of elementary geometry, as we now show.

Let f be any figure, M any motion, and let $M(f) = f'$. Then f and f' are congruent in the sense of Definition 3.1. If f is a segment, triangle, or circle, we showed earlier that f and f' are also congruent in the sense of elementary geometry and mentioned that this statement can be extended to the few other figures dealt with in that subject. Thus, if two figures of elementary geometry are congruent by the new definition, they are also congruent by the old. It remains to show the converse. If, for example, two triangles are congruent in the sense of elementary geometry, say by having the sides of one equal to those of the other, we would have to show that there is a motion that sends one of these triangles into the other. To handle problems of this sort requires that we know something about the specific types of motions which are discussed in the sections to follow. We therefore postpone the solution of these problems until later in the chapter.

In elementary geometry two congruent figures always have the same name. A circle, for example, is never congruent to anything but a circle. This is also true of the more general congruent figures covered by Definition 3.1. For example, in the preceding paragraph if f is an ellipse, then its image f' is also an ellipse. To show this, let A, B be the foci of f, and k the sum of the distances of each point of f from A and B. If $M(A, B) = A', B'$, then, since M preserves distance, f' must be the locus of all points the sum of whose distances from A' and B' is k, and hence an ellipse. Analogous results could be obtained for parabolas, hyperbolas, cycloids, and all the other curves of higher geometry which are defined in metric terms.

In advanced work the term *figure* is used broadly to mean not only a figure bearing a special name, like those referred to above, but *any* set of points whatever. The term is used in the same way in Definition 3.1. Regardless of the nature of a figure f, then, its image f' under a motion is to be called *congruent* to it by 3.1, and this simply means that there is a 1-1 correspondence between the points of f and f' such that the distance between each two points of f equals the distance between their correspondents in f'.

EXERCISES

1. Prove that motions leave the size, that is, absolute value, of every angle invariant, and hence preserve perpendicularity.

2. Prove that motions leave *ratio of division* invariant. (If P, P_1, P_2 are distinct points on a directed line, the signed number P_1P/PP_2, where P_1P and PP_2 are directed distances, is called *the ratio in which P divides* $\overline{P_1P_2}$, or *the ratio in which P divides the point-pair* P_1, P_2, and hence is known as a **ratio of division.**)

3. Prove that a motion sends each of the following figures into another of the same name:

   (a) a parabola; (b) a hyperbola; (c) a square.

4. Show that a motion which leaves each of two points fixed must leave every point of their line fixed.

5. Show that a motion which leaves each of three noncollinear points fixed is necessarily the identity.

6. Show that a motion which leaves each of two nonparallel lines fixed leaves their point of intersection fixed. What can be said about the fixed points of a motion which leaves each of three nonconcurrent lines fixed?

7. Prove that the property of being an asymptote of a hyperbola is preserved by a motion.

4. Translations. There are four distinct types of motions. We discuss first the type known as a *translation.**

Starting with an example, let us consider the transformation T with equations

$$x' = x + 3, \quad y' = y + 4. \tag{1}$$

If P is any point, and $T(P) = P'$, then P' is 3 units to the right of P and 4 units above it. This gives a clear picture of the effect of T (Fig. II, 2).

Another way of viewing T makes use of the directed segment, or **vector,** $\overrightarrow{PP'}$ (Fig. II, 2). Since $PP' = 5$ and $\theta = \arctan \frac{4}{3} = 53°$, we may say that P' is 5 units from P in the direction $\theta = 53°$, or, more explicitly, in the upward direction along the line through P with inclination 53°. Thus, if each point of the plane were joined to its image by a segment directed toward the image, all such segments would have the same length and direction. Fig. II, 2 shows several of these directed segments.

It is clear from our discussion that T maps the plane on itself in a 1-1 way. This can also be seen by noting that equations (1) represent a linear transformation. To conclude that T is a motion we must still show it preserves distance. Hence let A, B be any two points, and A', B' be their images. The vectors $\overrightarrow{AA'}$ and $\overrightarrow{BB'}$ are then *equal*, meaning that they have the same length and direction. If, as in the left part of Fig. II, 3, A and B are on a line of slope $\frac{4}{3}$, then A', B' are on this same line and clearly $AB = A'B'$. Otherwise, as in the right part of Fig. II, 3, the equality of $\overrightarrow{AA'}$ and $\overrightarrow{BB'}$ means that $ABB'A'$ is a parallelo-

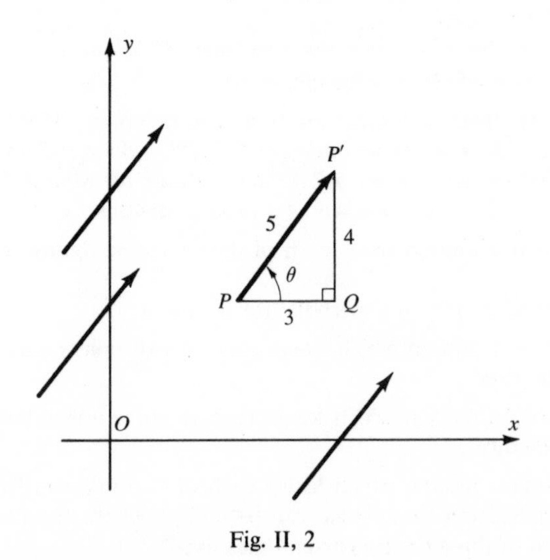

Fig. II, 2

* The term *translation* also occurs in analytic geometry, where it means a certain method of changing from one coordinate system to another.

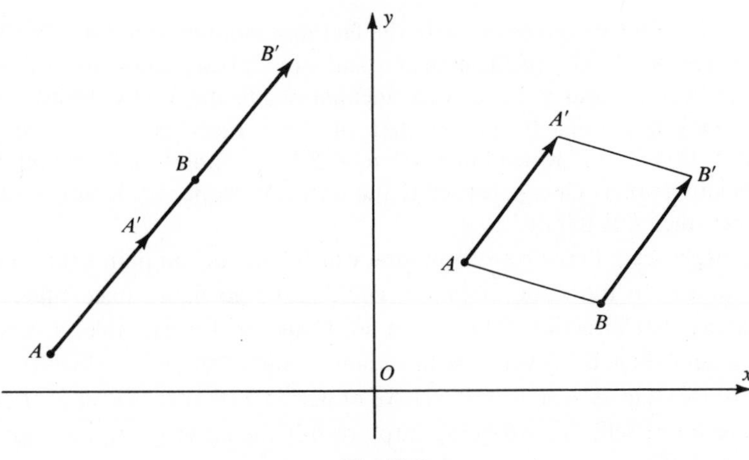

Fig. II, 3

gram, and hence that $AB = A'B'$. An algebraic proof could also be given (see Ex. 4).

The following additional properties of T are obvious from the discussion, but are worth stating since, as will be seen, they are characteristic of all translations. (1) *There are no fixed points.* (2) *Each line having the same inclination as $\overrightarrow{PP'}$ is fixed, whereas any other line goes into a parallel line.* Although there is no need to prove these properties for T, it will be instructive, nevertheless, to check them algebraically in order to illustrate procedures that are common when transformations are studied algebraically.

To check property (1) we substitute x for x', and y for y', in equation (1), obtaining $x = x + 3, \quad y = y + 4$. Since these equations have no solution, no point is its own image.

To check property (2) let g be any line, and

$$ax + by + c = 0 \tag{2}$$

be its equation (a, b are not both zero.). Substitution from equation (1) into this gives

$$a(x' - 3) + b(y' - 4) + c = 0, \tag{3}$$

or

$$ax' + by' - 3a - 4b + c = 0. \tag{4}$$

This is the equation of g', the image of g. For, whenever equation (2) is satisfied by a point (x, y), equation (3), and hence equation (4), is satisfied by the image point (x', y'); conversely, whenever (4), and hence (3), is satisfied by a point (x', y'), (2) is satisfied by (x, y), the original of that point. Since (4) is of the first degree, g' is a line, in agreement with the fact that motions send lines into lines, as we proved in 2.2. The coefficients of x and y in (2) being the same as those of x' and y' in (4), g and g' have equal inclinations to the x-axis. Hence, either $g = g'$, or g is parallel to g'. The first of these cases occurs if and only if $-3a - 4b + c = c$, which implies that $-a/b = \frac{4}{3}$ (unless a, b are both zero, which is impossible). Thus, g is fixed if and only if its slope is $\frac{4}{3}$. If this condition is not met, then g is parallel to g'.

It might seem that a transformation which "moves" all points through the same distance, in the same direction, could leave no figure fixed other than lines having that direction. This is not so. Consider, for example, a curve of infinite extent (Fig. II, 4) which winds about a line of slope $\frac{4}{3}$ and which, with respect to this line as axis, is a sine curve of period 5 (that is, $AB = 5$). T sends this curve into itself, arc AB going into arc BC, the latter going into arc CD, and so forth. Clearly, no figure of finite extent can go into itself.

A figure which is not left fixed by T goes into a congruent figure 5 units away in the direction $\theta = 53°$. Since the distance from the first of these figures to the other, measured from each point P to its image P', is constant, and the vectors $\overrightarrow{PP'}$ all have the same direction, the two figures are called *parallel*. Fig. II, 5 shows two parallel arcs and two parallel angles. We see, incidentally, that T preserves the *sense* of every angle (see Exs. 10, 11). Thus, $\angle BAC$ and its image $\angle B'A'C'$ are both positive.

We are now ready to generalize what has been done with T.

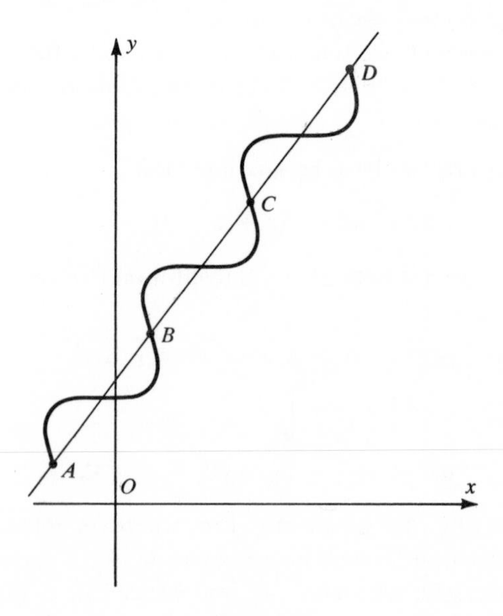

Fig. II, 4

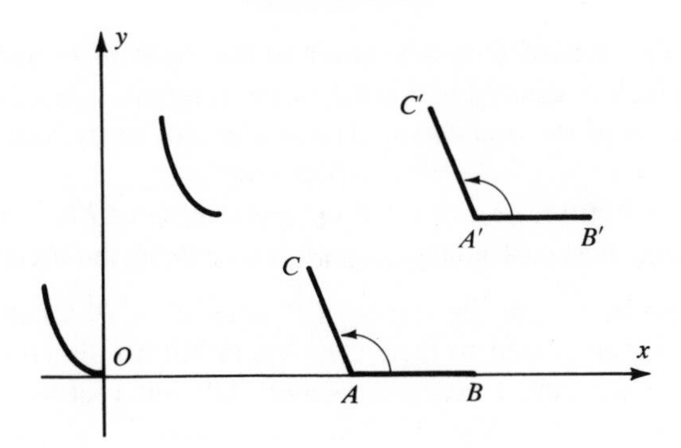

Fig. II, 5

4.1 Definition. *By a translation is meant a transformation with equations of the form*

(5) $$x' = x + a, \qquad y' = y + b,$$

where a and b are any numbers.

If $a = b = 0$, equations (5) reduce to $x' = x,\ \ y' = y$, the identity. Thus I is included among the translations. We note also that translations are linear transformations (I, §8).

Starting from any point P, we can reach its image P' under the translation (5) by going a units horizontally and b units vertically, the signs of a and b determining the sense of these directions. Fig. II, 6 illustrates the case in which $a < 0, b < 0$.

If the translation is not I, the position of P' relative to P can also be specified by the vector $\overrightarrow{PP'}$ (Fig. II, 6), which is called the *vector of the translation*. By definition, the *length* of $\overrightarrow{PP'}$ is $\sqrt{a^2 + b^2}$, and its *direction* (or *angle*) is the

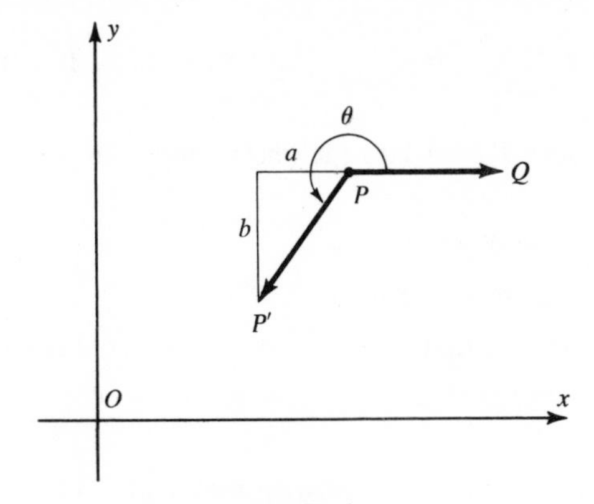

Fig. II, 6

angle $\theta = \angle QPP'$, where Q is any point to the right of P, and the ray PQ is horizontal. θ, which is usually restricted to the interval $-2\pi < \theta < 2\pi$, is also called the *direction of the translation*. The reader can verify that $\sin \theta$ and $\cos \theta$ are $b/\sqrt{a^2 + b^2}$ and $a/\sqrt{a^2 + b^2}$, respectively.

I cannot, of course, be exhibited by a vector since PP' is always zero for such a translation. It is customary, however, to extend the meaning of a vector to cover this case by calling the symbol $\overrightarrow{PP'}$ when $P = P'$ a **null vector** (or **zero vector**). Such a vector is said to have *zero length* but no direction. Vectors with a direction are often called *nonzero vectors*. All null vectors are regarded as equal.

The basic properties of the general translation (5) are exactly like those of the translation T of the illustrative example and are stated in the following theorem, whose proof is covered in exercises:

4.2 *Translations are sense-preserving motions. A translation other than the identity has no fixed point, leaves fixed each line having the same inclination as the vector of the translation, but no other line, and sends each figure not left fixed into a parallel figure.*

It is clear that in the translations we have all the transformations of the plane into itself in which points "move" a constant amount in a constant direction.

EXERCISES

T in Exercises 1–4 is the translation represented by equations (1).

1. Using T, find (a) the image of $(-5, -1)$, (b) the original of $(0, 0)$. In each case plot both points and draw the vector $\overrightarrow{PP'}$.

2. Find the equations of the images of the following loci under T and draw both graphs in each case:

(a) $2x + y - 5 = 0$; (b) $8x - 6y - 1 = 0$;

(c) $x^2 + y^2 = 1$; (d) $y = x^2$.

3. What line does T send into the line through the origin with slope 2? with slope $\frac{4}{3}$?

4. Prove algebraically, by using $\sqrt{(x_1 - x_2)^2 + (y_1 - y_2)^2}$, that T (a) preserves distance, (b) "moves" each point 5 units.

5. Prove algebraically that $x' = x + a, \quad y' = y + b$ is a motion.

6. Prove algebraically that $x' = x + a, \quad y' = y + b$ "moves" each point $\sqrt{a^2 + b^2}$ units.

7. Find the equations of the translation that sends $(-3, 1)$ into $(2, 0)$.

8. If A, A' are any points, not necessarily distinct, there is a unique translation in which A' is the image of A. Prove this (a) geometrically, (b) algebraically.

9. Draw several of the vectors $\overrightarrow{PP'}$ for the translation $x' = x + 3$, $y' = y - 4$, exhibit θ, and find a value for it (nearest degree).

10. That translations preserve the sense of angles can be proved as follows. Let $\angle AOB$ be a positive angle less than π. From analytic geometry, $\cot \angle AOB = (1 + m_2m_1)/(m_2 - m_1)$, where m_1, m_2 are the slopes of lines AO, BO, respectively, and neither line is vertical. Since translations preserve angle-size, $\angle AOB$ and its image, $\angle A'O'B'$, have the same absolute value. If $\angle A'O'B'$ were negative, $\angle B'O'A'$ would be positive. Using the above formula, show that this is impossible, and thus prove that $\angle A'O'B'$ is positive.

11. If one of the lines in Ex. 10 is vertical, the formula fails under the stated conditions, but holds if m_1, m_2 are interpreted as slopes with respect to the y-axis. Hence show that the sense of $\angle AOB$ is preserved.

12. Prove that the curve $y = \sin x$ is fixed for the translation $x' = x + 2\pi$, $y' = y$. Find several other fixed curves for this translation.

13. Prove that a translation always sends a nonzero vector into an equal vector.

14. Prove that the property of a line being tangent to a circle is preserved by translations.

15. Do for a secant and an exterior line what is asked in Ex. 14 for a tangent.

16. If $f(x)$ is defined for all values of x and has the period a, show that the curve $y = (b/a)x + f(x)$, where b has any value, is fixed for the translation $x' = x + a$, $y' = y + b$.

17. Find a translation that leaves the curve $y = x + \sin x$ fixed, and draw the curve (see Ex. 16).

18. If a curve other than a line is met by every line in only a finite number of points, show that it is not fixed for any translation. (Hence lines are the only algebraic curves which may be fixed in a translation.)

19. It is intuitively clear that the *sense of order of the vertices of a triangle* is preserved by translations. That is, if we traverse a triangle ABC so as to encounter A, B, C in that order, and then describe the image triangle $A'B'C'$ so as to encounter A', B', C' in that order, both traversals are clockwise or both are counterclockwise. This can be made more mathematical by noting that the angles AOB, BOC, COA, where O is any point inside triangle ABC, have the same sense, defining a sense of order of the vertices on the basis of this fact, and then proving that this sense is preserved by translations. Do this.

20. Show that a translation with vector of length p and direction θ has the equations $x' = x + p \cos \theta$, $y' = y + p \sin \theta$.

21. Show that $x' = x + a$, $y' = y + b$ is a linear transformation and find the value of its determinant (I, §8).

5. Inverses and Resultants. In §4 we used the translation

$$x' = x + 3, \qquad y' = y + 4 \tag{1}$$

for purposes of illustration and denoted it by T. Its inverse T^{-1} is the transformation $(x', y') \to (x, y)$ with equations

$$x = x' - 3, \qquad y = y' - 4. \tag{2}$$

T^{-1} is therefore a translation, and its vector $\overrightarrow{P'P}$ differs from that of T only in being oppositely directed (Fig. II, 7). Clearly, what we have done with T can be done with any other translation. In other words, *the inverse of a translation is a translation, and the vectors of the two translations differ only in being oppositely directed* (*assuming they are nonzero vectors*).

Let us now find the resultant T_1T_2 of any two translations T_1 and T_2:

$$T_1 \begin{cases} x' = x + a_1 \\ y' = y + b_1, \end{cases} \qquad T_2 \begin{cases} x' = x + a_2 \\ y' = y + b_2. \end{cases}$$

First we rewrite T_2 as

$$x'' = x' + a_2, \qquad y'' = y' + b_2, \tag{3}$$

that is, we call the initial point x', y' rather than x, y, and the image point x'', y'' rather than x', y'. If T_1 sends P into P', we can now think of T_2 as sending P' into P''. Then T_1T_2 sends P into P'' and its equations are found by combining those of T_1 with (3):

$$T_1T_2 \begin{cases} x'' = x + (a_1 + a_2) \\ y'' = y + (b_1 + b_2). \end{cases}$$

Thus, T_1T_2 is a translation. Similarly, by leaving T_2 as it was given originally

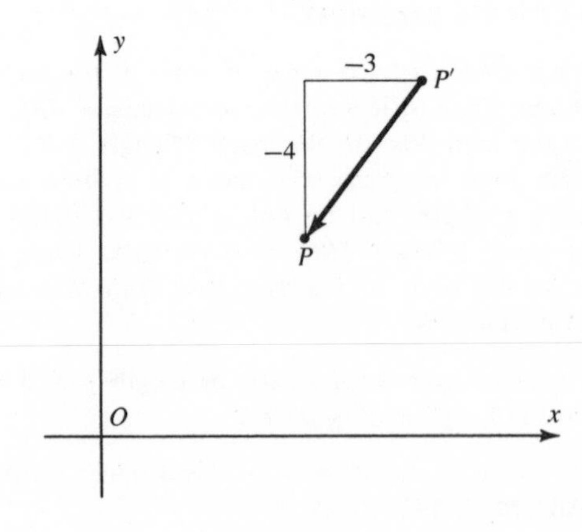

Fig. II, 7

and rewriting T_1, we can obtain the equations of T_2T_1. They will be the same as those of T_1T_2. Hence $T_1T_2 = T_2T_1$, that is, translations are *commutative*.

We have now proved:

5.1 *The set of all translations is a group. This group is commutative and is a subgroup of the group of motions.*

The compounding of two translations T_1, T_2 when neither is the identity or the inverse of the other is worth illustrating geometrically. We let $T_1(P) = P'$ and $T_2(P') = P''$, where P is any point, and draw the vectors $\overrightarrow{PP'}$, $\overrightarrow{P'P''}$ for T_1, T_2 (Fig. II, 8). Since T_1T_2 sends P into P'', its vector is $\overrightarrow{PP''}$. Although, for simplicity, we have taken a_1, b_1, a_2, b_2 positive, the relation among the three vectors is always the same: they form a triangle with sides directed as in Fig. II, 8. In the theory of vectors, the *sum* of the two vectors $\overrightarrow{AB}$, $\overrightarrow{BC}$ is defined to be the vector $\overrightarrow{AC}$; in symbols, $\overrightarrow{AB} + \overrightarrow{BC} = \overrightarrow{AC}$.* Accordingly, $\overrightarrow{PP''} = \overrightarrow{PP'} + \overrightarrow{P'P''}$, that is, *the vector of T_1T_2 is the sum of the vectors of T_1 and T_2.* Since $T_1T_2 = T_2T_1$, the vector $\overrightarrow{PP''}$ also represents T_2T_1. Hence if $T_2(P) = Q'$, then $T_1(Q') = P''$, and $\overrightarrow{PP''} = \overrightarrow{PQ'} + \overrightarrow{Q'P''}$ (Fig. II, 9). Thus, *the vector of T_2T_1 equals the sum of the vectors of T_2 and T_1.*

These results also hold for the translations excluded from the preceding

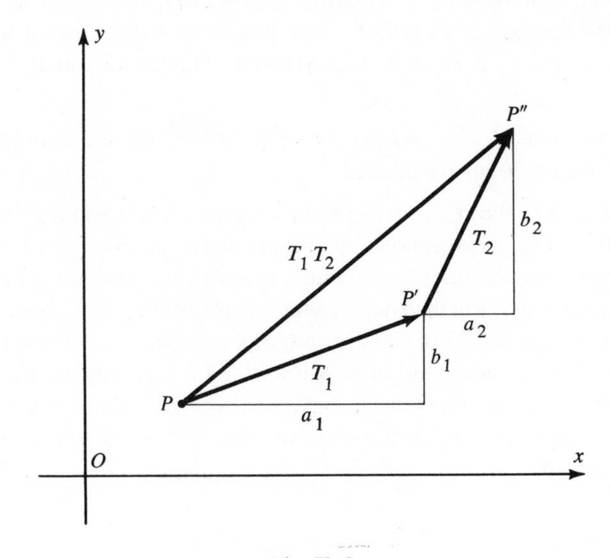

Fig. II, 8

* More generally, if $\overrightarrow{WX}$, $\overrightarrow{YZ}$ are any two vectors, and $\overrightarrow{AB}$, $\overrightarrow{BC}$ are equal to them, respectively, then $\overrightarrow{WX} + \overrightarrow{YZ}$ is defined to be $\overrightarrow{AC}$ or any vector equal to $\overrightarrow{AC}$.

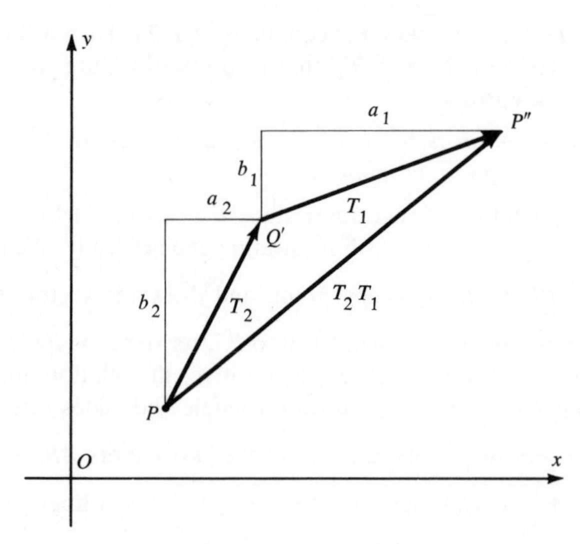

Fig. II, 9

discussion. For suppose T_1, T_2 are mutually inverse, with vectors $\overrightarrow{PP'}$, $\overrightarrow{P'P}$, respectively. Then $T_1T_2 = T_2T_1 = I$ and, by the above definition of the sum of vectors, $\overrightarrow{PP'} + \overrightarrow{P'P} = \overrightarrow{PP}$. Since the vector of I and the vector $\overrightarrow{PP}$ are null vectors, and all null vectors are equal, it follows that the vector of T_1T_2 equals the sum of the vectors of T_1 and T_2. The handling of the case in which one of the translations T_1, T_2 is I is left to the reader. This completes the proof of the theorem:

5.2 *The sum of the vectors, in either order, of two translations equals the vector of the resultant translation.*

It is clear that the relation between vectors and translations is very close. To each vector there corresponds a definite translation, in such a way that equal vectors always correspond to the same translation, and unequal vectors to different translations. Furthermore, when translations are compounded and their vectors are added, this correspondence persists, in the sense that if the vectors $\mathbf{v}_1$, $\mathbf{v}_2$ correspond to the translations T_1, T_2, respectively, then the vector $\mathbf{v}_1 + \mathbf{v}_2$ corresponds to the translation T_1T_2. Because of this close relation, the set of vectors and the set of translations are said to be *isomorphic*, that is, similar in structure.

EXERCISES

1. Given $T_1 \begin{cases} x' = x + 2 \\ y' = y + 3, \end{cases}$ and $T_2 \begin{cases} x' = x - 4 \\ y' = y + 1, \end{cases}$

(a) find T_1T_2 and T_2T_1, and illustrate them geometrically;

(b) find the length and direction θ (nearest degree) of the vector of T_1T_2.

2. Draw a vector which is the sum of $\overrightarrow{WX}$ and $\overrightarrow{YZ}$, where W, X, Y, Z are, respectively, the points $(0, 0)$, $(3, 2)$, $(0, 2)$, $(-1, 4)$.

3. Give the equations of the translations associated with the three vectors of Ex. 2.

4. Prove that a translation always sends the sum of two vectors into the sum of their images (see §4, Ex. 13).

5. Show algebraically that the resultant of the general translation and its inverse is the identity.

6. Show that the set of translations $x' = x + a, \quad y' = y$, where a has all possible values, is a group.

7. Verify that the translation I constitutes a group and is the only finite group of translations, that is, the only group consisting of a finite number of translations.

8. If T_1, T_2 are translations, prove that the inverse of T_1T_2 is $T_2^{-1}T_1^{-1}$.

9. If T_1, T_2, T_3 are translations, prove that $(T_1T_2)T_3 = (T_3T_2)T_1$.

10. If T_1, T_2, T_3 are translations and $T_1T_2 = T_3$, prove that $T_2 = T_1^{-1}T_3$.

11. Which of the statements requiring proof in Exs. 8–10 would be true if T_1, T_2, T_3 were any 1-1 mappings of the plane on itself?

6. Equations of Translations in Vector Form.*

From a numerical viewpoint a nonzero vector $\overrightarrow{PP'}$ that determines a translation is essentially a pair of numbers, its length and its direction θ. As we have seen, a translation is equally well determined by the two numbers a and b appearing in its equations. Moreover, like directed segments, these number-pairs a, b combine in a simple way when translations are compounded, as was seen in §5, when we obtained the equations of T_1T_2 from those of T_1 and T_2. For these reasons they, too, are called vectors, and more generally this term is applied to any ordered pair, triple, quadruple, etc., of numbers, such as the coordinates of a point, or the coefficients in a linear equation, and the like. In particular, an ordered number-pair is called a **two-dimensional vector**, or **vector with two components**, an ordered number-triple is called a **three-dimensional vector**, and so forth. In the remainder of this section we use the term *vector* to mean an ordered number-pair. Single numbers, as distinguished from ordered sets of numbers, are called **scalars.**

The vectors appearing in the equations of translations are (x, y), (x', y'), and (a, b). Since these equations are often written one above the other, these

* This section is not a prerequisite to anything in the book except §9 and IX, §§4, 5, 21.

vectors are also written $\begin{pmatrix} x \\ y \end{pmatrix}, \begin{pmatrix} x' \\ y' \end{pmatrix}, \begin{pmatrix} a \\ b \end{pmatrix}$, the upper number in each case being the *first component* of the vector, and the lower number the *second component*.

By the **sum of two vectors** is meant the vector whose first component is the sum of their first components, and whose second component is the sum of their second components. Thus we write, for example, $\begin{pmatrix} 1 \\ 2 \end{pmatrix} + \begin{pmatrix} 3 \\ 5 \end{pmatrix} = \begin{pmatrix} 4 \\ 7 \end{pmatrix}$. It is easily seen that the addition of vectors, as thus defined, is commutative and associative. Replacing "sum" by "difference" in the above definition gives the definition of the **difference of two vectors.** Then, for example, we have $\begin{pmatrix} 4 \\ 7 \end{pmatrix} - \begin{pmatrix} 3 \\ 5 \end{pmatrix} = \begin{pmatrix} 1 \\ 2 \end{pmatrix}$.

We can now write the translation

$$\begin{cases} x' = x + a \\ y' = y + b \end{cases} \tag{1}$$

as

$$\begin{pmatrix} x' \\ y' \end{pmatrix} = \begin{pmatrix} x \\ y \end{pmatrix} + \begin{pmatrix} a \\ b \end{pmatrix}, \tag{2}$$

and its inverse,

$$\begin{cases} x = x' - a \\ y = y' - b, \end{cases} \tag{3}$$

as

$$\begin{pmatrix} x \\ y \end{pmatrix} = \begin{pmatrix} x' \\ y' \end{pmatrix} - \begin{pmatrix} a \\ b \end{pmatrix}. \tag{4}$$

Equations (2) and (4) express the general translation and its inverse in **vector form,** while equations (1) and (3) express them in **scalar form.**

Equations (2) and (4) can be written more compactly if we denote each vector by a single letter, printed in bold type to distinguish it from a scalar. Symbolizing $\begin{pmatrix} x \\ y \end{pmatrix}$ by $\mathbf{z}$, $\begin{pmatrix} x' \\ y' \end{pmatrix}$ by $\mathbf{z}'$, and $\begin{pmatrix} a \\ b \end{pmatrix}$ by $\mathbf{h}$, we then obtain the concise vector equations

$$\mathbf{z}' = \mathbf{z} + \mathbf{h} \quad \text{and} \quad \mathbf{z} = \mathbf{z}' - \mathbf{h}. \tag{5}$$

It is to be noted that the second of these equations can be obtained from the first by subtracting $\mathbf{h}$ from both sides, as if we were dealing with scalars.

In §5 we compounded T_1 and T_2 to obtain T_1T_2, using scalar equations. This is done vectorially by writing the equations of T_1, T_2, T_1T_2, respectively, as

$$\mathbf{z}' = \mathbf{z} + \mathbf{h}_1, \qquad \mathbf{z}'' = \mathbf{z}' + \mathbf{h}_2, \qquad \mathbf{z}'' = \mathbf{z} + (\mathbf{h}_1 + \mathbf{h}_2), \tag{6}$$

where $\mathbf{h}_1$ is $\begin{pmatrix} a_1 \\ b_1 \end{pmatrix}$ and $\mathbf{h}_2$ is $\begin{pmatrix} a_2 \\ b_2 \end{pmatrix}$. We note that the last of these equations can be obtained by substituting the first in the second and using the associative property of the addition of vectors, just as if we were dealing with scalars.

EXERCISES

1. Express the following translations in vector form, showing the components of each vector:

(a) $x' = x + 3, \quad y' = y + 1$; (b) $x' = x - 2, \quad y' = y + 5$;
(c) $x' = x - 2, \quad y' = y$; (d) $x' = x, \quad y' = y$.

2. Working with vector equations, find the resultant of each of the pairs of translations given in the following parts of Ex. 1:

(a) a, b; (b) b, c; (c) a, d.

3. Express by an equation that the addition of vectors is commutative.

4. Prove that the addition of vectors is associative by showing that

$$\left[\begin{pmatrix} a_1 \\ b_1 \end{pmatrix} + \begin{pmatrix} a_2 \\ b_2 \end{pmatrix}\right] + \begin{pmatrix} a_3 \\ b_3 \end{pmatrix} = \begin{pmatrix} a_1 \\ b_1 \end{pmatrix} + \left[\begin{pmatrix} a_2 \\ b_2 \end{pmatrix} + \begin{pmatrix} a_3 \\ b_3 \end{pmatrix}\right].$$

5. If $\mathbf{z}' = \mathbf{z} + \mathbf{h}_1$, $\mathbf{z}'' = \mathbf{z}' + \mathbf{h}_2$, $\mathbf{z}''' = \mathbf{z}'' + \mathbf{h}_3$, prove that $\mathbf{z}''' = \mathbf{z} + (\mathbf{h}_1 + \mathbf{h}_2 + \mathbf{h}_3)$.

6. Express the translation $\begin{pmatrix} x' \\ y' \end{pmatrix} = \begin{pmatrix} x \\ y \end{pmatrix} - \begin{pmatrix} 2 \\ -3 \end{pmatrix}$ in scalar form.

7. The *negative* of the vector $\begin{pmatrix} c \\ d \end{pmatrix}$ is defined to be the vector $\begin{pmatrix} -c \\ -d \end{pmatrix}$. Prove that one can subtract a vector by adding its negative.

8. The vector $\begin{pmatrix} 0 \\ 0 \end{pmatrix}$ is called the *null vector* (or *zero vector*). Prove that the sum of any vector and its negative is the null vector.

7. Rotations. We turn now to a second type of motion, which we define as follows:

7.1 Definition. *Let A be any point and α any number of radians. The mapping in which A goes into itself, and any other point P of the plane goes into the point P' such that $AP = AP'$ and $\angle PAP' = \alpha$, is called a rotation or, more exactly, a rotation about A through the angle α. A is called the center of the rotation.*

Fig. II, 10 illustrates a case in which $0 < \alpha < \pi$.

There being only one angle α with vertex A and initial side AP, and only one point on the terminal side of this angle whose distance from A equals AP, a unique point P' is determined corresponding to any point P. Evidently, also, each given point of the plane is the image of a unique point. Thus, a rotation is a 1-1 mapping of the plane on itself. That distance is preserved can be seen with the aid of Fig. II, 11, where P', Q' are the images of P, Q. Since $AP = AP'$, $AQ = AQ'$, and $\angle PAQ = \angle P'AQ'$, triangles APQ, $AP'Q'$ are congruent and hence $PQ = P'Q'$. The proof offers no difficulty if A, P, Q are collinear.

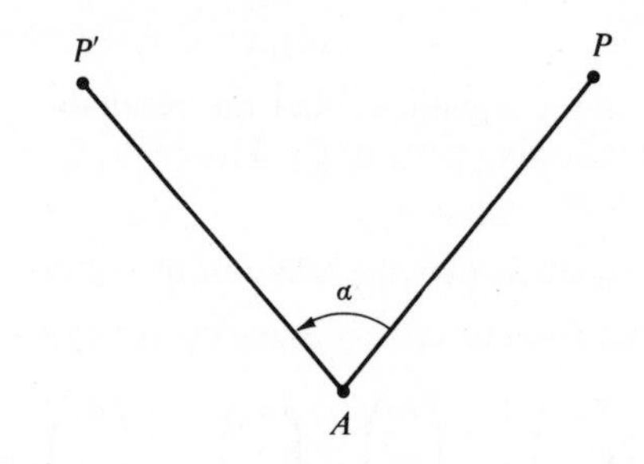

Fig. II, 10

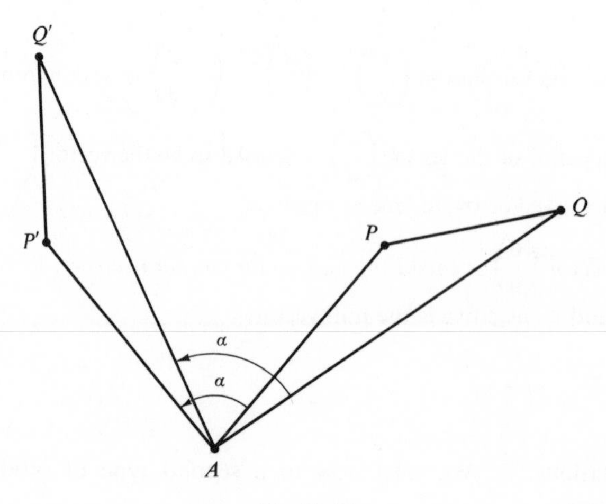

Fig. II, 11

In the discussion to follow we use R to denote a rotation about A through the angle α.

If our plane were a flat wooden platform and we rotated it *physically* about A through the angle α, each point or figure on it would move into the position previously occupied by its image. This gives one a good intuitive grasp of the *mathematical* rotation R, and also accounts for the terminology used in Definition 7.1. As this intuitive view clearly indicates, R leaves fixed each circle with center A, inducing in that circle a one-dimensional rotation through α, and sends each line g through A into a line through A making an angle α with g. As this view further suggests, the sense of every angle is preserved (the proof of this for rotations about the origin is covered in §8, Ex. 12).

If $\alpha = 0, 2\pi, 4\pi$, or any other integral multiple of 2π, R is the identity. Similarly, all the rotations corresponding to $\alpha = \pi + k(2\pi)$, where k is any integer, represent the same mapping. Thus, $\alpha = \pi$ when $k = 0$ and $\alpha = -\pi$ when $k = -1$, and the rotation through $-\pi$ is the same mapping as the rotation through π. This is illustrated in Fig. II, 12, where both rotations are seen to send a point P into the same point P'. In general, all the rotations $\alpha = \alpha_1 + 2k\pi$, where α_1 is constant and k is any integer, represent the same transformation.

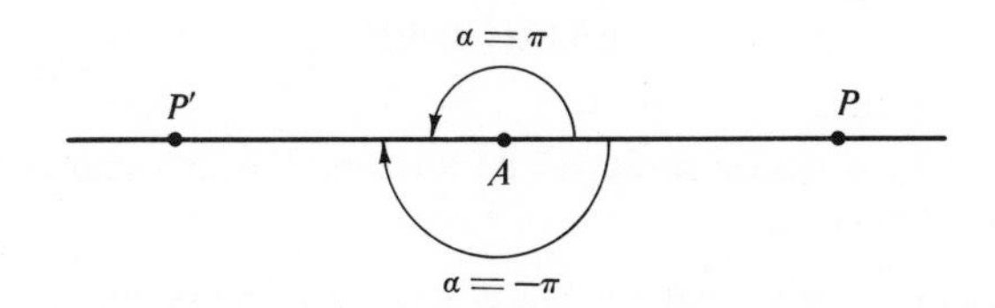

Fig. II, 12

Let us now see what points and lines are left fixed by R. If $R \neq I$, then A is the only fixed point. For the assumption that any other point is its own image clearly implies that α is an integral multiple of 2π, and hence that $R = I$. If R is the rotation through π, it leaves each line through A fixed, but no other line. For if g were another fixed line, it would meet some line h through A in a point B other than A. Then B, the intersection of two fixed lines, would be a fixed point, which contradicts that A is the only fixed point. If R is neither I nor the rotation through π, it leaves no line fixed. For, (1) it can clearly leave no line through A fixed without contradicting that R is not I or the rotation through π, and (2) if it left a line g not through A fixed, it would also have to leave fixed the line through A perpendicular to g since there is only one such perpendicular and perpendicularity is preserved (see §3, Ex. 1), and this would contradict (1).

The following theorem summarizes the discussion given in the foregoing paragraphs:

7.2 *Rotations are sense-preserving motions. A rotation other than I has no fixed point except its center, and has no fixed line unless it is a rotation through an odd multiple of π, in which case only the lines through its center are fixed.*

If R_1, R_2 are rotations about A through angles α_1, α_2, respectively, it is easily seen that their resultant R_1R_2 is a rotation about A through angle $\alpha_1 + \alpha_2$, and that the same is true of their resultant R_2R_1. In particular, α_2 may be the negative of α_1, in which case $\alpha_1 + \alpha_2 = 0$, $R_1R_2 = I$, and $R_2 = R_1^{-1}$. In other words, the inverse of a rotation about A through α is also a rotation about A, namely, through $-\alpha$. Thus we have:

7.3 *The set of all rotations about the same point is a commutative group. The latter is a subgroup of the group of motions.*

Nothing has been said thus far about the resultant of two rotations with different centers, nor about the set of all rotations in the plane. These matters will be considered later.

EXERCISES

1. Prove that a rotation preserves parallelism, that is, sends two parallel lines into two parallel lines.

2. It is intuitively clear that a rotation sends each figure into a congruent figure. What is the mathematical proof of this?

3. Show that a rotation through $\frac{1}{2}\pi$ sends each line into a line perpendicular to it.

4. Show that in a rotation through $\frac{1}{6}\pi$ the angle between a line and its image is $\frac{1}{6}\pi$.

5. State and prove the generalization of the facts stated in Exs. 3, 4.

6. Show that I constitutes a group of rotations and find another finite group of rotations.

7. A transformation ($\neq I$) which is its own inverse is said to be *involutory*. Show that a rotation through π is involutory. Are there any involutory translations?

8. Prove that a figure is symmetrical with respect to a point A if and only if it is left fixed by a rotation about A through π. (A figure is said to be *symmetrical with respect to a point A* if, whenever a point P belongs to the figure, so does the point Q such that A bisects $\overline{PQ}$.)

9. Show that the symmetry of a figure with respect to a point is preserved by all rotations about the point (see Ex. 8).

10. Show that the property of being a tangent, secant, or exterior line of a circle, respectively, is preserved by all rotations.

11. Show that an equilateral triangle is left fixed by a rotation through $\frac{2}{3}\pi$ about its center. Are there other rotations with this property?

12. Show that a square is left fixed by a rotation through $\frac{1}{2}\pi$ about its center. What other rotations have this same property?

13. Generalize Exs. 11, 12 for a regular polygon of n sides.

14. Show that in a rotation the distance from a point to its image varies directly as their distance from the center of rotation.

8. Equations of Rotations About the Origin. Each time we study a new type of transformation we shall want to find its equations with reference to a single rectangular coordinate system. This was done for translations and we now do it for rotations, first considering rotations about the origin O.

As a preliminary step, let us find the equations of these rotations with reference to the polar coordinate system which is superimposed in the usual way on our rectangular coordinate system. If α is the angle of rotation and $P'(r', \theta')$ denotes the image of the point $P(r, \theta)$, we obtain

$$(1) \qquad \begin{cases} r' = r \\ \theta' = \theta + \alpha. \end{cases}$$

This is illustrated in Fig. II, 13, where $\alpha > 0$. Denoting P by (x, y) and P' by

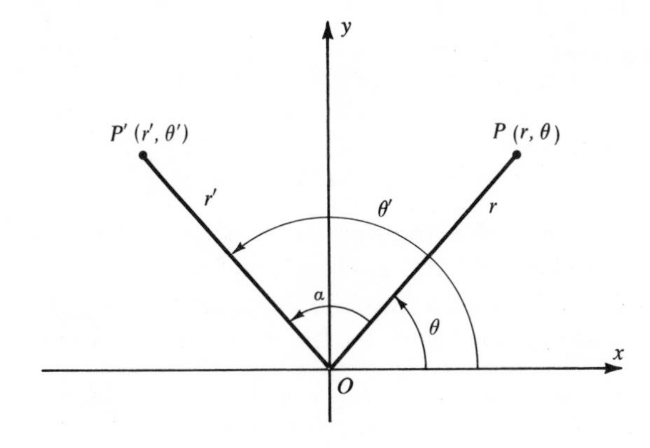

Fig. II, 13

(x', y'), we now combine equations (1) with the formulas relating rectangular and polar coordinates, and obtain

$$x' = r' \cos\theta' = r\cos(\theta + \alpha) = r\cos\theta\cos\alpha - r\sin\theta\sin\alpha = x\cos\alpha - y\sin\alpha,$$

$$y' = r'\sin\theta' = r\sin(\theta+\alpha) = r\cos\theta\sin\alpha + r\sin\theta\cos\alpha = x\sin\alpha + y\cos\alpha,$$

or

$$\begin{cases} x' = x\cos\alpha - y\sin\alpha \\ y' = x\sin\alpha + y\cos\alpha. \end{cases}$$

These are the equations of the rotation in rectangular coordinates. Although not as simple as (1), they serve our purposes better. We note that these equations represent a linear transformation whose determinant Δ is $\cos^2\alpha + \sin^2\alpha$, or 1 (see I, §8). Hence, to express the rotation in implicit form, that is, by equations solved for x and y, we need only apply I, 8.2 to the above equations. Our results are then as follows:

8.1 *A rotation about the origin through angle* α *has the equations*

$$\begin{cases} x' = x\cos\alpha - y\sin\alpha \\ y' = x\sin\alpha + y\cos\alpha, \end{cases} \tag{2}$$

or, in implicit form,

$$\begin{cases} x = x'\cos\alpha + y'\sin\alpha \\ y = -x'\sin\alpha + y'\cos\alpha. \end{cases} \tag{3}$$

Thus, the rotation about O through $\frac{1}{3}\pi$ has the equations

$$\begin{cases} x' = \dfrac{1}{2}x - \dfrac{\sqrt{3}}{2}y \\ y' = \dfrac{\sqrt{3}}{2}x + \dfrac{1}{2}y, \end{cases} \quad \text{or} \quad \begin{cases} x' = \frac{1}{2}(x - \sqrt{3}y) \\ y' = \frac{1}{2}(\sqrt{3}x + y), \end{cases}$$

and, in implicit form,

$$\begin{cases} x = \frac{1}{2}(x' + \sqrt{3}y') \\ y = \frac{1}{2}(-\sqrt{3}x' + y'). \end{cases}$$

The implicit form of a transformation is particularly useful in finding the equation of the image of a curve. This is done by substituting the expressions for x and y, given by the implicit form, in the equation of the given curve. Thus, in the above example, the image of the line $2x - 3y = 4$ is the line

$$\tfrac{2}{2}(x' + \sqrt{3}y') - \tfrac{3}{2}(-\sqrt{3}x' + y') = 4,$$

or

$$(2 + 3\sqrt{3})x' + (2\sqrt{3} - 3)y' = 8.$$

The reader can show that the image of the line $x + y = 0$ is the line $x' = (2 + \sqrt{3})y'$.

If R denotes the rotation (2), then R^{-1}, being a rotation about O through $-\alpha$, has the equations

$$\begin{cases} x' = x\cos(-\alpha) - y\sin(-\alpha) \\ y' = x\sin(-\alpha) + y\cos(-\alpha), \end{cases} \tag{4}$$

or

$$\begin{cases} x' = x\cos\alpha + y\sin\alpha \\ y' = -x\sin\alpha + y\cos\alpha, \end{cases} \tag{5}$$

since $\cos(-\alpha) = \cos\alpha$ and $\sin(-\alpha) = -\sin\alpha$. In these equations of R^{-1} the image point is denoted by (x', y'). Of course, equations (3) also represent R^{-1} if (x, y) is regarded as the image point.

Consider two rotations about O through α_1 and α_2:

$$\begin{cases} x' = x\cos\alpha_1 - y\sin\alpha_1 \\ y' = x\sin\alpha_1 + y\cos\alpha_1 \end{cases} \quad \text{and} \quad \begin{cases} x' = x\cos\alpha_2 - y\sin\alpha_2 \\ y' = x\sin\alpha_2 + y\cos\alpha_2. \end{cases} \tag{6}$$

Their resultant in either order, being a rotation about O through $\alpha_1 + \alpha_2$, must therefore have the equations

$$\begin{cases} x' = x\cos(\alpha_1 + \alpha_2) - y\sin(\alpha_1 + \alpha_2) \\ y' = x\sin(\alpha_1 + \alpha_2) + y\cos(\alpha_1 + \alpha_2). \end{cases} \tag{7}$$

These equations can also be obtained by combining the equations of the two rotations, after changing the notation of the variables suitably, just as we combined the equations of translations.

EXERCISES

1. A diagram readily shows that a rotation of $\frac{1}{2}\pi$ about O sends $(1, 0)$ into $(0, 1)$, the latter into $(-1, 0)$, and $(-1, 0)$ into $(0, -1)$. Show that the equations of the rotation meet these conditions.

2. Give the equations of a rotation of $\frac{1}{6}\pi$ about O and use them to find the point whose image is $(2, 1)$.

3. Find the equations of the images of the following curves under the rotation $x' = (x - y)/\sqrt{2}$, $y' = (x + y)/\sqrt{2}$, draw the original and image in each case, and find the value of α:

(a) $y = 0$; (b) $y = x$; (c) $y = x + 3$;
(d) $x = 3$; (e) $x^2 + y^2 = 1$; (f) $xy = 1$;
(g) $(x - 1)^2 + y^2 = 2$; (h) $y = x^2$.

4. Prove algebraically that the rotation (2) preserves distance.

5. Derive (3) by solving (2) for x and y.

6. Find the resultant of the two rotations (6) by combining them algebraically, and show that it is the rotation (7).

7. Prove algebraically that the rotation (2) leaves the origin fixed and has no other fixed point unless the rotation is I.

8. Express the rotation of Ex. 3 in implicit form (a) by solving for x and y, (b) by using I, 8.2.

9. In each part of this exercise find the equations of the rotation about O which sends (4, 3) into the given point and find a value of α:

(a) $(-4, -3)$; (b) $(-4, 3)$; (c) $(4, -3)$.

10. Find a value of A that makes the transformation $x' = Ax + 0.3y$, $y' = -0.3x + Ay$ a rotation about O, and find α (nearest degree).

11. Show that the rotation (2) sends the line $y = mx + b$ into a line whose slope m' (when it exists) is

$$m' = (m \cos \alpha + \sin \alpha)/(\cos \alpha - m \sin \alpha).$$

12. Using Ex. 11 prove that rotations about O preserve the sense of angles (see §4, Exs. 10, 11).

13. Show that a rotation about O through α changes the angle of every nonzero vector $\overrightarrow{AB}$ by α. (*Hint*. Note that the proposition is true when A is O, then use the fact that parallelograms go into parallelograms.)

14. Using a rotation prove the familiar test for symmetry: *a curve is symmetrical with respect to O if its equation is unchanged when x is replaced by $-x$, and y by $-y$.* (See §7, Ex. 8.)

15. Show that the set of all transformations of the form (2) is a group without using the fact that these transformations are rotations.

9. Equations of Rotations in Matrix Form.* Since all rotations about O have equations of the same general form, it is the coefficients of x and y in these equations that distinguish one rotation from another. Thus, the equations

$$\begin{cases} x' = \frac{1}{2}x - \frac{\sqrt{3}}{2}y \\ y' = \frac{\sqrt{3}}{2}x + \frac{1}{2}y \end{cases} \quad \text{and} \quad \begin{cases} x' = \frac{1}{2}x + \frac{\sqrt{3}}{2}y \\ y' = -\frac{\sqrt{3}}{2}x + \frac{1}{2}y \end{cases}$$

---

* This section presumes familiarity with §6, and is not a prerequisite to anything except IX, §21.

represent different rotations, their angles being $\frac{1}{3}\pi$ and $\frac{5}{3}\pi$, respectively.

When an ordered set of numbers, such as the coefficients in either of these pairs of equations, is arranged rectangularly in one or more rows and one or more columns, and enclosed by parentheses, it forms what is called a **matrix** (plural *matrices*), and the numbers are called the *elements* of the matrix. Thus, the two matrices formed from the coefficients of x and y in the above rotations are

$$\begin{pmatrix} \frac{1}{2} & -\frac{\sqrt{3}}{2} \\ \frac{\sqrt{3}}{2} & \frac{1}{2} \end{pmatrix} \quad \text{and} \quad \begin{pmatrix} \frac{1}{2} & \frac{\sqrt{3}}{2} \\ -\frac{\sqrt{3}}{2} & \frac{1}{2} \end{pmatrix},$$

and are called the *matrices of the rotations*, respectively. The matrix of the general rotation about O, as given by equations (2), §8, is

$$\begin{pmatrix} \cos\alpha & -\sin\alpha \\ \sin\alpha & \cos\alpha \end{pmatrix}.$$

The above matrices are of the special kind we consider in this section and are known as **2 × 2 matrices** since each has 2 rows and 2 columns. If there are m rows and n columns in a matrix, it is called an $\boldsymbol{m \times n}$ **matrix.** Thus, $\begin{pmatrix} 1 & 0 & 4 \\ 0 & -1 & 0 \end{pmatrix}$ is a 2×3 matrix since it has 2 rows and 3 columns, and $\begin{pmatrix} 5 \\ 0 \end{pmatrix}$ is a 2×1 matrix, with 2 rows and 1 column. We see, incidentally, that the 2-dimensional vectors considered in §6 are actually 2×1 matrices.

Although a 2×2 matrix and a determinant with 2 rows and 2 columns look much alike, they are essentially different things. Unlike the matrix, the determinant has a value. For example, the determinant of the coefficients of the first rotation above is

$$\begin{vmatrix} \frac{1}{2} & -\frac{\sqrt{3}}{2} \\ \frac{\sqrt{3}}{2} & \frac{1}{2} \end{vmatrix},$$

and its value is 1.

When $\alpha = 0$, the matrix of the general rotation about O reduces to $\begin{pmatrix} 1 & 0 \\ 0 & 1 \end{pmatrix}$. Since the rotation in this case is the identity, it is customary to denote $\begin{pmatrix} 1 & 0 \\ 0 & 1 \end{pmatrix}$ by I. No confusion will result from the double use of I.

To write the equations of a rotation in matrix form requires that the **product of a 2 × 2 matrix and a 2-component vector** be defined. We define this product to be another 2-component vector, as follows:

$$\begin{pmatrix} a & b \\ c & d \end{pmatrix}\begin{pmatrix} e \\ f \end{pmatrix} = \begin{pmatrix} ae + bf \\ ce + df \end{pmatrix}.$$

Thus,

$$\begin{pmatrix} 3 & 1 \\ -1 & 0 \end{pmatrix}\begin{pmatrix} -2 \\ 4 \end{pmatrix} = \begin{pmatrix} -6 + 4 \\ 2 + 0 \end{pmatrix} = \begin{pmatrix} -2 \\ 2 \end{pmatrix}.$$

The rotation through $\frac{1}{3}\pi$ given at the beginning of this section is then

$$\begin{pmatrix} x' \\ y' \end{pmatrix} = \begin{pmatrix} \frac{1}{2} & -\frac{\sqrt{3}}{2} \\ \frac{\sqrt{3}}{2} & \frac{1}{2} \end{pmatrix}\begin{pmatrix} x \\ y \end{pmatrix},$$

the rotation through $\frac{5}{3}\pi$ is

$$\begin{pmatrix} x' \\ y' \end{pmatrix} = \begin{pmatrix} \frac{1}{2} & \frac{\sqrt{3}}{2} \\ -\frac{\sqrt{3}}{2} & \frac{1}{2} \end{pmatrix}\begin{pmatrix} x \\ y \end{pmatrix},$$

and the general rotation about O is

$$\begin{pmatrix} x' \\ y' \end{pmatrix} = \begin{pmatrix} \cos\alpha & -\sin\alpha \\ \sin\alpha & \cos\alpha \end{pmatrix}\begin{pmatrix} x \\ y \end{pmatrix}.$$

These equations are called **matrix equations** of the corresponding rotations.

If, as in §6, we denote the vectors in any of these equations by $\mathbf{z}$, $\mathbf{z}'$, and the matrix by A (taking care not to regard this letter as a scalar), then the corresponding rotation can be expressed by the more concise matrix equation

$$\mathbf{z}' = A\,\mathbf{z}.$$

If the rotation is the identity, the above equation becomes $\mathbf{z}' = I\,\mathbf{z}$, where I is $\begin{pmatrix} 1 & 0 \\ 0 & 1 \end{pmatrix}$, and this simplifies to $\mathbf{z}' = \mathbf{z}$.

The compounding of rotations about O can be performed by use of their matrix equations. To do this in a significant way involves the multiplication of 2×2 matrices. The **product of two 2×2 matrices***

$$\begin{pmatrix} a & b \\ c & d \end{pmatrix} \quad \text{and} \quad \begin{pmatrix} e & f \\ g & h \end{pmatrix},$$

in that order, is defined to be another 2×2 matrix as given by the equation

$$\begin{pmatrix} a & b \\ c & d \end{pmatrix}\begin{pmatrix} e & f \\ g & h \end{pmatrix} = \begin{pmatrix} ae+bg & af+bh \\ ce+dg & cf+dh \end{pmatrix}.$$

(This is just the way in which second-order determinants are multiplied (see Appendix, 2.6).) If A_1, A_2 denote the given matrices, their product is denoted by A_1A_2.

If now we let

$$R_1 \begin{cases} x' = ax + by \\ y' = cx + dy, \end{cases} \qquad R_2 \begin{cases} x'' = ex' + fy' \\ y'' = gx' + hy', \end{cases}$$

be two rotations about O, with matrix equations

$$\mathbf{z}' = A_1\mathbf{z}, \qquad \mathbf{z}'' = A_2\mathbf{z}',$$

where A_1, A_2 denote the matrices of the rotations, then substitution gives

$$\mathbf{z}'' = A_2(A_1\mathbf{z}) \tag{1}$$

as a matrix equation of the resultant rotation R_1R_2.

To find the matrix of R_1R_2, let us note that since

$$A_1\mathbf{z} = \begin{pmatrix} a & b \\ c & d \end{pmatrix}\begin{pmatrix} x \\ y \end{pmatrix} = \begin{pmatrix} ax + by \\ cx + dy \end{pmatrix},$$

we have

$$A_2(A_1\mathbf{z}) = \begin{pmatrix} e & f \\ g & h \end{pmatrix}\begin{pmatrix} ax + by \\ cx + dy \end{pmatrix} = \begin{pmatrix} (ea + fc)x + (eb + fd)y \\ (ga + hc)x + (gb + hd)y \end{pmatrix}$$

$$= \begin{pmatrix} ea + fc & eb + fd \\ ga + hc & gb + hd \end{pmatrix}\begin{pmatrix} x \\ y \end{pmatrix} = (A_2A_1)\mathbf{z}.$$

---

* This definition is just a special case of the definition of the product of an $m \times n$ matrix by an $n \times k$ matrix (see H. Campbell, *An Introduction to Matrices, Vectors, and Linear Programming*). The definition given earlier for the product of a 2×2 matrix by a 2-component column vector is another special case of this general definition, for, as already noted, such a vector is a 2×1 matrix.

Hence (1) can be written

$$\mathbf{z}'' = (A_2A_1)\mathbf{z}, \tag{2}$$

from which we see that the matrix of R_1R_2 is A_2A_1, that is, the product of A_2 and A_1 in that order. Thus we have proved:

9.1 *The resultant of the rotations* $\mathbf{z}' = A_1\mathbf{z}$ *and* $\mathbf{z}'' = A_2\mathbf{z}'$ *is the rotation* $\mathbf{z}'' = (A_2A_1)\mathbf{z}$.

It is interesting to check this result in the case where R_2 is the inverse of R_1. Then

$$A_1 = \begin{pmatrix} \cos\alpha & -\sin\alpha \\ \sin\alpha & \cos\alpha \end{pmatrix}, \qquad A_2 = \begin{pmatrix} \cos\alpha & \sin\alpha \\ -\sin\alpha & \cos\alpha \end{pmatrix},$$

and hence

$$A_2A_1 = \begin{pmatrix} \cos\alpha & \sin\alpha \\ -\sin\alpha & \cos\alpha \end{pmatrix}\begin{pmatrix} \cos\alpha & -\sin\alpha \\ \sin\alpha & \cos\alpha \end{pmatrix} \tag{3}$$

$$= \begin{pmatrix} \cos^2\alpha + \sin^2\alpha & -\cos\alpha\sin\alpha + \sin\alpha\cos\alpha \\ -\sin\alpha\cos\alpha + \cos\alpha\sin\alpha & \sin^2\alpha + \cos^2\alpha \end{pmatrix} = \begin{pmatrix} 1 & 0 \\ 0 & 1 \end{pmatrix}.$$

This is as it should be since R_1R_2 is the identity.

It is easily verified that the product of $\begin{pmatrix} 1 & 0 \\ 0 & 1 \end{pmatrix}$ by any given 2×2 matrix in either order is the given matrix. In symbols, $AI = IA = A$, where A is any 2×2 matrix and I is $\begin{pmatrix} 1 & 0 \\ 0 & 1 \end{pmatrix}$, as previously agreed. Hence, $\begin{pmatrix} 1 & 0 \\ 0 & 1 \end{pmatrix}$ is called the **unit 2×2 matrix.**

By the *determinant of the matrix*

$$\begin{pmatrix} a & b \\ c & d \end{pmatrix} \tag{4}$$

is meant the determinant

$$\begin{vmatrix} a & b \\ c & d \end{vmatrix}.$$

If this is zero, the matrix is said to be *singular*. The matrix of a rotation is therefore nonsingular since its determinant is 1.

If the matrix (4) is nonsingular, and we denote it and its determinant by A and Δ, respectively, one easily verifies that the matrix

$$\begin{pmatrix} d/\Delta & -b/\Delta \\ -c/\Delta & a/\Delta \end{pmatrix} \tag{5}$$

is also nonsingular and has the property that its product by A in either order is the unit 2×2 matrix I. For this reason it is called the **inverse** of A, and is denoted by A^{-1}. We can thus write $AA^{-1} = A^{-1}A = I$. The equation

$$\begin{pmatrix} \cos\alpha & -\sin\alpha \\ \sin\alpha & \cos\alpha \end{pmatrix}\begin{pmatrix} \cos\alpha & \sin\alpha \\ -\sin\alpha & \cos\alpha \end{pmatrix} = \begin{pmatrix} 1 & 0 \\ 0 & 1 \end{pmatrix},$$

which comes from equation (3) above, is an illustration of this, the two matrices on the left-hand side being mutually inverse. Since they are the matrices of rotations which are mutually inverse, we have the following theorem:

9.2 *If two rotations about the origin are mutually inverse, so are their matrices. In symbols, if*

$$\mathbf{z}' = A\mathbf{z} \tag{6}$$

is the equation of a rotation about O, then

$$\mathbf{z} = A^{-1}\mathbf{z}' \tag{7}$$

is the equation of the inverse rotation $\mathbf{z}' \to \mathbf{z}$.

Equation (7) can also be deduced from (6) by formal manipulation, as we now show. Multiplying both sides of (6) by A^{-1}, we get

$$A^{-1}\mathbf{z}' = A^{-1}(A\mathbf{z}). \tag{8}$$

From the analysis used to obtain (2) we know that $A^{-1}(A\mathbf{z}) = (A^{-1}A)\mathbf{z}$. Since $A^{-1}A = I$ and $I\mathbf{z} = \mathbf{z}$, we see that (8) reduces to (7).

EXERCISES

1. Express the rotation about O through $\frac{1}{3}\pi$ by a matrix equation exhibiting the elements of the matrix. Do likewise for the rotation about O through $-\frac{1}{3}\pi$. Verify that the matrices of these rotations are mutually inverse.

2. Prove that $AI = IA = A$, where I is the unit 2×2 matrix and A is any 2×2 matrix.

3. Show by an example that the multiplication of matrices is not generally commutative.

4. Prove that the associative law holds in the multiplication of 2×2 matrices.

5. Find the inverse of the matrix $\begin{pmatrix} 2 & 3 \\ 4 & 5 \end{pmatrix}$. Then find the inverse of the inverse.

6. Find the product of the matrix of Ex. 5 and the matrix obtained by interchanging its rows and columns.

7. Which of the following are matrices of rotations about O?

(a) $\begin{pmatrix} 0 & 1 \\ -1 & 0 \end{pmatrix}$; (b) $\begin{pmatrix} 0.4 & -0.9 \\ 0.9 & 0.4 \end{pmatrix}$; (c) $\begin{pmatrix} 0.8 & 0.6 \\ -0.6 & 0.8 \end{pmatrix}$; (d) $\begin{pmatrix} 0.8 & -0.6 \\ 0.6 & 0.8 \end{pmatrix}$.

8. What relations must hold among the elements of the matrix (4) if it is to represent a rotation about O?

9. Prove that the product of the matrices (4) and (5) is the unit 2×2 matrix.

10. If $\mathbf{z}' = \begin{pmatrix} a & b \\ c & d \end{pmatrix} \mathbf{z}$ is a rotation about O, give a matrix equation of the inverse rotation which exhibits the elements of the matrix, and state which letter represents the image point.

11. Give an example of each of the following types of matrix:
(a) 3×2; (b) 2×1; (c) 1×4; (d) 3×3.

12. An $m \times n$ matrix in which $m = n$ is called a *square matrix*. How would you define the determinant of such a matrix? Use your definition to compute the determinant of the 3×3 matrix you gave in Ex. 11(d).

10. Equations of the General Rotation. Thus far we have found equations only for rotations about the origin. It remains to find the equations of rotations about an arbitrary point (h, k). Call this point O', the angle of rotation α, and let $P'(x', y')$, as usual, denote the image of $P(x, y)$.

It is convenient, first, to express the rotation in terms of an auxiliary ξ, η-coordinate system, with origin O', whose axes are parallel to the x, y axes and similarly directed (Fig. II, 14). If the coordinates of P and P' in the auxiliary

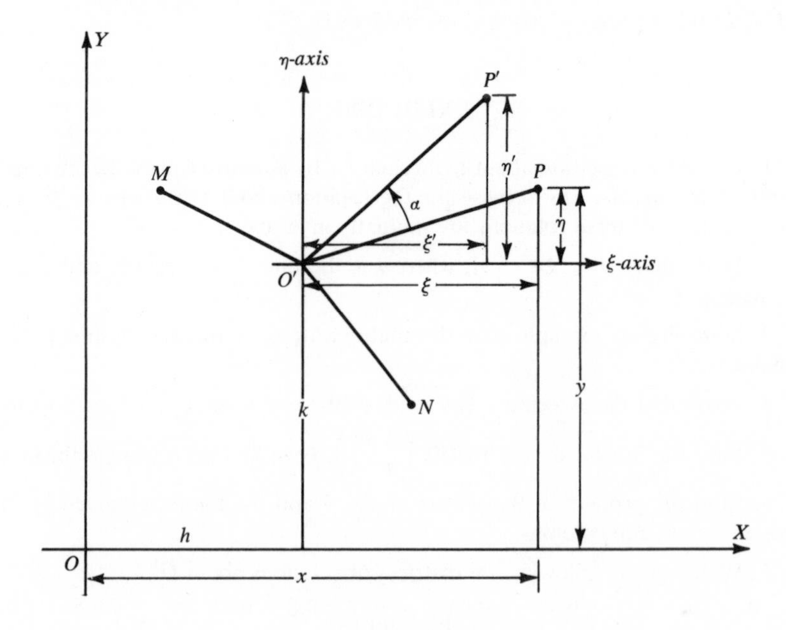

Fig. II, 14

system are (ξ, η) and (ξ', η'), then, by 8.1, we have

$$(1)\qquad \begin{cases} \xi' = \xi \cos \alpha - \eta \sin \alpha \\ \eta' = \xi \sin \alpha + \eta \cos \alpha. \end{cases}$$

It is easily seen from Fig. II, 14 that the relations between the original and the auxiliary coordinates of P, P' are

$$(2)\qquad \begin{cases} x = \xi + h \\ y = \eta + k \end{cases} \quad \text{and} \quad \begin{cases} x' = \xi' + h \\ y' = \eta' + k. \end{cases}$$

If we now solve (2) for ξ, η, ξ', η' and substitute into (1), we shall obtain the equations we have been seeking. We state the result as follows:

10.1 *The rotation through angle α about the point (h, k) has the equations*

$$(3)\qquad \begin{cases} x' - h = (x - h) \cos \alpha - (y - k) \sin \alpha \\ y' - k = (x - h) \sin \alpha + (y - k) \cos \alpha. \end{cases}$$

For example, the rotation through $\frac{1}{3}\pi$ about (4, 5) has the equations

$$\begin{cases} x' - 4 = \dfrac{1}{2}(x - 4) - \dfrac{\sqrt{3}}{2}(y - 5) \\ y' - 5 = \dfrac{\sqrt{3}}{2}(x - 4) + \dfrac{1}{2}(y - 5). \end{cases}$$

If (h, k) is (0, 0), equations (3) reduce to those of a rotation about O, and if $\alpha = 0$ they reduce to the identity I. Otherwise, they are generally complicated in appearance, and it would seem that compounding two such rotations about different points might present considerable difficulty. One can often simplify the problem, however, by using the fact that each such rotation is the resultant of a translation and a rotation about O. To show this, let us write (3) as

$$(4)\qquad \begin{cases} x' = x \cos \alpha - y \sin \alpha + (h - h \cos \alpha + k \sin \alpha) \\ y' = x \sin \alpha + y \cos \alpha + (k - k \cos \alpha - h \sin \alpha), \end{cases}$$

or

$$(5)\qquad \begin{cases} x' = x \cos \alpha - y \sin \alpha + a \\ y' = x \sin \alpha + y \cos \alpha + b, \end{cases}$$

where a and b denote the expressions in parentheses. Equations (5) are seen to be the resultant RT of the following rotation R about O and translation T:

$$R \begin{cases} x' = x \cos \alpha - y \sin \alpha \\ y' = x \sin \alpha + y \cos \alpha, \end{cases} \qquad T \begin{cases} x'' = x' + a \\ y'' = y' + b. \end{cases}$$

Hence the rotation represented by (3) is equal to RT, that is, (3) is the same transformation as RT.

Using this same R, we can also find a translation T' such that the rotation (3) is equal to $T'R$. To show this, we write

$$T' \begin{cases} x' = x + a' \\ y' = y + b' \end{cases} \quad \text{and} \quad R \begin{cases} x'' = x' \cos\alpha - y' \sin\alpha \\ y'' = x' \sin\alpha + y' \cos\alpha, \end{cases}$$

obtaining

$$T'R \begin{cases} x'' = (x + a') \cos\alpha - (y + b') \sin\alpha \\ y'' = (x + a') \sin\alpha + (y + b') \cos\alpha, \end{cases}$$

or

$$(6) \qquad T'R \begin{cases} x'' = x\cos\alpha - y\sin\alpha + (a'\cos\alpha - b'\sin\alpha) \\ y'' = x\sin\alpha + y\cos\alpha + (a'\sin\alpha + b'\cos\alpha). \end{cases}$$

We must now find a' and b' so that (6) is the same transformation as (3), and hence (4). This requires that we solve the equations

$$a'\cos\alpha - b'\sin\alpha = h(1 - \cos\alpha) + k\sin\alpha$$

$$a'\sin\alpha + b'\cos\alpha = k(1 - \cos\alpha) - h\sin\alpha$$

for a' and b'. The solution is

$$a' = h(\cos\alpha - 1) + k\sin\alpha$$
$$b' = k(\cos\alpha - 1) - h\sin\alpha.$$

It is left for the reader to verify this and to show that $T \neq T'$.

We can now state:

10.2 *Any given rotation, not the identity, through an angle α about a point other than the origin is the resultant of a rotation R through α about the origin followed by a translation T, or of a translation T' followed by R. The two translations are distinct, uniquely determined, and neither translation is I.*

The reader is referred to Ex. 4 for the geometric interpretation of T and T'.

For the rotation through $\frac{1}{3}\pi$ about (4, 5) mentioned earlier, the constants in the equations of T and T' are, respectively,

$$\begin{cases} a = 2 + \frac{5}{2}\sqrt{3} \\ b = \frac{5}{2} - 2\sqrt{3} \end{cases} \quad \text{and} \quad \begin{cases} a' = -2 + \frac{5}{2}\sqrt{3} \\ b' = -\frac{5}{2} - 2\sqrt{3}. \end{cases}$$

The following theorem is somewhat converse to 10.2:

10.3 *If T is a translation and R is a rotation through α about the origin,* *and neither transformation is I, then RT and TR are distinct rotations through* α *about points other than the origin.*

The proof is covered in the exercises.

EXERCISES

Draw a figure in each exercise when appropriate.

1. Using the equations of the rotation through $\frac{1}{3}\pi$ about (4, 5) given after equations (3), find the point whose image is O.

2. (a) Find a rotation about O and a translation whose resultant, in that order, is the rotation through $\frac{1}{2}\pi$ about (1, 0).
   (b) Solve (a) with the translation preceding the rotation about O.

3. Prove the last sentence in 10.2.

4. (a) With reference to Fig. II, 14 and the discussion of §10, if $R(O') = M$, show that $\overrightarrow{MO'}$ represents T. Similarly, if N is the point such that $R(N) = O'$, show that $\overrightarrow{O'N}$ represents T'. (b) Find M and N for the rotation through $\frac{1}{3}\pi$ about (4, 5) and draw $\overrightarrow{MO'}$, $\overrightarrow{O'N}$.

5. If $\overrightarrow{OA}$, $\overrightarrow{OA'}$ represent the translations T, T' mentioned in §10, then $\overrightarrow{A'A}$ has the same direction as $\overrightarrow{OO'}$ and is $2(1 - \cos\alpha)$ times as long. Verify this for the rotation through $\frac{1}{3}\pi$ about (4, 5).

6. Verify for the following rotation the statement made in Ex. 5:
$$\begin{cases} x' - 3 = \frac{3}{5}(x - 3) - \frac{4}{5}(y - 1) \\ y' - 1 = \frac{4}{5}(x - 3) + \frac{3}{5}(y - 1). \end{cases}$$

7. Prove the statement made in Ex. 5.

8. Prove 10.3.

9. Equations (3) represent a linear transformation. Find the value of its determinant and, using the fact that it is a rotation, give the equations of its inverse.

10. If $C_1(h_1, k_1)$, $C_2(h_2, k_2)$ are the centers of the rotations RT, TR of 10.3, and T has the equations $x' = x + a$, $y' = y + b$, show that $h_1 = h_2 + a$, $k_1 = k_2 + b$. (In other words, C_1 and C_2 are related by the fact that $\overrightarrow{C_2C_1}$ is the vector of T.)

11. Resultants of Translations and Rotations. We saw earlier that the resultant of two translations is a translation and that the resultant of two rotations about the same point is a rotation about that point. From 10.3 we know, further, that the resultant of a translation and a rotation about O (when neither is the identity) is a rotation about a point not O. It remains, then, to determine the nature of the resultant of (1) a translation and a rotation about a point not O, and (2) two rotations about different points. In doing this we suppose that none of these transformations is the identity.

(1) Let T be a translation, R a rotation about a point not O, and consider TR. By 10.2 we know that $R = T'R'$, where T' is a translation, and R' a rotation about O. Hence, using the associative property of transformations, we have

$$TR = T(T'R') = (TT')R' = T_1R',$$

where T_1 is a translation. Now, T_1R' is simply the rotation R' if $T_1 = I$, and, by 10.3, it is a rotation about a point not O if $T_1 \neq I$. Thus, TR is always a rotation. In the same way RT can be shown to be a rotation.

(2) Let R, R' be rotations about different points. If neither point is O, then, by 10.2, $R = TR_1$ and $R' = R_1'T'$, where R_1, R_1' are rotations about O and T, T' are translations. These equations are still true if one of the given rotations, say R, is about O, except that then $T = I$. In any case, then, we have

$$RR' = (TR_1)\,(R_1'T') = T(R_1R_1')T' = TR_2T' = (TR_2)T',$$

where R_2 is a rotation about O. If $R_2 = I$, which occurs if R_1, R_1' are mutually inverse, then RR' is clearly a translation. If $R_2 \neq I$, then TR_2 is a rotation (whose center is not O unless $T = I$). It follows, on using the result in (1) above, that $(TR_2)T'$, and hence RR', is a rotation.

We have thus proved:

11.1 *The resultant, in either order, of a translation ($\neq I$) and a rotation ($\neq I$) is a rotation. The resultant of two rotations about different points is a rotation or a translation.*

In view of what is stated at the beginning of this section, and of the fact that the inverses of translations and rotations are translations and rotations, respectively, it is clear that we have also proved:

11.2 *The set of all translations and rotations is a group. The set of all rotations is not a group.*

A translation and a rotation are generally not commutative. This was already apparent from 10.3. The same is true of two rotations, as the exercises readily show (see Ex. 1).

EXERCISES

1. Let R_1, R_2 be rotations through $\frac{1}{2}\pi$ about (0, 0), (0, 1), respectively. Show geometrically that $R_1R_2 \neq R_2R_1$ by considering their effect on (0, 0).

2. Let R be the rotation through $\frac{1}{2}\pi$ about (0, 0), and T the translation $x' = x + 2$, $y' = y + 4$. Show geometrically that $RT \neq TR$ by considering their effect on (0, 0).

3. Find the equations of the resultant of each of the following pairs of transformations in the given order, and determine whether the resultant is a translation or a rotation:

(a) a rotation through $\frac{1}{2}\pi$ about (1, 0) and a translation $x' = x + 3$, $y' = y - 1$;
(b) a rotation through $\frac{1}{2}\pi$ about (1, 0) and a rotation through $\frac{1}{2}\pi$ about (0, 1);
(c) a rotation through $\frac{1}{2}\pi$ about (1, 0) and a rotation through $\frac{3}{2}\pi$ about (0, 1).

4. If T is a translation ($\neq I$), and R is a rotation ($\neq I$) about a point not O, prove that RT is a rotation ($\neq I$).

5. If R_1, R_2 are rotations through π about different points A_1, A_2, prove that R_1R_2 is a translation whose vector has the same direction as $\overrightarrow{A_1A_2}$ and is twice as long.

6. Check the statement of Ex. 5 by taking A_1, A_2 to be (1, 0), (−1, 1), and finding the equations of R_1, R_2, R_1R_2.

7. Find two rotations whose resultant is the translation $x' = x + 2$, $y' = y - 3$ (see Ex. 5).

8. Prove that any given translation can be regarded as the resultant of two rotations (see Ex. 5).

12. The Group of Displacements. Taken collectively, translations and rotations are called **displacements**, or **rigid motions**. These terms serve to emphasize that the effect of such transformations, viewed intuitively, is to slide each figure, as if it were a rigid object, from one position to another without changing its size or shape.

We then have the theorem:

12.1 *The displacements constitute a group and have equations of the form*

$$\begin{cases} x' = x\cos\alpha - y\sin\alpha + a \\ y' = x\sin\alpha + y\cos\alpha + b. \end{cases} \tag{1}$$

Conversely, all the transformations represented by such equations are displacements.

Proof. The displacements form a group by 11.2. Equations (1) become those of any given translation if we take $\alpha = 0$ and choose a, b suitably, and those of any given rotation about O if we take $a = b = 0$ and choose α suitably. Furthermore, any given rotation about a point not O, being the resultant of a rotation about O and a translation by 10.2, has equations of the form (1). Thus, every displacement has equations of the type (1). The converse part of the theorem is likewise proved by use of the preceding facts.

For some purposes it is desirable to write equations (1) in a different notation. If we let

$$A = \cos\alpha, \quad B = -\sin\alpha, \quad C = a, \quad D = b,$$

then $A^2 + B^2 = 1$, and we can state the following theorem:

12.2 *Every displacement has equations of the form*

$$\begin{cases} x' = Ax + By + C \\ y' = -Bx + Ay + D, \end{cases} \tag{2}$$

where $A^2 + B^2 = 1$. Conversely, all the transformations represented by such equations, under the stated condition, are displacements.

The last sentence follows from the fact that if equations of the form (2), with $A^2 + B^2 = 1$, are given, they can be reduced to the form (1). For, knowing A and B, and the fact that $A^2 + B^2 = 1$, we can always find α so that $\cos \alpha = A$ and $\sin \alpha = -B$. Also, of course, we take $a = C$, $b = D$.

If P, P' are distinct points, there are infinitely many displacements that send P into P'. Just one of these is a translation, the others being rotations about points on the perpendicular bisector of $\overline{PP'}$.

If P, Q are distinct points, and P', Q' are another such pair, where $PQ = P'Q'$, there is at least one displacement sending P and Q into P' and Q', respectively. For suppose $P \neq P'$. There is then a translation T that sends P into P'. Let $T(Q) = Q_1$. Then $PQ = P'Q_1$, so that $P'Q_1 = P'Q'$. Some rotation R about P' sends Q_1 into Q'. Hence the displacement TR sends P, Q into P', Q'. If $P = P'$, there is a rotation about P which sends Q into Q', and this rotation is, of course, a displacement sending P, Q into P', Q'.

We can now prove the theorem:

12.3 *There is a unique displacement that sends given distinct points P and Q into given distinct points P' and Q', respectively, where $PQ = P'Q'$. In other words, a displacement is completely determined when two points and their images are specified.*

Proof. We have just seen that there is at least one displacement meeting the specified conditions. Call it D_1, and let D_2 be another such displacement. Let S be any point other than P or Q, and suppose that $D_1(S) = S_1'$, $D_2(S) = S_2'$. If P, Q, S are the vertices of a triangle, then P', Q', S_1' are the vertices of a congruent triangle, the pairs of equal angles of the triangles agreeing in sense (Fig. II, 15a). Likewise, triangles PQS and $P'Q'S_2'$ are congruent, with corresponding angles agreeing in sense. But there exists only one triangle with vertices P', Q' which is congruent to triangle PQS so that P' and Q' correspond to P and Q, respectively, and corresponding angles agree in sense. Hence $S_2' = S_1'$. If P, Q, S are collinear, suppose that S is between P and Q (Fig. II, 15b). Then S_1' is between P' and Q', with $P'S_1' = PS$. Likewise S_2' is between P' and Q', with $P'S_2' = PS$. Clearly, $S_2' = S_1'$. The same result is obtained if Q is between P and S, or if P is between S and Q. All possible positions of S having been considered, it follows that D_2 is the same transformation as D_1.

12.4 *Displacements are the only motions which preserve the sense of every angle.*

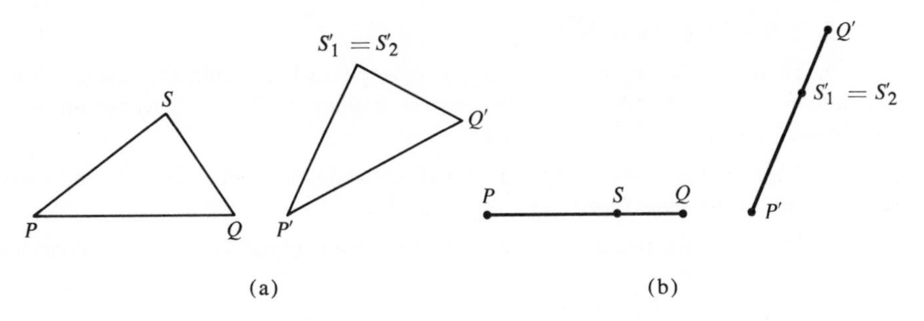

Fig. II, 15

Proof. We already know that every displacement possesses this property. It remains to show that any motion, M, which possesses this property is necessarily a displacement. Hence let $M(P, Q) = P', Q'$, where P, Q are any distinct points. Then $PQ = P'Q'$, so that P', Q' are also distinct. Let D be the displacement which, by 12.3, sends P, Q into P', Q', respectively. If S is any point other than P or Q, one can show exactly as in the proof of 12.3 that $M(S) = D(S)$. Hence M is the same transformation as D.

12.5 Definition. *Transformations which preserve the sense of every angle are called* ***direct****; those which reverse the sense of every angle are called* ***opposite****.*

Displacements are therefore the only direct motions.

EXERCISES

1. Which of the following are displacements? Find α (nearest degree) for those which are rotations.

(a) $\begin{cases} x' = \frac{3}{5}x + \frac{4}{5}y - 8 \\ y' = -\frac{4}{5}x + \frac{3}{5}y + 7; \end{cases}$ (b) $\begin{cases} x' = -\frac{3}{5}x - \frac{4}{5}y - 8 \\ y' = \frac{4}{5}x - \frac{3}{5}y + 7; \end{cases}$

(c) $\begin{cases} x' = y \\ y' = -x; \end{cases}$ (d) $\begin{cases} x' = y \\ y' = x; \end{cases}$ (e) $\begin{cases} x' = -y \\ y' = x. \end{cases}$

2. Find the equations of two displacements that send $(1, 2)$ into $(7, -6)$.

3. Using equations (2), prove algebraically that a displacement other than I has at most one fixed point.

4. Find the equations of the displacement which sends:

(a) $(0, 0)$, $(2, 0)$ into $(0, 3)$, $(0, 5)$, respectively;

(b) $(1, 2)$, $(4, 6)$ into $(-1, 5)$, $(-5, 8)$, respectively.

5. Finish the proof of 12.4.

6. Theorem 12.2 shows that a displacement is a linear transformation whose determinant is 1. Is a linear transformation a displacement if its determinant is 1? Justify your answer.

7. Find the equations of the transformation which is the inverse of (2) and show that they are of the same type as (2).

8. Show that the resultant of two transformations of the form (2) has equations of the same form.

9. Express equations (2) by a vector or matrix equation.

13. Reflections. We turn now to a third type of motion, called a *reflection.*

13.1 Definition. *If g is any line, the mapping in which each point of g goes into itself, and every other point of the plane goes into the symmetrical point with respect to g is called a reflection, or more exactly, a reflection in g, and g is called the axis of the reflection.*

Figure II, 16 shows three noncollinear points P, Q, R and their images P', Q', R' in a reflection whose axis is g. $\overline{PP'}$, $\overline{QQ'}$, $\overline{RR'}$ are bisected at right angles by g.

It is clear that a reflection is a 1-1 mapping of the plane onto itself. Also, it is easily shown by use of Fig. II, 16 that distance is preserved. Hence a reflection is a motion. Since the image of each figure is a figure symmetrical to it with

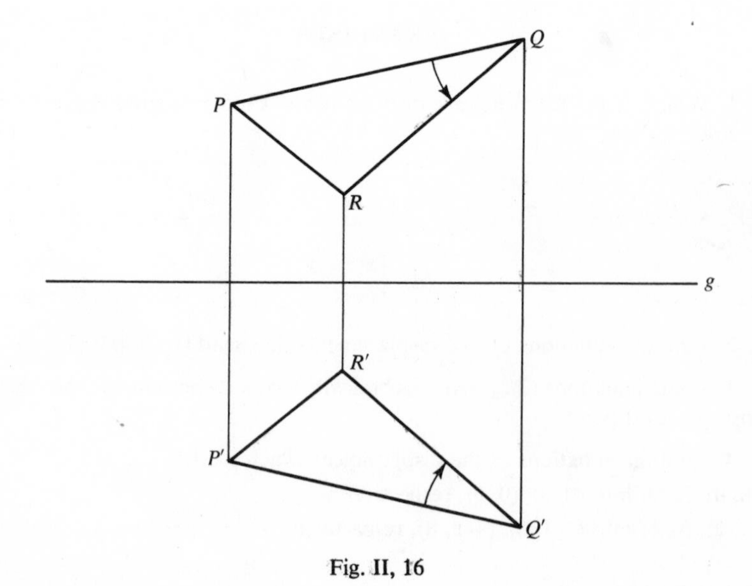

Fig. II, 16

respect to the axis of the reflection, the sense of every angle is reversed. Thus, if $\angle PQR$ is positive, as in Fig. II, 16, its image $\angle P'Q'R'$ is negative. A reflection is our first example of an *opposite* transformation. A method of proving that a reflection reverses the sense of angles is provided in §14, Exs. 5, 6.

Consider a reflection with axis g. The points of g are clearly the only fixed points for the reflection. In addition to g, each line perpendicular to g is left fixed. We distinguish g from these perpendiculars by calling it *pointwise fixed.* No other line is fixed, for no other line is symmetrical to g, and only figures symmetrical to g are fixed.

Summarizing, we have the following theorem:

13.2 *Reflections are opposite motions. A reflection leaves fixed each point of its axis but no other point. It leaves its axis fixed, and every line perpendicular to the axis, but no other line.*

Suppose now that a reflection R_g in g is followed by a reflection R_h in some other line h. Since R_g, R_h reverse sense, the resultant of the two reflections must preserve sense. Thus, R_gR_h is a direct motion, that is, a displacement. If g, h meet, their point of intersection is left fixed by R_g and R_h, and hence by R_gR_h, which is therefore a rotation about the point. If g, h are parallel, each line perpendicular to them is left fixed by R_g and R_h, and hence by R_gR_h, which is therefore a translation whose vector is perpendicular to g and h. The direction and length of this vector, and the angle of rotation in the case when R_gR_h is a rotation, are easily found by noting the effect of R_gR_h on a point of g. We state the results, leaving their verification to the reader:

13.3 *Let R_g, R_h be reflections in the distinct lines g, h. If g and h meet, R_gR_h is a rotation about their point of intersection through twice the (positive or negative) angle from g to h. If g, h are parallel, R_gR_h is a translation through twice the distance from g to h, and in the direction of the perpendicular from g to h.*

Thus, the resultant of successive reflections in the lines $y = 0$, $y = x$, in that order, is a rotation through $\frac{1}{2}\pi$ about O. Taking the reflections in the opposite order gives a rotation through $-\frac{1}{2}\pi$ about O. The resultant of successive reflections in the lines $y = 0$, $y = 1$ is the translation one of whose vectors extends from (0, 0) to (0, 2).

The following converse of 13.3 is also true and shows how basic reflections are. We leave the proof as an exercise.

13.4 *Every rotation is the resultant of reflections in two lines through its center. Every translation is the resultant of reflections in two lines perpendicular to its direction. In each case one of the lines can be chosen at will, the other then being uniquely determined.*

EXERCISES

1. Prove that a reflection preserves distance (do not assume that it is a motion).
2. Prove that a reflection is involutory, i.e., its own inverse (see §7, Ex. 7).

3. Prove that no set of reflections is a group.

4. Show that a reflection and the identity constitute a group.

5. What is the nature of a transformation which is the resultant of (a) an even number of reflections; (b) an odd number of reflections?

6. Does a reflection preserve (a) parallelism; (b) the size of an angle; (c) tangency to a circle? Justify your answers.

7. Verify by a diagram that a reflection reverses the order of the vertices in a triangle.

8. According to 13.3, $R_gR_h = D$, where D is a displacement. Prove that $R_h = R_gD$.

9. Prove that a motion (other than I) that leaves every point of a line fixed is necessarily a reflection in the line (see §3, Exs. 4, 5).

10. Prove that a reflection in g leaves fixed each figure that is symmetrical to g, but no other figure.

11. Complete the proof of 13.3.

12. Prove 13.4.

14. Equations of Reflections. The reflection R_x in the x-axis has the equations

$$(1) \qquad R_x \begin{cases} x' = x \\ y' = -y. \end{cases}$$

If h is any other line through O, and τ is its angle of inclination (Fig. II, 17), then R_xR_h is a rotation through 2τ about O by 13.3. In symbols,

$$R_xR_h = D,$$

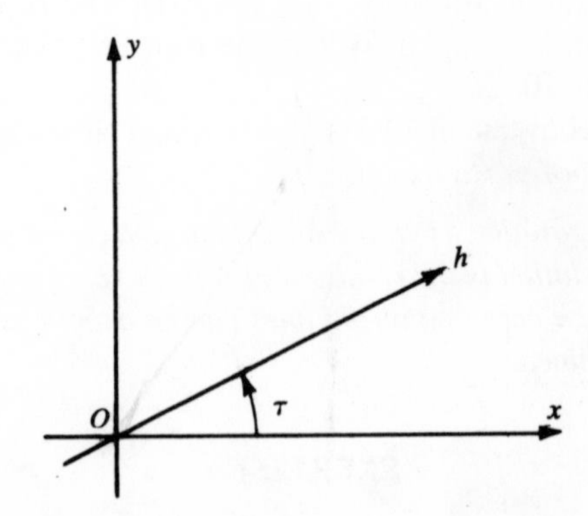

Fig. II, 17

where D represents the rotation. Hence, by §13, Ex. 8,

$$R_h = R_x D. \tag{2}$$

To obtain the equations of R_h we therefore combine equations (1) with those of a rotation through 2τ about O. The result is

$$R_h \begin{cases} x' = x \cos 2\tau + y \sin 2\tau \\ y' = x \sin 2\tau - y \cos 2\tau. \end{cases}$$

For example, the reflection in the line $y = x$ has the equations $x' = y, \quad y' = x$ since $\tau = \frac{1}{4}\pi$.

In deriving the equations of R_h we supposed that h is not the x-axis, and hence that $\tau \neq 0$. It is easily seen, however, that these equations reduce to equations (1) when $\tau = 0$, and hence include the reflection in the x-axis. We have thus proved:

14.1 *A reflection in the line through the origin with inclination τ has the equations*

$$\begin{cases} x' = x \cos 2\tau + y \sin 2\tau \\ y' = x \sin 2\tau - y \cos 2\tau. \end{cases} \tag{3}$$

To find the equations of the reflection R_k in a line k not through O, let h be the parallel to k through O, and let p be the distance between h and k (Fig. II, 18). If A is the foot of the perpendicular from O to k, then the length of $\overrightarrow{OA}$ is p. Let θ be the angle of $\overrightarrow{OA}$. By 13.3, $R_h R_k$ is the translation through distance $2p$

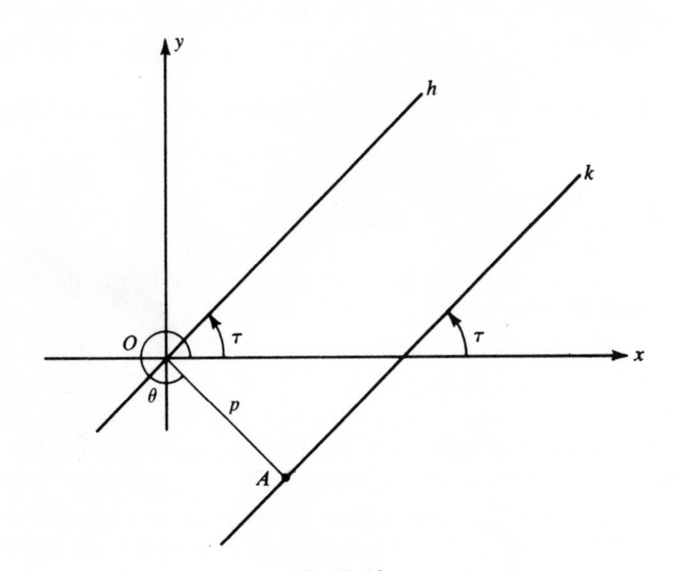

Fig. II, 18

in the direction of the perpendicular from h to k. This translation T then has a vector of length $2p$ and angle θ. Its equations are therefore

$$(4) \qquad T \begin{cases} x' = x + 2p \cos \theta \\ y' = y + 2p \sin \theta \end{cases}$$

by §4, Ex. 20. Since $R_h R_k = T$, it follows that

$$(5) \qquad R_k = R_h T,$$

and hence that the equations of R_k can be found by combining (3) and (4). The result is

$$R_k \begin{cases} x' = x \cos 2\tau + y \sin 2\tau + 2p \cos \theta \\ y' = x \sin 2\tau - y \cos 2\tau + 2p \sin \theta. \end{cases}$$

In obtaining these equations we supposed that k does not go through O, and hence that $p \neq 0$. But if we put 0 for p in these equations they reduce to (3). Hence they represent the reflections in *all* lines. We have therefore proved:

14.2 *A reflection in an arbitrary line has the equations*

$$(6) \qquad \begin{cases} x' = x \cos 2\tau + y \sin 2\tau + 2p \cos \theta \\ y' = x \sin 2\tau - y \cos 2\tau + 2p \sin \theta, \end{cases}$$

in which τ is the inclination of the line, p is the length of the perpendicular from the origin to the line, and, when $p \neq 0$, θ is the angle of the vector represented by this directed perpendicular.

For example, the reflection in the line whose intercept on each axis is 1 has the equations $x' = -y + 1$, $y' = -x + 1$ since $\tau = \frac{3}{4}\pi$, $p = \frac{1}{2}\sqrt{2}$, $\theta = \frac{1}{4}\pi$.

It is possible to simplify equations (6) by expressing τ in terms of θ (see Ex. 7).

EXERCISES

1. Find the equations of a reflection in (a) the y-axis; (b) the line bisecting quadrants I and III; (c) the line bisecting quadrants II and IV.

2. Find the equations of a reflection in (a) the line $y = 3$; (b) the line tangent to the circle $x^2 + y^2 = 25$ at $(3, -4)$.

3. A common test for symmetry states that *a curve is symmetrical to the x-axis if its equation is unchanged when y is replaced by $-y$, and to the y-axis if its equation is unchanged when x is replaced by $-x$.* Justify this by using the equations of R_x and R_y.

4. If the equation of a curve is given, state and prove a test for the symmetry of the curve with respect to (a) the line bisecting quadrants I and III, (b) the line bisecting quadrants II and IV.

5. Prove algebraically that a reflection in the x-axis reverses the sense of every angle (see §4, Exs. 10, 11).

6. Using Ex. 5 and equations (2) and (5), prove that all reflections are sense-reversing.

7. Verify that, for a line not through O, τ differs from θ by an odd multiple of $\frac{1}{2}\pi$. Then show that a reflection in such a line has the equations

$$\begin{cases} x' = -x \cos 2\theta - y \sin 2\theta + 2p \cos \theta \\ y' = -x \sin 2\theta + y \cos 2\theta + 2p \sin \theta, \end{cases}$$

where p, θ have the same meanings as in 14.2.

8. Find the axis of the reflection

$$\begin{cases} x' = -\frac{3}{5}x + \frac{4}{5}y + 1 \\ y' = \frac{4}{5}x + \frac{3}{5}y - \frac{1}{2}. \end{cases}$$

15. Other Opposite Motions. There are other sense-reversing motions besides reflections. Before considering them we write the equations of reflections differently than we did in §14. By making the substitutions

$$A = \cos 2\tau, \quad B = \sin 2\tau, \quad C = 2p \cos \theta, \quad D = 2p \sin \theta$$

in equations (6), §14, we obtain

$$\begin{cases} x' = Ax + By + C \\ y' = Bx - Ay + D, \end{cases} \tag{1}$$

where $A^2 + B^2 = 1$. Every reflection therefore has equations of this form.

These equations, however, are more general than equations (6), §14, since they represent other opposite motions besides reflections. For example, the resultant of the following translation and reflection in the x-axis,

$$\begin{cases} x' = x + 1 \\ y' = y \end{cases} \quad \text{and} \quad \begin{cases} x'' = x' \\ y'' = -y', \end{cases}$$

has the equations

$$\begin{cases} x'' = x + 1 \\ y'' = -y, \end{cases} \tag{2}$$

which are of the form (1), with $A = 1$, $B = 0$, $C = 1$, $D = 0$, and $A^2 + B^2 = 1$. Being the resultant of two motions, one direct and one opposite, (2) represents an opposite motion. This motion is not a reflection, for it clearly has no fixed point.

We now show that equations (1) represent every opposite motion and that any such motion, when not a reflection, is the resultant of a reflection and a displacement. As a step in this direction, we first prove:

15.1 *If P, Q are distinct points, and P', Q' are distinct points such that $PQ = P'Q'$, then there is a unique opposite motion that sends P, Q into P', Q', respectively. In other words, an opposite motion is completely determined when two points and their images are specified.*

Proof. Let D be the unique displacement which, by 12.3, sends P, Q into P', Q', and let R be the reflection in line $P'Q'$. Then DR is an opposite motion sending P, Q into P', Q'. Suppose M is another opposite motion with this property. One could now prove that $DR = M$ by showing that the images of an arbitrary point under DR and M are the same. The details are left to the reader since they differ from those in the proof of 12.3 only in that now the sense of every angle is reversed.

15.2 *Every opposite motion is the resultant of a displacement and a reflection, in that order.*

Proof. Let M be an opposite motion and P', Q' any distinct points. Since M is 1-1, these points are the images of two distinct points P, Q, respetively. Also, $PQ = P'Q'$. As shown in the proof of 15.1, there are a displacement D and a reflection R in line $P'Q'$ whose resultant DR sends P, Q into P', Q'. Since DR is an opposite motion, M must be identical with it by 15.1.

The reflection mentioned in 15.2 can always be chosen to be the reflection in the x-axis, for P', Q' in the above proof, being arbitrary, can be taken on that axis. It follows that the equations of all opposite motions can be obtained by finding the resultant of the displacements,

$$\text{(3)} \qquad \begin{cases} x' = Ax + By + C \\ y' = -Bx + Ay + D, \end{cases}$$

where $A^2 + B^2 = 1$, and the reflection in the x-axis,

$$\text{(4)} \qquad \begin{cases} x' = x \\ y' = -y, \end{cases}$$

in that order. Doing this, we obtain

$$\text{(5)} \qquad \begin{cases} x' = Ax + By + C \\ y' = Bx - Ay - D, \end{cases}$$

where $A^2 + B^2 = 1$. Thus, every opposite motion has equations of this form. Conversely, equations of this form always represent an opposite motion.*

To determine when the opposite motion (5) represents a reflection, and

* Lest a misunderstanding result from our use of equations (4), we wish to mention that even when an opposite motion is a reflection in a line other than the x-axis it will be the resultant of a displacement and a reflection in the x-axis.

when it does not, let us study the possible fixed points of (5). Replacing x' by x, and y' by y, and collecting terms, we get

$$(6) \qquad \begin{cases} (1 - A)x - By = C \\ -Bx + (1 + A)y = D. \end{cases}$$

These equations have a unique solution if and only if

$$\begin{vmatrix} 1 - A & -B \\ -B & 1 + A \end{vmatrix}$$

is not zero. Since $A^2 + B^2 = 1$ in (5), the above determinant *is* zero, being equal to $1 - A^2 - B^2$. Hence (5) cannot have exactly one fixed point. In other words, an opposite motion has either no fixed point or more than one fixed point. In the latter case there would have to be a line of fixed points (§3, Ex. 4), and this would imply that the motion is a reflection (§13, Ex. 9). From this we see that reflections are the only opposite motions having fixed points. Hence, (5) is a reflection if and only if (6) has solutions, and a necessary condition for this is that the determinant

$$\begin{vmatrix} -B & C \\ 1 + A & D \end{vmatrix}$$

be zero, that is, $AC + BD + C = 0$, or that

$$\begin{vmatrix} 1 - A & C \\ -B & D \end{vmatrix}$$

be zero, that is, $BC - AD + D = 0$. This condition is also sufficient.*

15.3 *The equations*

$$\begin{cases} x' = Ax + By + C \\ y' = Bx - Ay + D, \end{cases}$$

where $A^2 + B^2 = 1$ *and* C, D *are arbitrary, represent all the opposite motions and no other transformations. They represent reflections if and only if* $AC + BD + C = 0$ and $BC - AD + D = 0$.

The condition in 15.3, for example, is met by the reflection in the x-axis since $B = C = D = 0$. It is not met by the opposite motion (2), which is therefore not a reflection.

Although an opposite motion M which is not a reflection has no fixed point, as was shown above, it does have a fixed line (see Ex. 5), which we call g.

To see how the points of g move under M, let P_0 be any such point and

*For, since $1 - A$ and $1 + A$ are not both zero, the coefficient matrix and augmented matrix of (6) both have rank 1.

let $M(P_0) = P_1$ (Fig. II, 19). P_1 is on g since g is fixed. Let $P_0P_1 = k$. If $M(P_1) =$

P_{-4} P_{-3} P_{-2} P_{-1} P_0 P_1 P_2 P_3 P_4 g

Fig. II, 19

P_2, then $P_0P_1 = P_1P_2 = k$. There are two points on g at distance k from P_1, one of them being P_0. If P_2 coincided with P_0, M would send P_0, P_1 into P_1, P_0, respectively, and hence leave the midpoint of $\overline{P_0P_1}$ fixed, which is impossible. Hence P_2 is situated so that P_1 is the midpoint of $\overline{P_0P_2}$ (Fig. II, 19). Similarly, if $M(P_i) = P_{i+1}$, where i is any positive integer, P_i is the midpoint of $\overline{P_{i-1}P_{i+1}}$. Moreover, if $P_{-1}, P_{-2}, P_{-3}, \ldots$ are the points such that $P_0 = M(P_{-1})$, $P_{-1} = M(P_{-2})$, $P_{-2} = M(P_{-3})$, etc., then P_{-1} is the midpoint of $\overline{P_{-2}P_0}$, P_{-2} is the midpoint of $\overline{P_{-3}P_{-1}}$, etc. Thus M moves all the points P_i, where i is *any* integer, in the same direction along g, and through the same distance k. M must do likewise to the remaining points of g since each such point is between two of the points P_i, and M preserves distance.

Now consider any point P not on g. To find its image P' let Q be the foot of the perpendicular from P to g (Fig. II, 20). Let $M(Q) = Q'$ and $M(Q') = Q''$. Segment $\overline{PQ}$ must then go into a segment $\overline{P'Q'}$ of equal length perpendicular to g at Q'. Since $\angle PQQ'$ goes into $\angle P'Q'Q''$, and these angles have opposite senses, P' must be situated as shown in Fig. II, 20. Thus P and P' are on opposite sides of g.

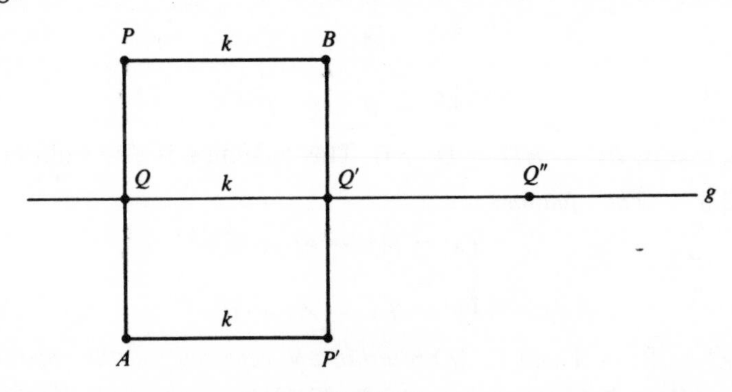

Fig. II, 20

The nature of M is now apparent: M is the resultant of a reflection in g and a translation parallel to g through distance k. Thus, in Fig. II, 20, the reflection sends P into A, and the translation sends A into P'. The resultant can also be taken in the opposite order. Then, the translation sends P into B and the reflection sends B into P'.

It is not difficult now to show that M has no fixed line other than g. This is left as an exercise.

To summarize, we have the following:

15.4 *An opposite motion which is not a reflection has no fixed point, but has a unique fixed line. The motion is the resultant, in either order, of a reflection in this line and a translation parallel to it.*

Because of this simple interpretation, an opposite motion which is not a reflection is called a **glide-reflection.** The fixed line is called the *axis* of the glide-reflection.

EXERCISES

1. Complete the proof of 15.1.

2. Verify that

$$x' = \tfrac{3}{5}x + \tfrac{4}{5}y - 1, \qquad y' = \tfrac{4}{5}x - \tfrac{3}{5}y + 2$$

is a reflection and find the equation of its axis.

3. Verify that

$$x' = -\tfrac{3}{5}x + \tfrac{4}{5}y + 2, \qquad y' = \tfrac{4}{5}x + \tfrac{3}{5}y + 1$$

is an opposite motion not a reflection and prove algebraically that it has no fixed point.

4. Show that equations (6), §14, which represent all the reflections, meet the condition $AC + BD + C = 0$ and $BC - AD + D = 0$.

5. Prove that every opposite motion not a reflection has a fixed line by showing that $x' = Ax + By + C$, $y' = Bx - Ay + D$, where $A^2 + B^2 = 1$, leaves the line (whenever it exists) $2\,Bx - 2(1 + A)y + AD - BC + D = 0$ fixed. (*Hint.* To simplify the algebra, regard the line as the image, that is, replace x, y by x', y', and find its original.)

6. Find the equation of the axis of the glide-reflection of Ex. 3. Plot the axis and several pairs of corresponding points not on it. (*Hint.* Either use Ex. 5 or the fact that the midpoint of the segment joining two corresponding points is necessarily on the axis.)

7. Given the glide-reflection $x' = x + 1$, $y' = -y + 6$, find the reflection and the translation which are its components in accordance with 15.4. (*Hint.* See Ex. 5 or 6.)

8. Prove that a glide-reflection cannot leave two lines fixed.

9. Equations (5) were derived by use of the reflection in the x-axis. Show that they can also be obtained by using the general reflection (1).

10. What kinds of transformations are the following: (a) the inverse of an opposite motion; (b) the resultant of two opposite motions; (c) the resultant of three opposite motions?

11. Can you give a set of opposite motions which is a group? Justify your answer.

16. Equations of the Group of Motions. As a first step in finding these equations we shall show, by a sequence of theorems, that there is no motion which preserves the sense of some angles and reverses the sense of others.

16.1 *A motion which leaves each of three noncollinear points fixed is necessarily the identity.*

Proof. Let A, B, C be distinct noncollinear points, and M be a motion that leaves each of them fixed (Fig. II, 21). By §3, Ex. 4, M leaves each point of lines AB, BC, AC fixed. Let P be a point not on any of these lines. If Q is a point of line AB distinct from A and B, line PQ cannot be parallel to both of the lines AC, BC, and hence meets one of them, say AC, in a point R. Since M leaves Q and R fixed, it leaves each point of line QR fixed, and hence P. Thus M leaves every point of the plane fixed.

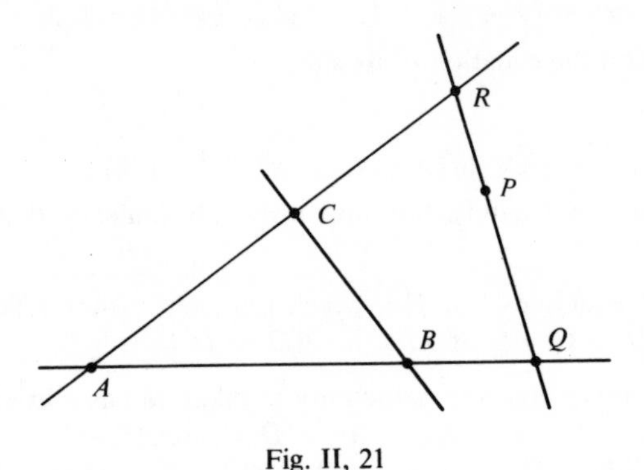

Fig. II, 21

To achieve our objective we also need the theorem:

16.2 *There cannot be two motions that send three specified noncollinear points into three specified noncollinear points, respectively.*

Proof. Assume M, N are distinct motions that send the noncollinear points A, B, C into the noncollinear points A', B', C', respectively. The motion MN^{-1} then leaves each of the points A, B, C fixed and must therefore be I by 16.1. From $MN^{-1} = I$ we obtain $(MN^{-1})N = IN$, $M(N^{-1}N) = N$, $MI = N$, and finally $M = N$. The truth of the theorem follows from this contradiction.

16.3 *A motion is either direct or opposite.*

Proof. Let M be a motion, and A, B, C three noncollinear points. Let $M(A, B) = A', B'$. Then M must send C into a point C' such that $AC = A'C'$ and $BC = B'C'$. An attempt to construct C' by straightedge and compass shows that C' must have one of two possible positions, C_1' and C_2', one on each side of line $A'B'$ (Fig. II, 22). For C' to be in the position C_1' would mean that M preserves the sense of the angles of triangle ABC, and for C' to occupy the

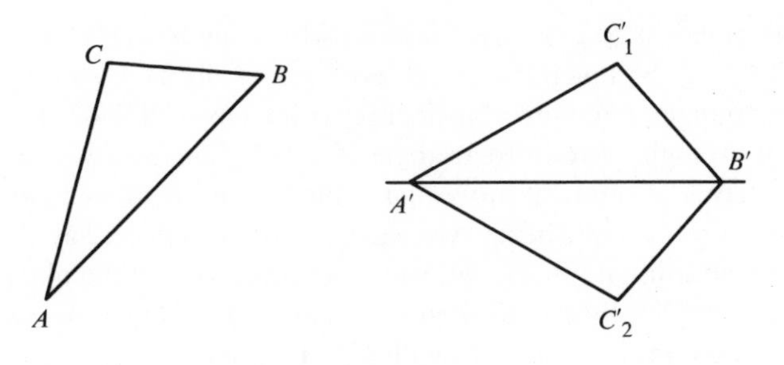

Fig. II, 22

position C'_2 would mean that M reverses the sense of these angles. Now, by 12.3 there is a unique direct motion that sends A, B into A', B', and it necessarily sends C into C'_1 since it preserves the sense of all angles. If $M(C) = C'_1$, then M is identical with this direct motion by 16.2. Similarly, by 15.1 there is a unique opposite motion that sends A, B into A', B' and it must send C into C'_2 since it reverses the sense of all angles. If $M(C) = C'_2$, then M is identical with this opposite motion by 16.2. Thus, M is either direct or opposite.

From 16.3 we see that there are no motions which preserve the sense of some angles and reverse that of others. Hence, motions are either translations, rotations, reflections, or glide-reflections, and nothing else. Also, from this theorem and the details used in its proof the following can be deduced. This is left as an exercise.

16.4 *There is a unique motion which sends three noncollinear points A, B, C into three noncollinear points A', B', C', respectively, if $AB = A'B'$, $AC = A'C'$, $BC = B'C'$.*

This theorem is sometimes expressed by saying that *a motion is completely determined by three noncollinear points and their images*, or that *there is a unique motion that maps one of two congruent triangles onto the other so that vertices go into vertices in a specified way.*

The equations representing the set of all motions can now be found by combining the equations of the direct motions and those of the opposite motions. From 12.2 and 15.3, then, we obtain the following:

16.5 *All motions have equations of the form*

$$\begin{cases} x' = Ax + By + C \\ y' = \pm(-Bx + Ay) + D, \end{cases}$$

where $A^2 + B^2 = 1$, the $+$ sign corresponding to a direct motion, the $-$ sign to an opposite motion. Conversely, all the transformations represented by such equations, where $A^2 + B^2 = 1$, are motions.

As was noted in §12, a direct motion which sends a triangle ABC into a triangle $A'B'C'$ can be imagined as achieved physically by a sliding movement which brings triangle ABC into coincidence with triangle $A'B'C'$. If an opposite motion sends triangle ABC into triangle $A''B''C''$, and we wish to view this intuitively in terms of physical movements, then a sliding movement of triangle ABC can do no more than bring two sides into coincidence, say side AB with side $A''B''$. An additional movement, which involves leaving the plane, is needed to bring the triangles into coincidence; triangle ABC must be swung about side AB as an axis until C coincides with C''. This may be regarded as illustrating the fact that an opposite motion is the resultant of a displacement and a reflection.

EXERCISES

1. Show that each direct motion is a linear transformation whose determinant is 1, and that each opposite motion is a linear transformation whose determinant is -1. Is the converse true? Justify your answer.

2. Which of the following transformations are direct motions, and which are opposite?

(a) $\begin{cases} x' = \frac{1}{3}x + \frac{2}{3}y + 4 \\ y' = -\frac{2}{3}x + \frac{1}{3}y - 5; \end{cases}$ (b) $\begin{cases} x' = -x + 2 \\ y' = y; \end{cases}$

(c) $\begin{cases} x' = -x + 2 \\ y' = -y; \end{cases}$ (d) $\begin{cases} x' = x - y + 2 \\ y' = x + y + 3; \end{cases}$

(e) $\begin{cases} x' = \dfrac{5}{6}x - \dfrac{\sqrt{11}}{4}y + 1 \\ y' = \dfrac{\sqrt{11}}{9}x + \dfrac{5}{6}y + 2; \end{cases}$ (f) $\begin{cases} x' = x - y + 4 \\ y' = -x + y - 5. \end{cases}$

3. Prove 16.4.

4. Name the motions which have (a) no fixed line; (b) exactly one fixed line; (c) more than one fixed line; (d) no fixed point; (e) exactly one fixed point; (f) more than one fixed point.

5. Name the motions which have (a) exactly one fixed circle; (b) more than one fixed circle; (c) no fixed circle.

17. Further Remarks on Congruence. Now that we are familiar with motions of various types we can complete the discussion of congruence begun in §3. It will be convenient to refer to the old, intuitive definition of congruence used in elementary geometry as Definition I, and to refer to the new definition (see 3.1) of congruence based on motions as Definition II. In §3 we showed that whenever two of the figures considered in elementary geometry are congruent by Definition II they are also congruent by Definition I. One of our remaining tasks is to show, conversely, that any two figures which are congruent by Definition I are also congruent by Definition II.

In elementary geometry we know by Definition I that any two points are congruent, that is, they can be made to coincide. Similarly, any two lines are congruent, or any two segments of equal length. To show that any two points, say P, P', are also congruent by Definition II, we simply note that there is a motion that sends P into P', for example, the translation with vector $\overrightarrow{PP'}$. Likewise, any two lines, say g, g', are congruent by Definition II. For, if they are parallel, and P, P' are any points on them, respectively, the translation $\overrightarrow{PP'}$ maps g on g', whereas, if they meet in Q some rotation about Q will map g on g'. Finally, any two segments $\overline{AB}$, $\overline{A'B'}$ of equal length are congruent according to the new definition. For, since $AB = A'B'$, by 12.3 there is a displacement that sends A, B into A', B' and hence $\overline{AB}$ into $\overline{A'B'}$.

A familiar fact of elementary geometry, often stated as an axiom and based on Definition I, is that all right angles are congruent, that is, they can be made to coincide. To show that they are also congruent by Definition II, consider any two right angles, one with vertex A and sides h, k, the other with vertex A' and sides h', k'. Take points B, B' on h, h', respectively, so that $AB = A'B'$. There is a displacement D that sends A, B into A', B', and hence ray h into ray h' by 2.3. Since D sends rays into rays and preserves perpendicularity it must send k into a ray with endpoint A' and perpendicular to h'. This ray is necessarily k' if the given angles have the same sense, and so, in this case, D would send angle h, k into angle h', k'. The reasoning is the same in case the angles have unlike senses, except that then, instead of D, we would use the opposite motion that sends A, B into A', B'.

Two triangles ABC, $A'B'C'$ whose sides are equal, respectively, are proved congruent in elementary geometry by superposing one triangle on the other and then using Definition I. The triangles are also congruent by Definition II, for the equalities $AB = A'B'$, $BC = B'C'$, $AC = A'C'$ imply that there is a motion which sends A, B, C into A', B', C', respectively (16.4), and this motion sends $\overline{AB}$, $\overline{BC}$, $\overline{AC}$ into $\overline{A'B'}$, $\overline{B'C'}$, $\overline{A'C'}$ (2.3), and hence triangle ABC into triangle $A'B'C'$.

We do not attempt to dispose of the remaining situations in elementary geometry where two figures are called congruent. Actually, the satisfactory handling of some of these situations cannot be achieved by use of Definition II alone, but requires also that certain changes be made in the axioms of elementary geometry.

Before ending our discussion of congruence let us consider one further matter. To say that two figures f, f' are congruent by Definition II means, as we have seen, that there is a motion which carries f into f'. This, in turn, implies that there is a 1-1 correspondence between the points of f and f' such that

$PQ = P'Q'$, where P, Q and P', Q' are any corresponding pairs of points. The question then arises: Is the converse true? That is, if the points of two figures f, f' are known to be in such a 1-1 correspondence, does a motion necessarily exist which sends f into f'? We mean, of course, a motion of the *entire* plane. The answer is in the affirmative, as we now prove.

17.1 *If f and f' are geometric figures whose points are in a* 1-1 *correspondence such that the distance between each two points in f equals the distance between their correspondents in f', then there exists a motion of the entire plane which maps f onto f'.**

Proof. Let A, B be any two points of f, and A', B' their correspondents in f'. Since $AB = A'B'$, there are two motions, one direct and one opposite, which send A, B into A', B' (12.3 and 15.1). If f has a point X collinear with A, B, its correspondent X' must be collinear with A', B' in view of the assumed distance relations. Also, because of these relations, each of the two specified motions send X into X'. If f has a point C not on line AB, its correspondent C' is not on line $A'B'$ because of the assumed distance relations. By 16.4, exactly one of the two specified motions will send C into C'. Call this motion M. Let D be any other point of f not on line AB. Then its correspondent D' is not on line $A'B'$. We now show that $M(D) = D'$.

If D is on line AC or line BC, it follows, by the same reasoning used for X, that $M(D) = D'$. If D is on neither line, consider the three circles with centers A, B, C and radii equal to AD, BD, CD, respectively (Fig. II, 23). (D is, of course, on these circles.) Then D' is on neither of the lines $A'C'$, $B'C'$, and is on the circles with centers A', B', C' and radii equal to $A'D'$, $B'D'$, $C'D'$, and hence to AD, BD, CD. Let $M(D) = D''$. Since M sends A, B, C into A', B', C' and preserves distance, the statement just made for D' also holds for D''. If D' and D'' did not coincide, the circles mentioned in the statement would meet in D' and D'', $\overline{D'D''}$ would be a common chord, and the perpendicular bisector of $\overline{D'D''}$ would

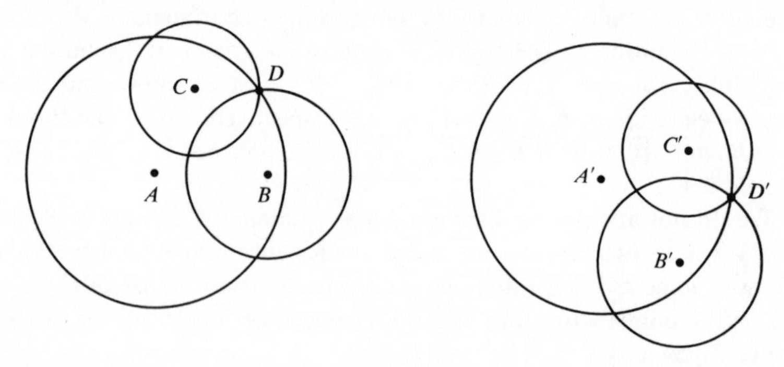

Fig. II, 23

* In stating this theorem we have ignored the case in which f and f' consist of a single point each.

contain each center. This contradicts that A', B', C' are noncollinear. Hence $D'' = D'$. Thus, $M(D) = D'$. We have now shown that M sends every point of f into its correspondent in f'. Conversely, since there is a 1-1 correspondence between the points of f and f', each point of f' is the image of a point of f. Thus, M maps f onto f'.

On combining the theorem just proved and Definition 3.1 we obtain another theorem:

17.2 *Two geometric figures are congruent if and only if there is a* 1-1 *correspondence between their points such that the distance between two points of one figure always equals the distance between their correspondents in the other.*

This characterization of congruent figures does not involve the term *motion* explicitly. Hence it can be used as an alternative definition of congruent figures. This is often done in parts of geometry not concerned with motions. Between them, Definition 3.1 and Theorem 17.2 go a long way toward putting the intuitive concept of congruence into precise mathematical language.

EXERCISES

In the following exercises interpret *congruent* according to Definition 3.1.

1. Show that two circles with equal radii are congruent.

2. Prove that two parabolas are congruent if the distance from the focus to the directrix of one equals the corresponding distance in the other.

3. Prove that two ellipses are congruent if the major and minor axes of one are equal, respectively, to the major and minor axes of the other.

4. Prove that two hyperbolas with equal transverse axes are congruent if the angles between their asymptotes are equal (see §3, Ex. 7).

5. Every conic section has a focus and a directrix, and can be defined as *the locus of a point, the ratio of whose distances from the focus and from the directrix is a constant* (called the *eccentricity* of the conic). Show that two conics with the same eccentricity are congruent if the distance between focus and directrix is the same for both.

6. Prove that the interiors of two circles with equal radii are congruent.

7. Prove that two arcs of the same circle are congruent if they subtend equal central angles.

8. Show that the figure consisting of the x- and y-axes is congruent to the figure consisting of the lines $x = 1$ and $y = 1$.

9. If f consists of the vertices of a square and the point of intersection of the diagonals, and f' consists of the corresponding five points of another square of the same size, prove that f and f' are congruent.

10. The figure f consisting of $(0, 0)$, $(0, 1)$, $(0, 3)$ is congruent to each of the

following figures f' by 17.2. Find a motion which maps f onto f' in each case (equations need not be given).

(a) (1, 1), (0, 1), (−2, 1); (b) (1, 0), (2, 0), (4, 0);
(c) (0, −1), (0, −2), (0, −4); (d) (1, 1), (1, 0), (1, −2).

11. For each of the following pairs of figures find the equations of a motion which maps the first figure onto the second:

(a) (1, 1), (0, 1), (−2, 1) and (0, 0), (0, 1), (0, 3);
(b) (1, 1), (0, 1), (−2, 1) and (0, −1), (0, −2), (0, −4);
(c) (0, 0), (1, 0), (0, 1) and (3, 0), (3, −1), (4, 0);
(d) (0, 0), (1, 0), (0, 1) and (5, 4), (5, 3), (4, 4).

12. If there is a 1-1 correspondence of the type mentioned in 17.1 between the points of two lines, show that the midpoints of the segments $\overline{PP'}$, where P' corresponds to P, are distinct and collinear, or else coincide. (This is called ***Hjelmslev's Theorem***, after J. T. Hjelmslev, 1873–1950.)

CHAPTER III

Transformations of Similarity

1. Introduction. Having considered all the transformations of the plane into itself which preserve distance, we turn now to those which preserve *angle-size*, or, to be more explicit, which always send an angle into another of the same absolute value. Such transformations are called *similarity transformations* (or *similarities*, or *similitudes*), since they send each figure into a similar figure, that is, speaking intuitively, one of the same shape, but not necessarily of the same size. Motions send each angle into an angle of equal size and are therefore included among the similarity transformations, being the special kind of similarity in which figures agree with their originals in size as well as shape.

2. General Properties of Similarity Transformations. The above description of similarity transformations gives one a quick grasp of them. The following description, however, will serve us better as a working definition:

2.1 Definition. *A similarity transformation is a* 1-1 *mapping of the plane onto itself in which each distance is multiplied by the same positive number* k. *We call* k *the ratio of the similarity transformation, or the similarity ratio.*

From this definition all the facts concerning similarity transformations can be deduced, including those mentioned in §1. Thus, since the definition does not exclude $k = 1$, and to multiply a distance by 1 is to leave it unchanged, we again see that motions are included among the similarity transformations. Until §3 is reached, we take it for granted that similarity transformations other than motions do exist.

A similarity transformation S in which $k = 2$ doubles all distances. S^{-1}, being a 1-1 mapping of the plane onto itself which cuts all distances in half, is therefore a similarity transformation with $k = \frac{1}{2}$. The resultant of two similarity transformations with $k = 2$ and $k = 3$, respectively, is a 1-1 mapping of the plane onto itself which multiplies all distances by 6, and hence is a similarity transformation with $k = 6$. Since these examples can clearly be generalized, we have the theorem:

2.2 *The set of all similarity transformations is a group. (This group is called the* **similarity group**.*) The set of all motions is a subgroup of the similarity group.*

Suppose that B is a point between points A and C:

$$AB + BC = AC. \tag{1}$$

If $S(A, B, C) = A', B', C'$, where S is a similarity transformation with ratio k, then $A'B' = k \cdot AB$, $B'C' = k \cdot BC$, and $A'C' = k \cdot AC$. Multiplying (1) by k gives

$$k \cdot AB + k \cdot BC = k \cdot AC,$$

or

$$A'B' + B'C' = A'C'.$$

Thus, since A', B', C' are distinct, B' is between A' and C'. S therefore preserves the between-relation. The following theorem is then immediate, as in the case of motions (see II, §2):

2.3 *Similarity transformations preserve the between-relation and hence preserve collinearity and noncollinearity. They send lines, segments, rays, angles, and n-sided polygons into figures of the same kind, respectively.*

In particular, the image of a triangle is a triangle. Since the sides of the latter are k times as long as those of the original triangle, the sides of the triangles are in proportion. It follows from elementary geometry that the corresponding angles of the triangles are equal, and that the triangles are similar. This proves the first assertion in the following theorem. The rest of the proof is left as an exercise.

2.4 *Similarity transformations send triangles into similar triangles and, more generally, n-sided polygons into similar n-sided polygons. The image of each angle is therefore an angle of equal size. In particular, perpendicularity is preserved.*

As for nonrectilinear figures, it is left as an exercise to show that the image of a conic is always a conic of the same eccentricity, and hence, speaking intuitively, of the same shape. Thus, a circle always goes into a circle.

Similarity transformations are the only 1-1 transformations of the plane into itself which preserve angle-size, or more exactly, which send each angle into an angle of the same absolute value, vertex going into vertex, and each side into a side. This is proved in the next chapter (IV, 6.5, 10.5).

EXERCISES

Prove the property of similarity transformations stated in each exercise.

1. The parallelism of lines is preserved.

2. An n-sided polygon goes into a similar n-sided polygon.

3. Angle-size is preserved.

4. Each conic goes into another of the same eccentricity (see II, §17, Ex. 5).

5. The image of the exterior of a circle is the exterior of the image circle.

6. The image of the interior of a triangle is the interior of the image triangle.

7. If line g goes into g', the two half-planes determined by g go, respectively, into the two half-planes determined by g' (see II, §2, Ex. 3).

8. No similarity transformation other than a motion has more than one fixed point.

9. Ratio of division is preserved (see II, §3, Ex. 2).

10. A similarity transformation with ratio k multiplies the area of every triangle and every circle by k^2.

3. A Key Similarity Transformation. If A is any point, consider the mapping in which A goes into itself, and any other point P of the plane goes into the point P' on ray AP such that $AP' = 2 \cdot AP$ (Fig. III, 1). This is clearly a 1-1 mapping of the plane onto itself. Also, it is easily seen that all distances are doubled. For example, if Q is not on line AP, and Q goes into Q', then $P'Q' = 2 \cdot PQ$ since triangles APQ, $AP'Q'$ are similar. Thus the transformation is a similarity. We see incidentally that similarity transformations other than motions do exist.

If, instead of 2 in the above example, we use k, representing any positive number, we obtain what is called a **radial transformation with ratio k and center A.** This transformation is clearly a similarity with ratio k which "moves" the points of the plane straight away from A if $k > 1$, as in the above example, and straight toward A if $k < 1$. The terms *expansion of the plane* and *contraction of the plane* are sometimes convenient for distinguishing these two cases. If $k = \frac{1}{2}$, for example, P' is midway between P and A, and we have a contraction of the

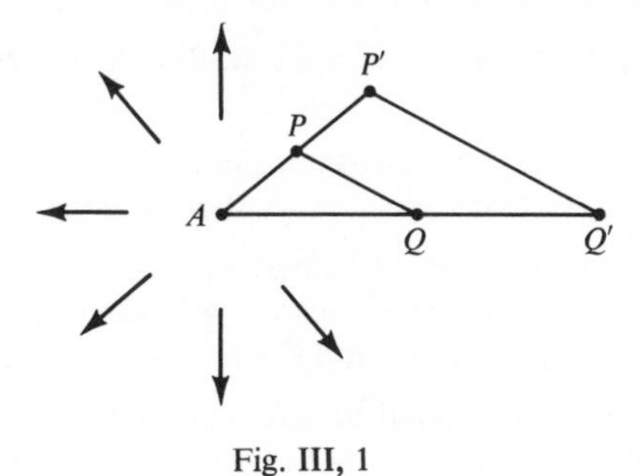

Fig. III, 1

plane toward A. A radial transformation in which $k = 1$ is, of course, the identity.

As diagrams correctly suggest, a radial transformation preserves the sense of every angle and hence is a direct similarity transformation (II, 12.5). This is illustrated in Fig. III, 1, where $\angle APQ$ and its image $\angle AP'Q'$ are both positive. The resultant of a reflection and a radial transformation that doubles distance is therefore a 1-1 mapping of the plane on itself which doubles distance and reverses the sense of every angle. Thus we see that *opposite similarity transformations* which are not motions do exist.

It is easily verified that a radial transformation with ratio k and center at the origin of coordinates has the equations

$$x' = kx, \quad y' = ky, \quad k > 0. \tag{1}$$

Later we shall see that by combining these equations with those representing the group of motions one obtains the equations of all similarity transformations.

EXERCISES

1. Prove that a radial transformation with ratio k is a similarity which multiplies all distances by k.

2. Take a triangle in whose interior lies the center of a radial transformation with $k = \frac{1}{3}$, draw its image, and prove that the triangles are similar.

3. Do Exercise 2 after replacing "interior" by "exterior."

4. Prove algebraically that the transformation (1) when not the identity, (a) is a similarity which multiplies distances by k; (b) is direct (see II, §4, Exs. 10, 11); (c) leaves each line through O fixed, but has no other fixed line; (d) has no fixed point other than O; (e) sends each circle with center O into another such circle; (f) sends each line not through O into a parallel line; (g) sends a conic into another conic with the same eccentricity.

5. Does the set of all transformations of the form (1), where $k > 0$, form a group? Justify your answer.

6. Determine some properties of the transformation $x' = 2x$, $y' = -2y$. Is it a similarity? Is it direct? Generalize your results.

7. Determine some properties of the transformations represented by equations (1) when $k < 0$.

8. Find the equations of the radial transformation with ratio k and center (a, b). Use the equations to prove that the transformation is direct.

9. Draw the image of the quadrilateral with consecutive vertices (1, 0), (2, 0), (2, 1), (1, 1) in: (a) the radial transformation with center (0, 0) and $k = \frac{1}{2}$; (b) the resultant of the transformation in (a) and the reflection in the y-axis.

10. Show that there is a radial transformation with center O which maps the parabola $y = ax^2$ onto the parabola $y = bx^2$, where a and b are positive.

11. Show that there is a radial transformation with center O which maps the ellipse $b^2x^2 + a^2y^2 = a^2b^2$ onto the ellipse $d^2x^2 + c^2y^2 = c^2d^2$ if a, b, c, d are positive and $a/b = c/d$. (The ellipses then have the same eccentricity.)

12. Show that there is a radial transformation with center O which maps the hyperbola $b^2x^2 - a^2y^2 = a^2b^2$ onto the hyperbola $d^2x^2 - c^2y^2 = c^2d^2$ if a, b, c, d are positive and $a/b = c/d$. (The hyperbolas then have the same eccentricity.)

4. Equations of the Similarity Group. As a first step in obtaining these equations we prove:

4.1 *There cannot be more than one similarity transformation that sends three given noncollinear points into three given noncollinear points, respectively.*

Proof. Let S and T be distinct similarity transformations that send the noncollinear points A, B, C into the noncollinear points A', B', C', respectively. Then ST^{-1} is a similarity transformation which leaves each of the points A, B, C fixed. It can only do this by preserving distance, and hence it is a motion. Being a motion with three noncollinear fixed points, it is necessarily the identity (II, 16.1). Thus $ST^{-1} = I$, so that $S = T$. The truth of the theorem follows from this contradiction.

Also, we need the following fact:

4.2 *The resultant of a motion and a radial transformation is a similarity transformation. Conversely, any given similarity transformation is the resultant of a motion and a radial transformation, and hence is either direct or opposite.*

Proof. The resultant of a motion and a radial transformation with ratio k is a 1-1 transformation of the plane into itself that multiplies distance by k, and hence is a similarity. Conversely, let S be a similarity with ratio k, and let A', B', C' be any noncollinear points. Since S is 1-1, points A, B, C exist such that $S(A, B, C) = A', B', C'$, and they are noncollinear by 2.3. Also, triangles ABC, $A'B'C'$ are similar by 2.4. Take B'' (Fig. III, 2) on side $\overline{A'B'}$ of triangle $A'B'C'$ so that $AB = A'B''$, and C'' on side $\overline{A'C'}$ so that $AC = A'C''$ (the sides must be extended to do this if $k > 1$). Then triangles ABC, $A'B''C''$ are congruent, and there is a motion M that maps the first on the second so that $M(A, B, C) = A', B'', C''$ (II, 16.4). If R is the radial transformation with center A' which has the same ratio as S, then $R(A', B'', C'') = A', B', C'$. Thus, MR maps A, B, C on A', B', C', respectively, and by the first part of the proof is a similarity. Using 4.1 we infer that $S = MR$. Since R is direct, S is direct if M is direct (Fig. III, 2a), and opposite if M is opposite (Fig. III, 2b).

We leave to the reader the proof of the following:

4.3 *There is a unique similarity transformation that sends a triangle ABC into a similar triangle $A'B'C'$ so that A, B, C go into A', B', C', respectively.*

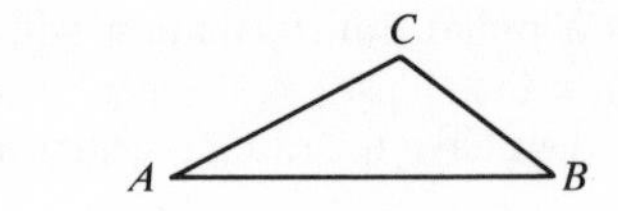

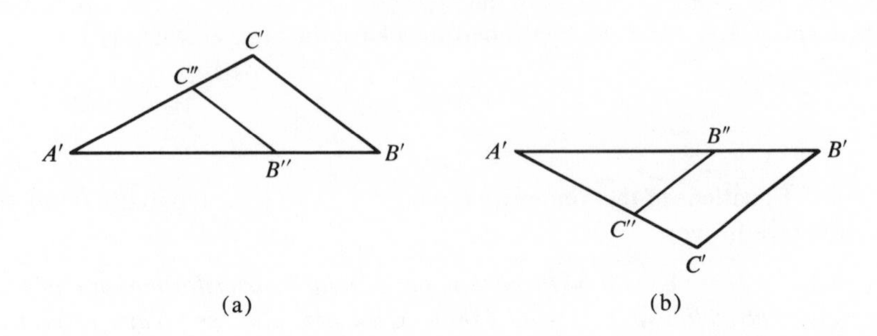

Fig. III, 2

In view of 4.2, the equations of the similarity group can be found by combining the equations of motions and those of radial transformations. Actually, it suffices to use only radial transformations with center at the origin for this purpose, for, A', B', C' in the proof of 4.2 being arbitrary, A' can be taken at the origin.

Thus, the direct similarities are the resultant of the displacements

$$\text{(1)} \qquad \begin{cases} x' = Ax + By + C \\ y' = -Bx + Ay + D, \end{cases} \qquad A^2 + B^2 = 1,$$

and the radial transformations

$$\text{(2)} \qquad \begin{cases} x' = kx \\ y' = ky, \end{cases} \qquad k > 0,$$

in that order, and hence have the equations

$$\text{(3)} \qquad \begin{cases} x' = k(\ \ Ax + By + C) & A^2 + B^2 = 1, \\ y' = k(-Bx + Ay + D), & k > 0. \end{cases}$$

Conversely, any given equations of this form necessarily represent a similarity since the procedure used to obtain them can be reversed.

If we let

$$\text{(4)} \qquad kA = a, \quad kB = b, \quad kC = m, \quad kD = n,$$

then (3) can be rewritten as

$$(5)\qquad \begin{cases} x' = \quad ax + by + m \\ y' = -bx + ay + n. \end{cases}$$

These have the same form as the displacements, but now $a^2 + b^2$ can have any positive value, for

$$a^2 + b^2 = (kA)^2 + (kB)^2 = k^2(A^2 + B^2) = k^2.$$

Conversely, any equations of the form (5), with a, b not both zero, represent a similarity. For if we take $k = \sqrt{a^2 + b^2}$ and then compute A, B, C, D from (4), it follows that $k > 0$, $A^2 + B^2 = 1$, and hence that (5) can be reduced to (3).

In like manner, by finding the resultant of the opposite motions and the radial transformations (2), we can show that the equations of the opposite similarities are

$$(6)\qquad \begin{cases} x' = ax + by + m \\ y' = bx - ay + n, \end{cases}$$

where $a^2 + b^2 = k^2$. The converse statement holds just as above.

Combining (5) and (6), then, we have:

4.4 *Every similarity transformation with ratio k has the equations*

$$\begin{cases} x' = \qquad ax + by + m \\ y' = \pm(-bx + ay) + n, \end{cases}$$

where $\sqrt{a^2 + b^2} = k$. Conversely, any equations of this form, with a, b not both zero, represent a similarity transformation with ratio $k = \sqrt{a^2 + b^2}$. The + and − signs correspond to direct and opposite transformations, respectively.

It is to be noted that similarity transformations are linear transformations of a special kind. In the next chapter we shall see that similarities are the only linear transformations which preserve angle-size.

EXERCISES

1. Prove 4.3.

2. Derive equations (6).

3. Verify that the transformations (5) and (6) are linear transformations. Interpret the absolute values of their determinants geometrically.

4. Show that $x' = 2x + 3y + 1$, $y' = 3x - 2y + 4$ is a similarity transformation, determine whether it is direct or opposite, and find its ratio.

5. Find a motion and a radial transformation whose resultant is the transformation of Ex. 4.

6. Find the inverse of the transformation in Ex. 4. Is it a similarity? If so, find its ratio.

7. If S is the transformation in Ex. 4, find S^2. Is it a similarity? If so, find its ratio.

5. Extension of the Notion of Similar Figures. In elementary geometry the range of figures to which the term *similar* is applied is quite limited, only polygons (including triangles) being designated in this way. Out of this restricted concept, however, has grown the broader view that any two geometric figures, regardless of type, ought to be called similar if they have the same shape. According to this view, which is intuitive since "shape" is not defined in mathematics, all circles are similar, all ellipses with the same eccentricity are similar, and, to take a practical illustration, the figures on a snapshot are similar to the corresponding figures on its enlargement.

To make this view more mathematical we proceed as we did in the preceding chapter in connection with congruence. Hence, having shown that similarity transformations send polygons into similar polygons (2.4), and conics into conics of the same eccentricity (§2, Ex. 4), we now adopt the following definition:

5.1 Definition. *Two geometric figures are called similar if and only if there is a similarity transformation which maps one on the other.*

An immediate consequence of this definition is that the range of figures to which the term *similar* can be applied is without restriction, for every geometric figure has an image under a similarity transformation. Another immediate consequence is that two polygons which are similar in the sense of this definition are also similar in the sense of elementary geometry. This follows from 2.4, in which "similar" is used in the sense of elementary geometry. To be sure, however, that this definition is entirely consistent with what is done in elementary geometry, we must still show, conversely, that two polygons which are called similar in elementary geometry are also similar in the sense of 5.1. That is, given two polygons whose sides are proportional and whose corresponding angles are equal, we must prove that there is a similarity transformation which maps one on the other. We do this by first proving the following theorem:

5.2 *If f and f' are any geometric figures whose points are in a* 1-1 *correspondence such that the distance between any two points of f' is the same positive constant k times the distance between the corresponding points of f, then f and f' are similar in the sense of Definition* 5.1.

Proof. Let P, Q be any distinct points of f, and P', Q' the corresponding points of f'. Then $P'Q' = k \cdot PQ$. Let S be a similarity transformation whose ratio has this same value k. S sends f into a figure f'', establishing between the points of f and f'' a 1-1 correspondence in which $P''Q'' = k \cdot PQ$, where P'', Q'' correspond to P, Q. The points of f' and f'' are then in a 1-1 correspondence in which $P'Q' = P''Q''$, where P'', Q'' correspond to P', Q'. The figures f' and f'' are therefore congruent (II, 17.2) and a motion M exists which maps f'' on f' (II, 3.1). Since SM, a similarity transformation, maps f on f', these figures are similar in the sense of Definition 5.1.

Returning now to the postponed problem let us consider two polygons (including triangles) which are similar in the sense of elementary geometry. Denote by f and f' the two sets of corresponding vertices $A, B, C, D, \ldots$, and $A', B', C', D', \ldots$, of these polygons (Fig. III, 3). Then f, f' are figures of the type described in 5.2. For, considering pairs of consecutive vertices first, we have, by hypothesis, $A'B' = k \cdot AB$, $B'C' = k \cdot BC$, etc., where $k > 0$. Such relations also hold for nonconsecutive vertices. Take A, C and A', C', for example. Since $A'B' = k \cdot AB$, $B'C' = k \cdot BC$, and $\angle A'B'C' = \angle ABC$, we know from elementary geometry that $A'C' = k \cdot AC$. Likewise, from this last equation, the equation $C'D' = k \cdot CD$, and the easily verified equation $\angle A'C'D' = \angle ACD$, we obtain $A'D' = k \cdot AD$. And so forth. It follows from 5.2, then, that there is a similarity transformation S which maps f on f'. Since S sends segments into segments, $\overline{AB}, \overline{BC}, \overline{CD}, \ldots$ go into $\overline{A'B'}, \overline{B'C'}, \overline{C'D'}, \ldots$, respectively. Thus, S maps one polygon on the other, and the polygons are therefore similar in the sense of Definition 5.1.

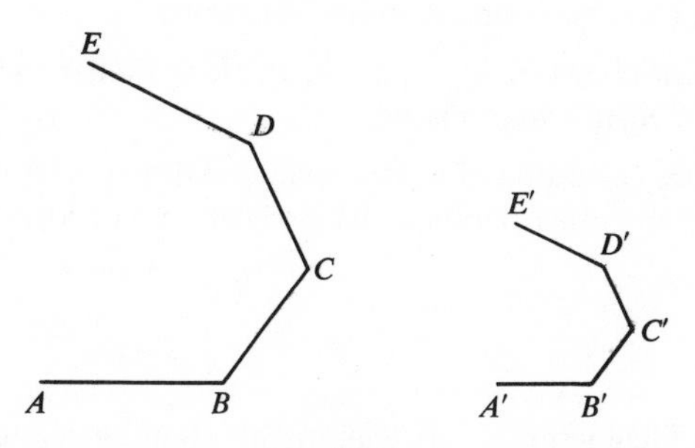

Fig. III, 3

To illustrate the broadened concept of similarity achieved by this definition, let us consider some figures other than polygons. Taking circles first, we shall show that all circles are similar. Suppose c_1, c_2 are circles with centers O_1, O_2 and radii r_1, r_2, respectively. Let T be the translation $O_1 \to O_2$. Then $T(c_1) = c$, where c is the circle with center O_2 and radius r_1. Let D be the radial transformation with center O_2 and ratio $k = r_2/r_1$. Then $D(c) = c_2$. Since TD, a similarity transformation, thus maps c_1 on c_2, these circles are similar by 5.1.

As another example let us prove that all parabolas are similar. Let p_1, p_2 be parabolas with vertices V_1, V_2, respectively. The translation $V_1 \to O$ (the origin of coordinates) sends p_1 into a parabola with vertex O. Some rotation about O sends the latter parabola into a parabola p_1' of the type $y = ax^2$, $a > 0$. The resultant of these transformations is a motion M_1 that sends p_1 into p_1'. In like manner, p_2 can be sent by a motion M_2 into a parabola p_2' of the type $y = ax^2$, $a > 0$. There is a radial transformation D with center O which maps p_1' on p_2' (§3, Ex. 10). It follows that $M_1DM_2^{-1}$, a similarity transformation, sends p_1 into p_2. These parabolas are therefore similar by 5.1.

Circles and parabolas are conics with eccentricity 0 and 1, respectively. In Exercise 2 the student will show that all ellipses with the same eccentricity are similar, and in Exercise 3 that all hyperbolas with the same eccentricity are similar. Since parabolas, ellipses, and hyperbolas are known collectively as *nondegenerate conics*, we can state the theorem:

5.3 *All nondegenerate conics with the same eccentricity are similar.*

EXERCISES

1. Prove that all segments are similar.

2. Prove that all ellipses with the same eccentricity are similar. (*Hint.* See §3 Ex. 11.)

3. Prove that all hyperbolas with the same eccentricity are similar. (*Hint.* See §3, Ex. 12.)

4. Prove that any two pairs of points are similar.

5. Is the set of points (0, 0), (1, 0), (0, 1), (1, 2) similar to the set (0, 0), (−2, 0), (0, −2), (−2, −4)? Justify your answer.

6. Is the figure consisting of a parabola, its directrix, and its focus similar to the figure consisting of any other parabola, its directrix, and its focus? Justify your answer.

6. Metric Geometry. It was in the opening section of this book that we first mentioned, in a general way, the Kleinian idea of using groups of transformations as a means of classifying geometric properties and distinguishing different kinds of geometry. Later, in I, §6, when it was possible to be more specific, we indicated the role which geometric invariants play in that idea. Now we can elaborate further.

Corresponding to each property encountered in Euclidean plane geometry there usually are certain 1-1 transformations of the plane into itself which preserve the property. Since it is characteristic of this set of transformations, and no others, to preserve the property, we may reasonably associate the property with this particular set of transformations. Thus, we associate *distance* with *the set of all motions*, for the latter, as we know, are the only 1-1 transformations of the

plane into itself which preserve distance. In like manner we associate *angle-size* **with** *the set of all similarity transformations*, for, as was mentioned at the end of §2, they are the only 1-1 transformations of the plane into itself which always send an angle into another of equal size. Since distance and angle-size are properties involving measurement and these two sets are groups, we refer to the latter as *metric groups*. **Area-size**, by which we mean the absolute value of the measure of an area, is another property in plane geometry which involves measurement. As we know, it is invariant for the group of motions. However, as we shall see in Chapter IV, motions are not the only 1-1 transformations of the plane into itself which preserve area-size. We therefore do not associate area-size with the group of motions, but with a larger, or more inclusive, group, known as the *equiareal group*, or *equiaffine group*. We include this group among the metric groups. No other group besides the three already designated will be called a metric group. By the **metric groups**, then, we mean the group of motions, the similarity group, and the equiareal group.

We can now define a **metric property** to be any property which is preserved by a metric group, but by no larger set of 1-1 transformations of the plane into itself. Thus, *distance, angle-size, and area-size are metric properties*. They are not the only ones, though they are the most basic in plane geometry. The property of being a square, for example, is also metric, for every square goes into a square under a similarity transformation, whereas no larger set of 1-1 transformations of the plane into itself can achieve this, as will be seen in the next chapter. For the same reason, the property of being an equilateral triangle is metric, as is the property of being a circle. The congruence of two figures and the similarity of two figures, likewise, are metric properties. Other examples could easily be given.

As our definition of a metric property suggests, some properties preserved by the metric groups are not metric. The property of being a line, for example, is one of these, for, as we shall see in Chapter IV, the metric groups do not represent all the 1-1 transformations of the plane into itself which send lines into lines. Parallelism of lines is not a metric property either, for it is easily verified that any 1-1 transformation of the plane into itself which sends lines into lines will preserve parallelism. Also nonmetric are the property of a point being between two others and the property of a figure being a parabola. Hearing such properties called nonmetric perhaps comes as a surprise to the student, for he is used to thinking of them in metric terms. However, it is not necessary to think of them in this way. He may be accustomed, for example, to think of parallel lines as lines which are everywhere equidistant, but the definition of parallel lines is simply that they are (coplanar) lines which do not meet, a statement involving no measurement.

Having defined metric properties in terms of the metric groups, we may now describe as **metric geometry** that portion of plane geometry concerned primarily with metric properties. The remaining portion may then be called **nonmetric geometry**. For the most part, the statements comprising metric geometry are easily recognized since they are the axioms, definitions, and theorems dealing directly or indirectly with distance, angle (meaning angle-size), and area (meaning area-size). Among the axioms of this type are two which were stated by Euclid: *all right angles are congruent*, and *a circle can be drawn with any center and any radius.* Among the definitions belonging to metric geometry are those of a circle, equilateral triangle, isosceles triangle, and angle-bisector. The various area formulas and the propositions on congruent and similar triangles may serve to illustrate the theorems belonging to metric geometry.

Some of the statements belonging to nonmetric geometry, too, are easy to recognize. We may mention, for example, the *axioms* which state that two points determine a line and that there is not more than one parallel to a line through a point outside the line, the *theorem* stating that two lines parallel to the same line are parallel to each other, and the *definitions* of parallel lines, of a triangle, and of a polygon. On the other hand, there are quite a few statements of importance whose nonmetric character can be appreciated only after a study of nonmetric geometry. We begin this study in the next chapter.

EXERCISES

1. Mention a property which is preserved by only one of the metric groups.

2. Mention several properties each of which is preserved by only two of the metric groups.

3. If possible, mention a property that is preserved by all three metric groups.

4. Prove that a 1-1 mapping of the plane onto itself which sends lines into lines necessarily preserves parallelism.

5. Prove that a 1-1 mapping of the plane onto itself which sends lines into lines necessarily sends two intersecting lines into two intersecting lines. Hence show that the property of being a figure consisting of three lines meeting so as to form a triangle is not metric.

6. Show that the transformation $x' = 2x, \quad y' = y$ maps the plane onto itself in a 1-1 way and sends lines into lines, but is not a similarity or an equiareal transformation. Hence prove that the property of being a line is not metric.

7. Show that the transformation of Ex. 6 always sends a parabola into a parabola. Hence prove that the property of being a parabola is not metric.

CHAPTER IV

Affine Transformations

1. Introduction. As we have seen, some of the properties preserved by the metric groups are also preserved by still larger sets of transformations, and hence are called nonmetric rather than metric. Since there are a number of these more inclusive sets of transformations, we may expect to find that nonmetric properties are of various types. This was already noted in Chapter I, where we referred in a general way to affine, projective, and topological properties and to the three nonmetric geometries which are concerned with them, respectively (I, §§1, 6).

The road to take in starting the exploration of nonmetric properties lies straight ahead, for we have only to continue with the study of linear transformations begun in the preceding two chapters. There we were concerned only with linear transformations of a special kind, namely,

$$\begin{cases} x' = ax + by + m \\ y' = -bx + ay + n \end{cases} \quad \text{and} \quad \begin{cases} x' = ax + by + m \\ y' = bx - ay + n, \end{cases}$$

representing the direct and opposite similarities, respectively, and, in particular, the direct and opposite motions when $a^2 + b^2 = 1$. Now our concern is with the set of *all* linear transformations

$$\begin{cases} x' = ax + by + m \\ y' = cx + dy + n, \end{cases} \qquad \begin{vmatrix} a & b \\ c & d \end{vmatrix} \neq 0,$$

where the coefficients are any numbers meeting the condition stated at the right. Among other things, we shall determine the geometric properties that are associated with this set of transformations. Since those properties are commonly called *affine properties*, we shall also refer to the transformations themselves as **affine transformations.**

When coordinates are used for points designated as $P_1, P_2, P_3, \ldots$ or $P_1', P_2', P_3', \ldots$, we shall understand that (x_i, y_i) are the coordinates of P_i, and (x_i', y_i') are the coordinates of P_i' for the various values of i.

2. The Affine Group. We have already shown that an affine transformation maps the plane on itself in a 1-1 way and that its inverse is also affine (I, §7). If

$$\begin{cases} x' = ax + by + m \\ y' = cx + dy + n \end{cases} \quad \text{and} \quad \begin{cases} x'' = \alpha x' + \beta y' + \mu \\ y'' = \gamma x' + \delta y' + \nu \end{cases}$$

are two affine transformations, then their resultant is

$$\begin{cases} x'' = (a\alpha + c\beta)x + (b\alpha + d\beta)y + (m\alpha + n\beta + \mu) \\ y'' = (a\gamma + c\delta)x + (b\gamma + d\delta)y + (m\gamma + n\delta + \nu). \end{cases}$$

This transformation is also affine. For it has the form of an affine transformation, and the determinant

$$\begin{vmatrix} a\alpha + c\beta & b\alpha + d\beta \\ a\gamma + c\delta & b\gamma + d\delta \end{vmatrix}$$

is not zero, being the product of the nonzero determinants,

$$\begin{vmatrix} \alpha & \beta \\ \gamma & \delta \end{vmatrix} \quad \text{and} \quad \begin{vmatrix} a & b \\ c & d \end{vmatrix},$$

of the given transformations (see Appendix, 2.6). This proves the theorem:

2.1 *The resultant of two affine transformations is an affine transformation whose determinant is the product of their determinants.*

Combining 2.1 with the fact that the inverse of an affine transformation is an affine transformation, we obtain:

2.2 *The set of all affine transformations is a group.*

This group is called the **affine group**. Among its important subgroups are the three metric groups, of which we have already encountered the motions and the similarities. The equiareal group is considered later in the chapter.

In the discussion of the transformation

$$(1) \qquad \begin{cases} x' = ax + by + m \\ y' = cx + dy + n \end{cases}$$

given in I, §7, we saw that the condition $ad - bc \neq 0$ guarantees that the transformation maps the plane on itself in a 1-1 way. Examples readily show what happens when this condition is not met, and this knowledge is useful at times. Consider

$$\begin{cases} x' = 2x + 3y + 1 \\ y' = 6x + 9y + 4. \end{cases} \tag{2}$$

Since the coefficients of x and y are in proportion, if we multiply the first equation by 3, and subtract the result from the second, x and y are eliminated together, giving

$$y' - 3x' = 1. \tag{3}$$

All image points therefore have coordinates satisfying this equation. The locus of the latter being a line, we see that the transformation sends the entire plane into points of a single line. Also, the transformation is not 1-1. The point (1, 4), for example, is the image of (0, 0), but it is also the image of every point on the line $2x + 3y = 0$. The proof of the following general statement can now be left as an exercise:

2.3 *If $ad - bc = 0$, the transformation* (1) *does not map the plane on itself and is not* 1-1. *In particular, if a, b, c, d are zero, all the points of the plane go into a single point; otherwise they go into points of a single line.*

The transformation (1) is often called an affine transformation regardless of the value of $ad - bc$. If the latter is zero, the transformation is then said to be *singular*; if it is not zero, the transformation is called *nonsingular*. In this book we shall continue to use "affine transformation" to mean a nonsingular transformation.

We now use 2.3 in proving the important theorem:

2.4 *There is a unique affine transformation that sends three noncollinear points P_1, P_2, P_3 into three noncollinear points P'_1, P'_2, P'_3, respectively.*

Proof. We must show that a unique set of values of a, b, c, d, m, n, where $ad - bc \neq 0$, can be found such that the resulting transformation (1) sends P_1, P_2, P_3 into P'_1, P'_2, P'_3, respectively. Substituting the coordinates of these points in (1) gives

$$\begin{aligned}
x_1a + y_1b + m \qquad\qquad\qquad &= x'_1 \\
x_2a + y_2b + m \qquad\qquad\qquad &= x'_2 \\
x_3a + y_3b + m \qquad\qquad\qquad &= x'_3 \\
x_1c + y_1d + n &= y'_1 \\
x_2c + y_2d + n &= y'_2 \\
x_3c + y_3d + n &= y'_3 .
\end{aligned}$$

This is a system of six first-degree equations in the six unknowns a, b, c, d, m, n, and its determinant is

$$\begin{vmatrix} x_1 & y_1 & 1 & 0 & 0 & 0 \\ x_2 & y_2 & 1 & 0 & 0 & 0 \\ x_3 & y_3 & 1 & 0 & 0 & 0 \\ 0 & 0 & 0 & x_1 & y_1 & 1 \\ 0 & 0 & 0 & x_2 & y_2 & 1 \\ 0 & 0 & 0 & x_3 & y_3 & 1 \end{vmatrix}.$$

This determinant is not zero, for according to Laplace's development of a determinant it is equal to

$$\begin{vmatrix} x_1 & y_1 & 1 \\ x_2 & y_2 & 1 \\ x_3 & y_3 & 1 \end{vmatrix} \quad \begin{vmatrix} x_1 & y_1 & 1 \\ x_2 & y_2 & 1 \\ x_3 & y_3 & 1 \end{vmatrix},$$

whose factors are not zero since P_1, P_2, P_3 are noncollinear (IV, 8.1). The system therefore has a unique solution (see Appendix, 5.1). Also, $ad - bc$ is not 0. For if it were, 2.3 shows that we could not have noncollinear image points, and this would contradict the assumption that P_1', P_2', P_3' are noncollinear.

EXERCISES

1. Find the inverse of the transformation $x' = 2x - y + 1, \quad y' = x + 3y - 2$ in explicit form and show that it is affine.

2. Find the resultant of the transformation $x' = 3x + y - 4, \; y' = -2x + y + 5$ and the given transformation of Ex. 1, in that order, and show that it is affine.

3. Do affine transformations obey the commutative law? the associative law? Justify your answers.

4. Show that $x' = 2x, \quad y' = y$ is an affine transformation other than a similarity and that it does not preserve angle-size.

5. Show that the transformation (2) maps all points of the line $2x + 3y + a = 0$, where a is any constant, on the point $(1-a, 4-3a)$.

6. Prove 2.3.

7. Deduce from 2.2 and 2.4 that affine transformations preserve collinearity.

8. Verify that there are exactly six affine transformations which map the vertices of a given triangle on those of another.

3. A Key Affine Transformation. Just as we were able to understand all similarities from an acquaintance with the special kind we called radial transformations, so we can understand all affine transformations from a knowledge of certain simple ones, which we call **primitive transformations**. The equations of the latter are either $x' = x$, $y' = cx + dy$, where $d \neq 0$, or $x' = ax + by$, $y' = y$, where $a \neq 0$. Clearly the properties of one of these two types of primitive transformations differ from those of the other only in that the roles of x and y are interchanged. We therefore study only the first, which we denote by T, and state our results in such a way that they also hold for the second. It will be helpful to exhibit T more clearly as

$$\begin{cases} x' = x \\ y' = cx + dy, \end{cases} \qquad d \neq 0, \tag{1}$$

and to give its implicit form,

$$\begin{cases} x = x' \\ y = (-cx' + y')/d. \end{cases} \tag{2}$$

One should note that T has a determinant of value d, and that it is not a similarity except when it is the identity or the reflection in the x-axis. We exclude these cases from the following discussion by assuming that $c \neq 0$ whenever $d = \pm 1$.

Lines. Let g be any line. If it is vertical (that is, perpendicular to the x-axis), T sends each point of g into a point of g since $x' = x$. Conversely, each point of g is the image of a point of g. Thus each vertical line is its own image.

If g is not vertical, its equation is

$$y = Ax + B. \tag{3}$$

Using (2) we find that T transforms (3) into

$$y' = (Ad + c)x' + Bd, \tag{4}$$

which represents a nonvertical line g'. Hence each point of g goes into a point of g'. Conversely, let $P'(x', y')$ be a point of g'. Then x', y' satisfy (4). If we substitute for x' and y' in (4), using (1), we obtain an equation in x, y, the coordinates of the point P whose image is P'. This equation is precisely (3). Hence P is on g, that is, each point of g' is the image of a point of g. Thus, the image of each nonvertical line is a nonvertical line.

Two distinct lines g, h cannot have the same image, otherwise each point of the latter would be the image of a point of g and of a point of h, which is impossible. If g, h meet in P, and $T(P) = P'$, the image lines must meet in P'. If g, h are parallel, so are their images; for if the latter met in Q', then g, h would meet in the original of Q', which is impossible. It follows that the images of the vertices of a parallelogram are themselves the vertices of a parallelogram.

We have now proved:

3.1 *In a primitive transformation the image of every line is a line, distinct lines having distinct images, and a* 1-1 *correspondence is thus established* *among the lines of the plane. Intersecting lines go into intersecting lines, and parallel lines into parallel lines. The property of being the vertices of a parallelogram is therefore preserved.*

We have, of course, also proved certain additional facts about vertical and nonvertical lines in connection with the particular primitive transformation T.

Sense on a Line. It is customary in analytic geometry to direct vertical lines upward, that is, to make the upward direction the positive one. There being no fixed procedure concerning nonvertical lines, let us agree that their positive directions are toward the right. The *sense of an ordered point-pair* P_1, P_2 on a directed line means the sense of the direction from P_1 to P_2 on the line. We now prove:

3.2 *If two pairs of points on the same line or on parallel lines agree in sense, so do their image pairs under primitive transformations, and if they differ in sense, so do the image pairs.*

Let P_1, P_2 and P_3, P_4 be the original pairs. Suppose, first, that each pair is on a vertical line. Then each image pair is also on a vertical line. In this case the sense of the ordered point-pair P_1, P_2 is given by the sign of $y_2 - y_1$, that of the ordered pair P_3, P_4 by the sign of $y_4 - y_3$, and similarly for the image pairs. Thus, if $y_2 - y_1 > 0$, the sense of P_1, P_2 is positive, or upward. To check this we note that P_2 is above P_1 since $y_2 > y_1$. Denoting the image points and their ordinates by primes, we obtain, on using (1) and the fact that $x_1 = x_2 = x_3 = x_4$,

$$\frac{y_2' - y_1'}{y_4' - y_3'} = \frac{c(x_2 - x_1) + d(y_2 - y_1)}{c(x_4 - x_3) + d(y_4 - y_3)} = \frac{y_2 - y_1}{y_4 - y_3}.$$

If the senses of P_1, P_2 and P_3, P_4 are the same, the ratio on the right-hand side is positive. Then the ratio on the left-hand side is positive, and hence P_1', P_2' and P_3', P_4' have the same sense. Similar reasoning disposes of the case in which P_1, P_2 and P_3, P_4 have opposite senses.

Suppose now that the original pairs are on a nonvertical line or are on two such parallel lines. Then the same is true, respectively, of the image pairs. The sense of P_1, P_2 is given by the sign of $x_2 - x_1$, that of P_3, P_4 by the sign of $x_4 - x_3$, and similarly for the image pairs. Thus if $x_2 - x_1 > 0$, the sense of P_1, P_2 is positive. To check this we note that P_2 is to the right of P_1 since $x_2 > x_1$. On using (1) we get

$$\frac{x_2' - x_1'}{x_4' - x_3'} = \frac{x_2 - x_1}{x_4 - x_3}.$$

Reasoning as above we again substantiate 3.2.

Distance. Not being a similarity, T does not multiply all distances by the same amount, but changes them in some less uniform way. To determine what this is, let P_1, P_2 be two points, with images P_1', P_2'. By the distance formula and equations (1),

$$(5) \qquad (P_1P_2)^2 = (x_1 - x_2)^2 + (y_1 - y_2)^2,$$

$$(6) \qquad (P_1'P_2')^2 = (x_1' - x_2')^2 + (y_1' - y_2')^2$$
$$= (x_1 - x_2)^2 + [c(x_1 - x_2) + d(y_1 - y_2)]^2.$$

If P_1, P_2 are on a vertical line, then P_1', P_2' are on this line (since each vertical line is its own image), and the four abscissas are equal. Hence (5), (6) give

$$(P_1P_2)^2 = (y_1 - y_2)^2,$$

$$(P_1'P_2')^2 = d^2(y_1 - y_2)^2 = d^2 \cdot (P_1P_2)^2,$$

$$P_1'P_2' = |d| \cdot P_1P_2.$$

Thus, all distances on vertical lines are multiplied by the same constant $|d|$.

If P_1, P_2 are on a nonvertical line of slope m, then $m = (y_1 - y_2)/(x_1 - x_2)$. Equations (5), (6) now give

$$(P_1P_2)^2 = (x_1 - x_2)^2(1 + m^2),$$

$$(P_1'P_2')^2 = (x_1 - x_2)^2\,[1 + (c + dm)^2],$$

$$\frac{(P_1'P_2')^2}{(P_1P_2)^2} = \frac{1 + (c + dm)^2}{1 + m^2},$$

$$P_1'P_2' = \sqrt{\frac{1 + (c + dm)^2}{1 + m^2}} \cdot P_1P_2.$$

Thus, all distances on lines of slope m are multiplied by a constant whose value depends only on m.

We have therefore proved the theorem:

3.3 *A primitive transformation multiplies all distances on the same line or on parallel lines by the same amount, that is, by the same positive constant.*

For example, on vertical lines the constant is $|d|$, as we saw above; on horizontal lines it is $\sqrt{1 + c^2}$, since $m = 0$; and on lines inclined at 45° it is $\sqrt{\tfrac{1}{2}[1 + (c + d)^2]}$.

Ratios of Distances. Several invariants can be noted as a result of the preceding discussions. First, it is clear from 3.3 that the ratio of two distances on the same line or on parallel lines is preserved. In particular, then, the equality of two such distances is preserved. This agrees with the previously noted fact that the vertices of a parallelogram go into the vertices of a parallelogram.

On applying 3.2 we see that it is also true that the ratio of two *directed* distances on the same line or on parallel lines is invariant. In particular, *ratio of division* (II, §3, Ex. 2) is preserved. For let P, P_1, P_2 be distinct collinear points. Then

$$r = \frac{P_1P}{PP_2}$$

is the ratio in which P divides the point-pair P_1, P_2 (or the segment $\overline{P_1P_2}$). The image points P', P_1', P_2' are then also distinct and collinear, and

$$r' = \frac{P_1'P'}{P'P_2'}$$

is the ratio in which P' divides $\overline{P_1'P_2'}$. Since the distances involved here are directed distances, P_1P is positive or negative according as the sense of P_1, P is positive or negative, and r is positive or negative according as P_1, P and P, P_2 agree or disagree in sense. It is understood, of course, that P, P_1, P_2 are distinct. Since the ratios of the corresponding *undirected* distances are equal, as seen above, we have

$$\frac{|P_1P|}{|PP_2|} = \frac{|P_1'P'|}{|P'P_2'|},$$

or

$$\left|\frac{P_1P}{PP_2}\right| = \left|\frac{P_1'P'}{P'P_2'}\right|.$$

Thus the absolute values of r and r' are the same. The signs of r and r' are also the same in view of 3.2. Hence $r = r'$.

If, for example, P is the midpoint of $\overline{P_1P_2}$, in which case $r = 1$, then $r' = 1$ and P' is the midpoint of $\overline{P_1'P_2'}$.

3.4 *Primitive transformations preserve the ratio of any two directed distances, and also of any two undirected distances, on the same line or on parallel lines. In particular, they send equal distances into equal distances and preserve ratio of division, thus sending midpoints into midpoints.*

EXERCISES

1. Find the ratio in which P divides $\overline{P_1P_2}$ if (a) P_2 is midway between P_1 and P; (b) P_1 is midway between P and P_2; (c) P is $\frac{1}{3}$ of the way from P_1 to P_2.

2. What can be said about the sign and absolute value of the ratio r in which P divides $\overline{P_1P_2}$ if (a) (P_1PP_2); (b) (P_1P_2P); (c) (PP_1P_2)?

3. If P recedes indefinitely far from P_1, P_2, what happens to r in Ex. 2(b)? in Ex. 2(c)?

4. Find the ratio in which $P(1, -1)$ divides the point-pair $P_1(0, -3)$, $P_2(4, 5)$.

5. Subject the points $A(0, -6)$, $B(3, 0)$, $C(3, 6)$, $D(0, 0)$ to the transformation $x' = x$, $y' = 2x + 4y$ and verify 3.2, 3.3, 3.4.

6. It is shown in analytic geometry, and easily verified, that the formula for the ratio r in which the point $P(x, y)$ divides the point-pair P_1, P_2 is $r = (x - x_1)/(x_2 - x)$ if the points are on a nonvertical line, and $r = (y - y_1)/(y_2 - y)$ if they are on a nonhorizontal line. Show that r is invariant for primitive transformations.

7. Using the formulas of Ex. 6, show that there is a unique point P which divides a point-pair P_1, P_2 in a specified ratio r, not 0, 1, or -1, by showing that the coordinates (x, y) of P are

$$x = \frac{x_1 + rx_2}{1 + r}, \qquad y = \frac{y_1 + ry_2}{1 + r}.$$

Also verify that the point with coordinates (x, y) as given by these formulas is collinear with the points (x_1, y_1) and (x_2, y_2).

8. Prove each of the following for the primitive transformation $x' = ax + by$, $y' = y$, where $a \neq 0$:

(a) 3.1; (b) 3.2; (c) 3.3.

9. Prove that $x' = x$, $y' = 2x + 4y$ leaves the line $2x + 3y = 0$ pointwise fixed. Generalize this by showing that every primitive transformation leaves a certain line pointwise fixed.

10. How do points "move" under the special primitive transformation $x' = x$, $y' = ky$, where $k > 0$? (This transformation is called a *one-dimensional strain* by analogy with the behavior of the cross sections of elastic objects when the latter are subjected to certain forces. More generally, a one-dimensional strain is defined as follows. Let g be any line, and k any positive constant. Each point of g goes into itself, and any other point P of the plane goes into the point P' on the same side of g as P such that line PP' is perpendicular to g, and $OP' = k \cdot OP$, where O is the intersection of g and line PP'.)

11. Prove that the primitive transformation T given by (1) preserves distance in just one direction when $c = 0$, and in just two directions when $c \neq 0$. More generally, show that distance is multiplied by the same constant in at most two directions. (As in §3, assume that T is not a similarity.)

4. A Key Affine Transformation (continuation). For convenient reference we restate the equations of the primitive transformation T:

$$(1) \qquad \begin{cases} x' = x \\ y' = cx + dy, \end{cases} \qquad d \neq 0,$$

$$(2) \qquad \begin{cases} x = x' \\ y = (-cx' + y')/d. \end{cases}$$

The Between-relation. The invariance of the ratio of division $r = P_1P/PP_2$ has important consequences. When P is between P_1 and P_2, P_1P and PP_2 have the same sign, so that $r > 0$. For no other position of P is this true. Hence, if $T(P_1, P_2) = P_1', P_2'$, each point between P_1, P_2 goes into a point between P_1', P_2'. Thus T preserves the between-relation.

Conversely, each point P' between P_1', P_2' is the image of a point between P_1, P_2. For let $T(P) = P'$. Then $T^{-1}(P') = P$. Inspection of (2), which is the explicit form of T^{-1}, shows that T^{-1} is a primitive transformation of the same type as T. Hence T^{-1} preserves the between-relation, too, so that P is between P_1, P_2. We may now conclude that T maps the complete segment $\overline{P_1P_2}$ onto the complete segment $\overline{P_1'P_2'}$. Using these results, the student should show that ray $P_1'P_2'$ is the image of ray P_1P_2.

It now follows that the image of an angle is an angle, that the image of a triangle is a triangle (vertices going into vertices, sides into sides, interior into interior), and, more generally, that an n-sided polygon always goes into an n-sided polygon. Also, on using 3.2 and 3.4 we see that two segments which are on the same line or on parallel lines, and which agree in length and direction, must go into two segments which possess these same properties. In other words, the image of a pair of equal vectors is a pair of equal vectors.

4.1 *Primitive transformations preserve the between-relation, and hence the properties of being a segment, ray, angle, half-plane, n-sided polygon, or pair of equal vectors.*

Area. It is shown in analytic geometry that the area A of a triangle with vertices P_1, P_2, P_3 is given by the formula

$$A = \frac{1}{2}\begin{vmatrix} x_1 & y_1 & 1 \\ x_2 & y_2 & 1 \\ x_3 & y_3 & 1 \end{vmatrix},$$

and that A is positive or negative according as the order of the vertices P_1, P_2, P_3 is counterclockwise or clockwise. If $T(P_1, P_2, P_3) = P_1', P_2', P_3'$, then for the area, A', of triangle $P_1'P_2'P_3'$ we have

$$A' = \frac{1}{2}\begin{vmatrix} x_1' & y_1' & 1 \\ x_2' & y_2' & 1 \\ x_3' & y_3' & 1 \end{vmatrix}.$$

Using (1) we get

$$A' = \frac{1}{2}\begin{vmatrix} x_1 & cx_1 + dy_1 & 1 \\ x_2 & cx_2 + dy_2 & 1 \\ x_3 & cx_3 + dy_3 & 1 \end{vmatrix}.$$

Consequently (see Appendix, 2.5)

$$A' = \frac{1}{2}\begin{vmatrix} x_1 & cx_1 & 1 \\ x_2 & cx_2 & 1 \\ x_3 & cx_3 & 1 \end{vmatrix} + \frac{1}{2}\begin{vmatrix} x_1 & dy_1 & 1 \\ x_2 & dy_2 & 1 \\ x_3 & dy_3 & 1 \end{vmatrix} = \frac{1}{2}c\begin{vmatrix} x_1 & x_1 & 1 \\ x_2 & x_2 & 1 \\ x_3 & x_3 & 1 \end{vmatrix} + \frac{1}{2}d\begin{vmatrix} x_1 & y_1 & 1 \\ x_2 & y_2 & 1 \\ x_3 & y_3 & 1 \end{vmatrix}.$$

Since the next to last of these determinants has two identical columns, its value is 0. Hence

$$A' = \tfrac{1}{2}d \begin{vmatrix} x_1 & y_1 & 1 \\ x_2 & y_2 & 1 \\ x_3 & y_3 & 1 \end{vmatrix} = dA. \tag{3}$$

Since d is the value of the determinant of T, we have proved:

4.2 *A primitive transformation multiplies the areas of all triangles by a constant factor, which is equal to the determinant of the transformation. It therefore preserves the ratio of the areas of any two triangles.*

Thus we see that similarity transformations are not the only transformations which multiply all triangular areas by the same amount. It is to be noted, in particular, that a primitive transformation with determinant 1 or -1 preserves the absolute value of the area of every triangle. Thus, motions are not the only transformations which leave area-size invariant.

From (3) we see that A and A' have the same signs if $d > 0$, and opposite signs if $d < 0$. Hence the sense of order of the vertices of a triangle is preserved if $d > 0$, and reversed if $d < 0$. Suppose, for example, that $A = 3$, in which case the order of P_1, P_2, P_3 is counterclockwise. If $d = 2$, then $A' = 6$ and the order of P_1', P_2', P_3' is counterclockwise; if $d = -2$, then $A' = -6$ and the order of these points is clockwise. Now, as diagrams readily suggest, the preservation or reversal of the order of the vertices of a triangle goes hand in hand with the preservation or reversal, respectively, of the sense of its angles (see II, §4, Ex. 19). Thus, we have the theorem:

4.3 *A primitive transformation is direct or opposite according as its determinant is positive or negative.*

Conic Sections. Every conic section has an equation of the form

$$Ax^2 + Bxy + Cy^2 + Dx + Ey + F = 0, \tag{4}$$

where A, B, C are not all zero. Conversely, the locus of this equation (whenever there is a locus) is always a conic section, the particular type depending on the value of the so-called *indicator*, $B^2 - 4AC$. Thus the conic is hyperbolic (a hyperbola or two intersecting lines), parabolic (a parabola or two parallel lines, which may coincide), or elliptic (an ellipse or a single point) according to whether the indicator is positive, zero, or negative, respectively.

Substituting equations (2) into (4) gives

$$A'x'^2 + B'x'y' + C'y'^2 + D'x' + E'y' + F' = 0, \tag{5}$$

where

$$A' = Ad^2 - Bcd + Cc^2, \qquad B' = Bd - 2Cc, \qquad C' = C,$$
$$D' = Dd^2 - Ecd, \qquad E' = Ed, \qquad F' = Fd^2.$$

Hence each point on the locus of (4) goes into a point on the locus of (5).

Conversely, each point on the latter locus is the image of a point on the locus of (4), for it can be verified that substitution of equations (1) into (5) reduces (5) to (4). Thus, the locus of (5) is the image of the locus of (4). This algebraic work also shows that the locus of (4) is the image of the locus of (5) under the inverse transformation T^{-1}, for (1) and (2) are, of course, also the equations of T^{-1}, with (x', y') representing the original point. It follows that A', B', C' are not all zero. For if they were, (5) would be a first-degree equation, and it could not then be transformed by the affine transformation T^{-1} into (4), which is a second-degree equation. Hence the locus of (5) is a conic section.

The indicator of this conic section is $B'^2 - 4A'C'$. If we substitute in it the expressions for A', B', C' given above, and simplify, we obtain

$$B'^2 - 4A'C' = d^2(B^2 - 4AC).$$

Since $d \neq 0$, we know that $d^2 > 0$. It follows that $B'^2 - 4A'C'$ is positive, negative, or zero when $B^2 - 4AC$ is positive, negative, or zero, respectively. Hence (5) is the same type of conic as (4). Also, since T is 1-1 and sends lines into lines, (5) is degenerate (consisting of two lines, which may coincide, or a single point) only if (4) is. Hence T sends parabolas, ellipses, and hyperbolas into parabolas, ellipses, and hyperbolas, respectively. Eccentricity is usually not preserved, as is to be expected from a transformation which is not a similarity. The image of a circle, for example, is always an ellipse other than a circle. These statements will be verified in exercises.

4.4 *The image of a conic section under a primitive transformation is a conic section of the same type, but not necessarily of the same eccentricity, and is degenerate only if the original conic is.*

Curves of Higher Degree. Having seen that lines go into lines, and conic sections into conic sections, one may wonder whether primitive transformations always preserve the degree of a curve. This is the case, as we now show.

Let c be a curve of degree n, that is, the locus of an equation

$$f(x, y) = 0, \tag{6}$$

where $f(x, y)$ is a polynomial of degree n in x, y. We need only consider the case $n > 2$. Exactly analogous to the situation for lines and conic sections, the image of c is the locus c' of the equation

$$F(x', y') = 0 \tag{7}$$

obtained from (6) by substitution of (2), the implicit form of T. Since (6) is of degree n and equations (2) are of the first degree, $F(x', y')$ is clearly a polynomial in x', y' of degree not greater than n. On the other hand, suppose that the degree of $F(x', y')$ were less than n. Since c is the image of c' under T^{-1}, substitution of the equations of T^{-1} into (7) changes the latter to (6). Under our supposition, this means that an equation of lower degree than n is changed into one of degree n. This is impossible since the equations of T^{-1} in implicit form are equations (1), which are of the first degree. It follows that $F(x', y')$ is of degree n.

4.5 *The image of a curve of degree n (≥ 1) under a primitive transformation is a curve of degree n.*

EXERCISES

1. In §3, Ex. 5, certain points were subjected to the transformation $x' = x$, $y' = 2x + 4y$. (a) Verify that the image of each pair of equal vectors determined by these points is a pair of equal vectors; (b) compare the sense and area of the various image triangles with those of the originals; (c) compare the sense of each image angle with that of its original.

2. Do what is asked in Ex. 1, but use the transformation $x' = x$, $y' = 2x - 4y$.

3. Prove that in a primitive transformation: (a) the image of a ray is a ray; (b) the image of a half-plane is a half-plane.

4. How do primitive transformations affect the area (a) of a quadrilateral? (b) of an n-sided polygon? (c) of a circle?

5. Show that the primitive transformation $x' = x$, $y' = \frac{1}{2}y$ sends the ellipse $b^2x^2 + a^2y^2 = a^2b^2$ into an ellipse and the hyperbola $b^2x^2 - a^2y^2 = a^2b^2$ into a hyperbola, but changes the eccentricity in each case.

6. Apply the transformation T to the following conics and verify 4.4:
(a) $x^2 + y^2 = 1$; (b) $x^2 - y^2 = 0$; (c) $x^2 + y^2 = 0$; (d) $y^2 - 1 = 0$.

7. Prove the following for the primitive transformation $x' = ax + by$, $y' = y$, where $a \neq 0$:
(a) 4.2; (b) 4.4.

8. Find the image of the curve $y = x^3$ under the transformation $x' = x$, $y' = 4x - y$ and draw both curves.

9. Find the image of the curve $xy^2 - x^2 - 5xy + 6x = 0$ in the transformation of Ex. 8, and draw both curves.

10. Prove that a primitive transformation which is not a similarity (a) never sends a circle into a circle; (b) never sends a triangle into a similar triangle. (*Hint.* See §3, Ex. 11.)

11. Show that the set of all primitive transformations is not a group.

12. Is the set of all primitive transformations of the form (1), where $d \neq 0$, a group? Justify your answer.

5. Resolution of the General Affine Transformation. Let us now see why it is that primitive transformations are a key to all affine transformations.

An affine transformation

$$(1)\qquad \begin{cases} x' = ax + by + m \\ y' = cx + dy + n \end{cases}$$

is the resultant of the transformations

$$(2)\qquad \begin{cases} x' = ax + by \\ y' = cx + dy \end{cases} \quad \text{and} \quad \begin{cases} x'' = x' + m \\ y'' = y' + n, \end{cases}$$

in that order. The second of these is a translation. The first can be resolved into the primitive transformations

$$(3)\qquad \begin{cases} x' = ax + by \\ y' = y \end{cases} \quad \text{and} \quad \begin{cases} x'' = x' \\ y'' = \dfrac{c}{a}x' + \dfrac{ad - bc}{a}y' \end{cases}$$

if $a \neq 0$, and into the primitive transformations

$$(4)\qquad \begin{cases} x' = x - y \\ y' = y \end{cases}, \quad \begin{cases} x'' = x' \\ y'' = x' + y', \end{cases}$$

$$\begin{cases} x''' = -x'' + y'' \\ y''' = y'' \end{cases}, \quad \begin{cases} x^{iv} = bx''' \\ y^{iv} = y''' \end{cases}, \quad \begin{cases} x^{v} = x^{iv} \\ y^{v} = \dfrac{d}{b}x^{iv} + cy^{iv} \end{cases}$$

if $a = 0$ (in which case $b \neq 0$). This proves the theorem:

5.1 *Every affine transformation can be resolved into primitive transformations and a translation, in that order.*

The preceding analysis being entirely general, one or more of the transformations mentioned in 5.1 can be the identity. Thus, in the extreme case in which $b = c = m = n = 0$ and $a = d = 1$, every transformation in (1), (2), (3) is the identity.

There are other ways of analyzing the general affine transformation besides that described in 5.1. It can be shown*, for example, that every affine transformation is the resultant of similarities and very simple affine transformations known as *one-dimensional strains* (see §3, Ex. 10). This is analogous to the situation encountered in Chapter III, where we saw that every similarity is the resultant of motions and the simple similarities we called radial transformations.

The following is a corollary of 5.1. We leave its proof as an exercise.

5.2 *An affine transformation multiplies the areas of all triangles by the value of its determinant.*

---

* See W. F. Osgood and W. C. Graustein, *Plane and Solid Analytic Geometry*, p. 343.

EXERCISES

1. Resolve $x' = 2x - 3y$, $y' = 5x + 4y$ into primitive transformations.

2. Resolve $x' = 3y$, $y' = 5x + 4y$ into primitive transformations.

3. Verify that the transformations (3) and (4) are all affine.

4. Find the product of the determinants (a) of the transformations (2); (b) of the transformations (3); (c) of the transformations (4). Hence prove 5.2.

5. When a plane section of a solid object is deformed so that each point (x, y) of the section moves into a new position (x', y') of the section in accordance with the affine equations $x' = ax + by$, $y' = cx + dy$, which are the first equations in (2), the deformation is called a *homogeneous strain*. (Slight deformations of elastic objects are approximately of this type.) Describe the images of the following under a homogeneous strain: (a) a rectangle with one vertex at O; (b) a square with center O; (c) a circle with center O; (d) a triangle and its medians if the latter meet in O.

6. Affine Properties. All the properties of primitive transformations mentioned in §3 and §4 are possessed also by the resultant of two or more such transformations. (The resultant is, of course, affine, but not necessarily a primitive transformation. See §4, Ex. 11.) This is obvious for the properties of collinearity, parallelism, ratio of division, betweenness, and, in fact, for all the properties we have studied except perhaps those stated in 4.2 and 4.3. An example will clarify these two cases. Suppose that T_1, T_2 are primitive transformations with determinants 2 and -3. By 4.2, then, T_1 and T_2 multiply areas by 2 and -3, respectively. Hence their resultant T_1T_2 multiplies areas by -6. Also, the determinant of T_1T_2, being the product of the determinants of T_1 and T_2, is -6. Thus T_1T_2, though not necessarily a primitive transformation, has the property stated in 4.2. It also has the property stated in 4.3, for it is opposite (since T_1 is direct and T_2 is opposite) and has a negative determinant.

Translations, too, possess all the properties of primitive transformations stated in the theorems of §3 and §4. To verify this one need only replace "primitive transformation" in each of these theorems by "translation" and note that the resulting statement is still correct. Some of these statements, however, may sound strange. From 4.3, for example, we obtain: a translation is direct or opposite according as its determinant is positive or negative. Since every translation has a positive determinant, of value 1, and is therefore direct, this statement is correct.

Since every affine transformation is the resultant of primitive transformations and a translation, we have the theorem:

6.1 *All affine transformations have the properties of primitive trans-*

formations as stated in Theorems 3.1 *to* 3.4 *and* 4.1 *to* 4.4, *that is, these theorems remain correct if "primitive transformation" is replaced by "affine transformation."*

Thus, for example, every affine transformation sends a conic into a conic of the same type, is direct or opposite according as its determinant is positive or negative, and multiplies the areas of all triangles by the value of its determinant.

It will be recalled that when metric properties were mentioned in III, §6, several of the properties preserved by metric groups were not included among them because these properties are also preserved by more extensive groups. In the light of 6.1 it is now clear that parallelism, betweenness, collinearity, ratio of division, and the property of being a parabola, just to mention a few examples, were not included because they are preserved by the affine group.

Another thing done in III, §6, can also be checked now. We stated that the area-size of a triangle is preserved by a more extensive group than the group of motions, namely, by the equiareal or equiaffine group, and that it is with this group that the property of area-size is associated. By the **equiaffine group** is meant the set of all affine transformations whose determinants have the value 1 or -1. These transformations are called **equiaffine transformations.** That they constitute a group is easily verified; that they preserve the area-size of a triangle follows from 4.2 and 6.1. The motions are, then, a subgroup of the equiaffine group. A simple example of an equiaffine transformation that is not a motion is $x' = 2x, \quad y' = \frac{1}{2}y$. Equiaffine transformations are, by definition, the only affine transformations which preserve the area-size of a triangle. Also, affine transformations are the only 1-1 mappings of the plane onto itself which carry a triangle into a triangle so that vertices go into vertices, and sides into sides (see §10, Ex. 1c). Hence, *equiaffine transformations are the only* 1-1 *transformations of the plane into itself which send each triangle into a triangle of the same area-size.* This explains why it is reasonable to associate area-size of a triangle with the equiaffine group.

In turning our attention now to *affine properties* let us emphasize that this term will not mean *all* properties preserved by the affine group, just as "metric properties" does not mean *all* properties preserved by the metric groups. To be called "affine" a property must, first of all, be preserved by the affine group. (Hence the ways in which affine transformations change distance and area, interesting and useful though they may be, are not affine properties.) Beyond this we should expect, from remarks made earlier in the book, that the property should not be preserved by any larger set of 1-1 transformations of the plane into itself. It is customary, however, to replace this by the condition that the property should not be preserved by the *projective group*, that is, by the set of all *projective transformations*. These transformations are discussed in Chapter VI, where it is shown that they include the affine transformations as a subgroup.

6.2 Definition. *An affine property is a property that is preserved by all affine transformations, but not by all projective transformations.*

The degree of a curve, for example, is preserved by the affine group, but it is not an affine property since it is also preserved by the projective group, as we shall see later on. The property of being a conic of some special type, such as a parabola, ellipse, or hyperbola, however, is affine, for it is preserved by the affine group but not by the projective group. There are projective transformations, for example, which do not always send an ellipse into an ellipse, and which may, in fact, send each type of conic into a different type. Most of the properties of primitive transformations mentioned in §3 and §4 are affine. To be precise, we have the theorem:

6.3 *The property of being a segment, vector, ray, line, angle, triangle, n-sided polygon, or conic section of some special type is affine. Also affine are the parallelism and the concurrence of lines, the betweenness of points, the ratio of triangular areas, and the ratio of distances (directed or undirected) on the same line or on parallel lines. In particular, ratio of division and the equality of two vectors are affine.*

Another way of stating the first sentence in 6.3 is to say that segments, vectors, lines, parabolas, ellipses, hyperbolas, and so forth, are **affine figures.** Such figures may be thought of as having structures which are preserved by all affine transformations, but not by all projective transformations.

Affine figures, of course, do have metric aspects. Segments, for example, have length, angles have size, triangles have area, a pair of parallel lines are equidistant, and each type of conic has its special distance relations.

Also, it should be noted that special kinds of affine figures are usually not affine, but metric. A right triangle, for example, is not sent into a right triangle by all affine transformations, but only by all similarities, and hence is a metric figure. Isosceles triangles, equilateral triangles, scalene triangles, regular polygons, circles, and rectangular hyperbolas are other examples of metric figures.

As we know, an affine transformation which is a similarity sends all triangles into similar triangles and multiplies distances in all directions by the same amount. What can be said, in this regard, about affine transformations that are not similarities? Theorems 6.1 and 6.3 are helpful in obtaining the following answer to this question.

6.4 *An affine transformation that is not a similarity sends no triangle into a similar triangle (by which is meant that the image sides are not proportional to their originals),* and hence cannot multiply distances in three different directions by the same amount.*

Proof. Suppose that T, an affine transformation not a similarity, sends triangle ABC into the similar triangle $A'B'C'$. Then

$$\frac{A'B'}{AB} = \frac{B'C'}{BC} = \frac{A'C'}{AC} = k, \tag{1}$$

---

* See Ex. 6.

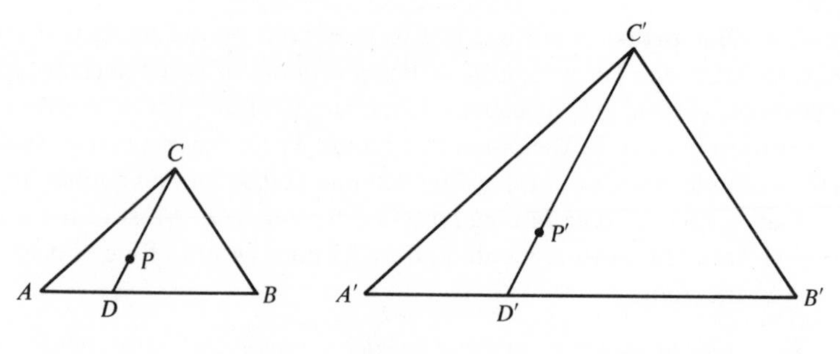

Fig. IV, 1

where $k > 0$. It follows from 6.3 that T multiplies by k all distances in the directions of lines AB, BC, and AC, that is, distances on these lines or on lines parallel to them. Take any point P inside triangle ABC. Its image P' is then inside triangle $A'B'C'$ since T preserves betweenness (II, §2). Extend $\overline{CP}$, $\overline{C'P'}$ to meet $\overline{AB}$, $\overline{A'B'}$ in D, D' (Fig. IV, 1). Then $\overline{C'D'}$, $\overline{A'D'}$, $\overline{D'B'}$ are the images of segments $\overline{CD}$, $\overline{AD}$, $\overline{DB}$ by 6.3. Also, since ratio of division is preserved,

$$\frac{A'D'}{D'B'} = \frac{AD}{DB}.$$

Hence

$$\frac{A'D' + D'B'}{D'B'} = \frac{AD + DB}{DB},$$

or, since $(A'D'B')$ and (ADB),

$$\frac{A'B'}{D'B'} = \frac{AB}{DB},$$

from which we get

$$\frac{A'B'}{AB} = \frac{D'B'}{DB}.$$

Combining the last equation with (1) gives $B'C'/BC = D'B'/DB = k$. Since $\angle B = \angle B'$, it follows that triangles BCD, $B'C'D'$ are similar. Thus, $C'D'/CD = k$, that is, distances in the direction of line PC are multiplied by k. In like manner, distances in the directions of lines PA, PB can be shown to be multiplied by k.

It then follows that if Q is an arbitrary point on triangle ABC, say on $\overline{AB}$, distance in the direction of line PQ must also be multiplied by k since Q', the image of Q, is on $\overline{A'B'}$ and triangles APQ, $A'P'Q'$ are similar. Now, any given direction in the plane is the direction of some such line PQ. It follows that T multiplies all distances in the plane by k, and hence is a similarity. Since this contradicts our original assumption, the proof is complete.

An affine transformation that preserves angle-size sends each triangle into a similar triangle, and hence is a similarity transformation by 6.4. Thus we have the theorem:

6.5 *Similarities are the only affine transformations that preserve angle-size.*

EXERCISES

1. Classify the following figures as affine, metric, or neither: (a) three points of which one is midway between the others; (b) a triangle and its angle-bisectors; (c) a triangle and its medians; (d) a triangle and its altitudes; (e) an ellipse and its center; (f) a circle and its diameters; (g) a central conic; (h) a hyperbola of eccentricity 2.

2. Prove that in an affine transformation the image of a half-plane determined by a line is a half-plane determined by the image of the line.

3. Prove that the set of affine transformations whose determinants are 1 or -1 forms a group. Give three examples of such transformations other than motions.

4. Show that in an affine transformation the image of the sum of two vectors is the sum of the images of the vectors.

5. A plane figure is called *convex* if, when two points belong to it, so do all points of their segment. (a) Verify that a line is convex and a circle not convex, that the interior of a circle is convex and the exterior not convex. (b) Show that the property of being a convex figure is preserved by the affine group.

6. Find the images of $A(0, 0)$, $B(1, 0)$, $C(0, 2)$ under the transformation $x' = 2x$, $y' = \frac{1}{2}y$. Show that triangles ABC, $A'B'C'$ are not similar and that triangles ABC, $A'C'B'$ are similar, but that this does not contradict 6.4.

7. Affine Geometry. We may now designate as **affine geometry**, or, more exactly, **affine geometry of the Euclidean plane,** that large part of Euclidean plane geometry which is primarily concerned with affine properties. Besides many formal statements (definitions, theorems, axioms), it includes all those steps in proofs where a point is taken between two others, or a midpoint is chosen, or a point is taken inside or outside of a triangle, or a point is taken on a certain side of a line (that is, in a certain half-plane determined by the line), or a parallel is drawn, etc.

All definitions of affine properties, including the property of being an affine figure, of course, belong to affine geometry. Among them, for example, are the definition of a triangle as the figure consisting of three noncollinear points and the segments determined by them, and the definition of parallel lines as lines having no common point. Although distance enters into some affine definitions, it does so only in an affine way, which means that these definitions deal only with *distances on the same line or on parallel lines*, sometimes comparing them by considering their ratios. The definition of ratio of division clearly does this. In particular, the affine definition of the midpoint of $\overline{AB}$ is that it is the point C on $\overline{AB}$ such that $AC = CB$ or $AC/CB = 1$. The definition of an isosceles triangle, however, is not affine, for although it, too, deals with two equal distances, they are not on the same line or on parallel lines.

Among the affine theorems, in some of which distance again enters, but only in an affine way, are the following: (1) *two lines parallel to the same line are parallel to each other*, (2) *the segment joining the midpoints of two sides of a triangle is parallel to the third side and half as long* (that is, the ratio of their lengths is $\frac{1}{2}$), (3) *the medians of a triangle are concurrent*, (4) *the opposite sides of a parallelogram are equal*, (5) *the diagonals of a parallelogram bisect each other*, and (6) *two lines which are not parallel have exactly one common point.* In addition to these theorems, all familiar, we mention the following two, which are not usually encountered in high school geometry. The proofs will be given in the next section.

7.1 Theorem of Menelaus. *If r_1, r_2, r_3 are the ratios in which three points Q_1, Q_2, Q_3 divide the sides $\overline{P_2P_3}$, $\overline{P_3P_1}$, $\overline{P_1P_2}$ of a triangle, respectively, the points are collinear if and only if $r_1r_2r_3 = -1$.*

As the presence of -1 shows, negative ratios are permitted. Hence one or more of the sides may be divided externally by Q_1, Q_2, Q_3.

7.2 Theorem of Ceva. *If r_1, r_2, r_3 are the ratios in which three points Q_1, Q_2, Q_3 divide the sides $\overline{P_2P_3}$, $\overline{P_3P_1}$, $\overline{P_1P_2}$ of a triangle, respectively, the lines P_1Q_1, P_2Q_2, P_3Q_3, joining these points to the opposite vertices, are parallel or concurrent if and only if $r_1r_2r_3 = 1$.*

Here, too, negative ratios are allowed. This theorem is a generalization of the affine theorem (3) mentioned above.

Menelaus (Alexandria, around 100 A.D.) and Ceva (Italy, 1647–1734) did not state their theorems in terms of ratio of division, but in a form easily converted to the propositions we have given.

Of the axioms used in most high school plane geometry texts just the following are affine: (1) *one, and only one, line goes through two distinct points*, and (2) *through a point outside of a line there passes exactly one line which is parallel to it.* In more rigorous presentations of geometry, however, the number of affine axioms is usually much greater, partly because of the presence of axioms dealing with the between-relation. Hilbert uses five axioms on this relation alone, and Veblen uses six.* One of Hilbert's axioms, for example, is: *of any three points situated on a line there is one and only one which is between the other two.*

---

* See D. Hilbert, *The Foundations of Geometry* (Open Court Publ. Co., 1902), and O. Veblen, *The Foundations of Geometry*, an article in *Monographs on Topics of Modern Mathematics*, edited by J. W. A. Young (Longman's, Green and Co., 1932).

EXERCISES

1. Classify each of the following figures as affine or metric: (a) trapezoid; (b) rhombus; (c) rectangle; (d) pentagon; (e) regular hexagon; (f) ellipse with eccentricity $\frac{1}{2}$; (g) two intersecting lines; (h) two parallel lines and a common perpendicular; (i) two parallel lines and a transversal; (j) a line and a point not on it; (k) a line and a point on it.

2. Classify each of the following theorems as affine or metric: (a) the angle-bisectors of a triangle are concurrent; (b) the altitudes of a triangle are concurrent; (c) the midpoints of the sides of a quadrilateral are the vertices of a parallelogram; (d) the medians drawn to the equal sides of an isosceles triangle are equal; (e) the segment joining the midpoints of the nonparallel sides of a trapezoid bisects each of the diagonals.

3. Show that the affine theorem (3) in our listing can be deduced from the theorem of Ceva.

4. Deduce the following from the theorem of Menelaus: *A line which goes through no vertex of a triangle, but meets the triangle in a point, can meet it in at most one other point.*

8. Collinearity and Concurrence. When affine geometry is developed by means of coordinates and equations, it is desirable to have simple ways of determining whether given points are collinear and given lines are concurrent or parallel. We shall develop such tests and then use them to prove the theorems of Menelaus and Ceva.

Let P_1, P_2, P_3 be distinct points. The line P_1P_2 then has the equation

$$y - y_1 = \frac{y_2 - y_1}{x_2 - x_1}(x - x_1) \tag{1}$$

if $x_1 \neq x_2$, and the equation $x = x_1$ if $x_1 = x_2$. Hence if (1) is rewritten as

$$(y_1 - y_2)(x - x_1) - (x_1 - x_2)(y - y_1) = 0, \tag{2}$$

this equation represents line P_1P_2 in both cases.

The point P_3 is on this line if and only if its coordinates satisfy (2), that is, if and only if

$$(y_1 - y_2)(x_3 - x_1) - (x_1 - x_2)(y_3 - y_1) = 0.$$

On multiplying this out one finds that the result can be written as

$$\begin{vmatrix} x_1 & y_1 & 1 \\ x_2 & y_2 & 1 \\ x_3 & y_3 & 1 \end{vmatrix} = 0. \tag{3}$$

Only when this condition is met, then, are the three points collinear. When the condition is not met, the points are the vertices of a triangle and, as was noted in §4, the value of the determinant is twice the area of the triangle.

Now consider three distinct lines

$$A_1x + B_1y + C_1 = 0, \quad A_2x + B_2y + C_2 = 0, \quad A_3x + B_3y + C_3 = 0.$$

If they are not parallel, let us suppose that the first two meet. Their point of intersection then has the coordinates

$$x = \frac{\begin{vmatrix} B_1 & C_1 \\ B_2 & C_2 \end{vmatrix}}{\begin{vmatrix} A_1 & B_1 \\ A_2 & B_2 \end{vmatrix}}, \qquad y = \frac{\begin{vmatrix} C_1 & A_1 \\ C_2 & A_2 \end{vmatrix}}{\begin{vmatrix} A_1 & B_1 \\ A_2 & B_2 \end{vmatrix}}.$$

This point is on the third line if and only if its coordinates satisfy the equation of the line. Substituting these coordinates and clearing of fractions gives

$$A_3 \begin{vmatrix} B_1 & C_1 \\ B_2 & C_2 \end{vmatrix} + B_3 \begin{vmatrix} C_1 & A_1 \\ C_2 & A_2 \end{vmatrix} + C_3 \begin{vmatrix} A_1 & B_1 \\ A_2 & B_2 \end{vmatrix} = 0,$$

or

$$\begin{vmatrix} A_1 & B_1 & C_1 \\ A_2 & B_2 & C_2 \\ A_3 & B_3 & C_3 \end{vmatrix} = 0. \tag{4}$$

Thus, when not parallel, the three lines are concurrent if and only if (4) holds. But (4) holds also when the lines are parallel since A_1, B_1 and A_2, B_2 are then both proportional to A_3, B_3. Hence (4) is a necessary and sufficient condition for the lines to be parallel or concurrent.

To summarize:

8.1 *Three distinct points (x_1, y_1), (x_2, y_2), (x_3, y_3) are collinear if and only if equation* (3) *holds, and three distinct lines*

$$A_1x + B_1y + C_1 = 0, \quad A_2x + B_2y + C_2 = 0, \quad A_3x + B_3y + C_3 = 0$$

are concurrent if and only if equation (4) *holds.*

In proving the theorem of Menelaus, 7.1, let us first consider the triangle with vertices $P_1(0, 0)$, $P_2(1, 0)$, $P_3(0, 1)$ (Fig. IV, 2). Using the fact that the point which divides the point-pair (x_1, y_1), (x_2, y_2) in the ratio r has the coordinates

$$\frac{x_1 + rx_2}{1 + r}, \quad \frac{y_1 + ry_2}{1 + r}$$

by §3, Ex. 7, we find the following coordinates for Q_1, Q_2, Q_3:

$$Q_1\left(\frac{1}{1 + r_1}, \frac{r_1}{1 + r_1}\right), \quad Q_2\left(0, \frac{1}{1 + r_2}\right), \quad Q_3\left(\frac{r_3}{1 + r_3}, 0\right).$$

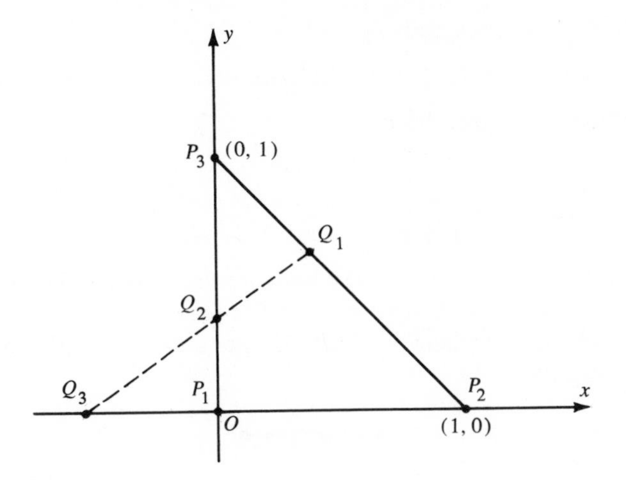

Fig. IV, 2

Denoting the determinant in (3) by D, and substituting the above coordinates in D, we get (see Appendix, 2.2)

$$D = \begin{vmatrix} \dfrac{1}{1+r_1} & \dfrac{r_1}{1+r_1} & 1 \\ 0 & \dfrac{1}{1+r_2} & 1 \\ \dfrac{r_3}{1+r_3} & 0 & 1 \end{vmatrix} = \frac{1}{(1+r_1)(1+r_2)(1+r_3)} \begin{vmatrix} 1 & r_1 & 1+r_1 \\ 0 & 1 & 1+r_2 \\ r_3 & 0 & 1+r_3 \end{vmatrix}$$

$$= \frac{1 + r_1r_2r_3}{(1+r_1)(1+r_2)(1+r_3)}.$$

Q_1, Q_2, Q_3 are collinear if and only if $D = 0$, and the condition for this is seen to be that $1 + r_1r_2r_3 = 0$, or $r_1r_2r_3 = -1$.

Now let $P_1'P_2'P_3'$ be any other triangle, and Q_1', Q_2', Q_3' be points dividing sides $\overline{P_2'P_3'}$, $\overline{P_3'P_1'}$, $\overline{P_1'P_2'}$ in the ratios r_1', r_2', r_3'. By 2.4 there is an affine transformation T that sends P_1, P_2, P_3 into P_1', P_2', P_3', respectively. Let Q_1, Q_2, Q_3 be the points which T sends into Q_1', Q_2', Q_3', respectively. Since ratio of division is preserved, Q_1, Q_2, Q_3 divide sides $\overline{P_2P_3}$, $\overline{P_3P_1}$, $\overline{P_1P_2}$ in ratios r_1, r_2, r_3 such that $r_1 = r_1'$, $r_2 = r_2'$, $r_3 = r_3'$. Also, Q_1', Q_2', Q_3' are collinear if and only if Q_1, Q_2, Q_3 are. The latter are collinear if $r_1r_2r_3 = -1$, and only then, as shown above. It follows that Q_1', Q_2', Q_3' are collinear if and only if $r_1'r_2'r_3' = -1$.

Turning now to the theorem of Ceva, 7.2, let us first prove it for the triangle $P_1P_2P_3$ which was used in the preceding proof. The equations of lines

P_1Q_1, P_2Q_2, P_3Q_3 are, respectively,

$$r_1x - y = 0, \quad x + (1 + r_2)y - 1 = 0, \quad (1 + r_3)x + r_3y - r_3 = 0.$$

The determinant in (4) then becomes

$$\begin{vmatrix} r_1 & -1 & 0 \\ 1 & 1 + r_2 & -1 \\ 1 + r_3 & r_3 & -r_3 \end{vmatrix},$$

and the value of this is $1 - r_1r_2r_3$. The lines are therefore parallel or concurrent if and only if $1 - r_1r_2r_3 = 0$, or $r_1r_2r_3 = 1$. The reasoning by which this result is extended to any other triangle $P_1'P_2'P_3'$ is the same as in the proof of the Theorem of Menelaus.

EXERCISES

1. Use 8.1 to test the following sets of points for collinearity:
(a) $(1, -3), (4, 3), (7, 8)$; (b) $(-1, 5), (4, 0), (9, -5)$.

2. Use 8.1 to test the following sets of lines for concurrence:
(a) $x + y - 1 = 0,\ 3x + 2y - 2 = 0,\ 2x + 2y - 3 = 0$;
(b) $2x - y + 3 = 0, 3x - y + 4 = 0,\ 2x + y + 1 = 0$;
(c) $x + 2y - 2 = 0, x - 2y - 2 = 0,\ 2x + 3y - 3 = 0$;
(d) $2x - y + 1 = 0, -2x + y + 1 = 0, 6x - 3y + 9 = 0$.

3. Use 8.1 to verify that the points (x_1, y_1), (x_2, y_2), $\left(\frac{x_1 + rx_2}{1 + r}, \frac{y_1 + ry_2}{1 + r}\right)$ are collinear.

4. The coordinates $(hx_1 + kx_2, hy_1 + ky_2)$, where h, k are any numbers not both zero, are said to be a *linear combination* of the coordinates (x_1, y_1) and (x_2, y_2). Show that the coordinates of a point that divides $\overline{P_1P_2}$ in the ratio r are a linear combination of the coordinates of P_1 and P_2 in which $h + k = 1$, neither h nor k is 0, and $k/h = r$.

5. If $A_1x + B_1y + C_1 = 0$, $A_2x + B_2y + C_2 = 0$, $A_3x + B_3y + C_3 = 0$ are distinct lines, no two of which are parallel, show that they are concurrent if the coefficients of one are a *linear combination* of the coefficients of the others, that is, if constants h, k exist so that, for example, $A_3 = hA_1 + kA_2$, $B_3 = hB_1 + kB_2$, $C_3 = hC_1 + kC_2$.

6. Prove the converse of the proposition in Ex. 5, that is, if the given lines are concurrent, then the coefficients of one are a linear combination of the coefficients of the others.

9. Affine Equivalence. Two geometric figures are said to be **equivalent with respect to a group of transformations** if and only if the group contains a transformation which maps one figure on the other.

Consider, for example, the group of motions. According to the above definition two figures are equivalent with respect to this group if and only if there is a motion which maps one onto the other. Now, according to II, 3.1, two figures are called congruent if and only if there is a motion sending one into the other. Thus, "equivalent with respect to the group of motions" and "congruent" mean the same thing. In like manner, reference to III, 5.1 shows that the two expressions, "figures equivalent with respect to the group of similarity transformations" and "similar figures", mean the same thing.

When we turn to the affine group we find no familiar word analogous to "congruent" or "similar" for expressing equivalence with respect to this group. The term **affinely equivalent** is therefore used for this purpose. Any two triangles, for example, are affinely equivalent. For, according to 2.4, there is an affine transformation which maps the vertices of one of these triangles on those of the other, and this transformation, according to 6.3, sends the sides of the first triangle, respectively, into those of the second.

If two figures are similar, they are, of course, also affinely equivalent. Hence, in view of the discussion and exercises of III, §5, any two segments, or any two lines, or any two parabolas are affinely equivalent. Not all ellipses are similar, but they are all affinely equivalent, and the same is true of hyperbolas. To prove this, let E_1, E_2 be any two ellipses. If they have the same eccentricity, they are similar by III, 5.3, and hence affinely equivalent. If they have different eccentricities, consider two ellipses E_1', E_2' with centers at O, foci on the x-axis, and having the same eccentricity as E_1, E_2, respectively. Also, let E_1', E_2' have equal major axes. The equations of E_1', E_2' are then, respectively,

$$\frac{x^2}{a^2} + \frac{y^2}{b_1^2} = 1 \qquad \text{and} \qquad \frac{x^2}{a^2} + \frac{y^2}{b_2^2} = 1,$$

where $b_1 \neq b_2$. Since E_1, E_1' are similar, there is a similarity transformation T_1 that maps E_1 on E_1' (III, 5.1). Likewise, there is a similarity transformation T_2 that maps E_2 on E_2'. Furthermore, it is easily verified that the affine transformation T_3 whose equations are

$$x' = x, \qquad y' = \frac{b_2}{b_1}y,$$

sends E_1' into E_2'. We can now write $T_1(E_1) = E_1'$, $T_2(E_2) = E_2'$, $T_3(E_1') = E_2'$, from which it follows that $T_1T_3T_2^{-1}$, an affine transformation, sends E_1 into E_2. Thus, E_1 and E_2 are affinely equivalent. The corresponding proof for hyperbolas is left as an exercise. We also leave as exercises the other unproved parts of the following summary:

9.1 *Any two segments, angles, triangles, or parallelograms are affinely equivalent, and so are any two lines, parabolas, ellipses, or hyperbolas.*

The notion of affine equivalence can often be used to advantage in proving affine theorems. We give two examples of this. As our first example let us prove that *the medians of a triangle meet in a point interior to the triangle.* Consider the special triangle whose vertices are $O(0, 0)$, $A(2, 0)$, $B(0, 2)$. The medians

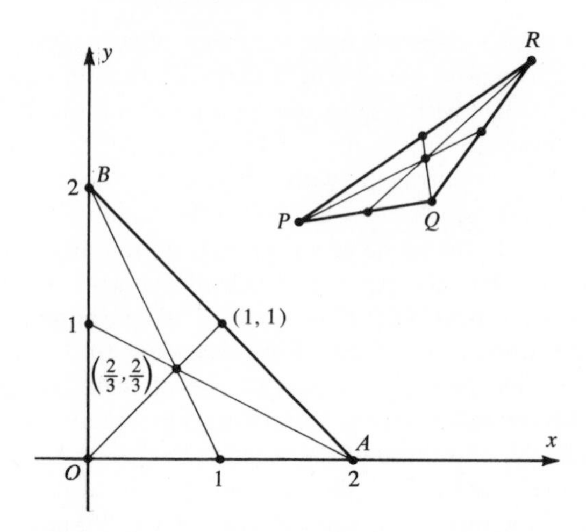

Fig. IV, 3

through O, A, B have the equations $x = y$, $x + 2y = 2$, $y + 2x = 2$, respectively. These equations have the common solution $x = \frac{2}{3}$, $y = \frac{2}{3}$. The medians therefore meet in the point $(\frac{2}{3}, \frac{2}{3})$, which is clearly inside the triangle (Fig. IV, 3). If PQR is any other triangle, triangle OAB can be mapped on it by the affine transformation that sends O, A, B into P, Q, R, respectively. Since medians go into medians, concurrent lines into concurrent lines, and interior points into interior points, the medians of triangle PQR must meet in a point interior to this triangle.

As our second example we prove the theorem: *the midpoints of each set of parallel chords of an ellipse are collinear, and lie on a diameter of the ellipse.* (A diameter of an ellipse is a chord through its center.) Just as we proved the theorem on triangles by first considering a special triangle, so now we first consider a special kind of ellipse, a circle. The theorem is obviously true for circles. Next, consider any ellipse E, not a circle, and let O_E be its center. Since all ellipses are affinely equivalent, if C is any circle, with center O_C, there is an affine transformation T that sends E into C. Each chord of E goes into a chord of C, and, conversely, each chord of C is the image of a unique chord of E. In particular, the chords through O_E go into the chords through the image of O_E. Since O_E bisects all the chords of E through it, its image has this same property regarding the chords of C. Its image must therefore be O_C. Thus, the diameters of E go into those of C in a 1-1 way. We note next that each family f of parallel chords of E goes into a family f' of parallel chords of C. Since the midpoints of the chords of f' lie on a diameter of C, so must the midpoints of the chords of f lie on the corresponding diameter of E.

Theorems analogous to the one just proved can also be proved for parabolas and hyperbolas (see Exs. 4, 5).

Our two examples serve to illustrate a *method* in which affine equivalence is used to prove an affine property for a general class of figures. In this method of proof we select a particular figure which is affinely equivalent to each figure of the class in question, and for which, because of its special position or nature, it is easy to prove the specified affine property. The affine equivalence then implies that the property is possessed by all the figures of the class.

EXERCISES

1. Prove that any two figures in each of the following categories are affinely equivalent: (a) angles; (b) parallelograms; (c) lines; (d) hyperbolas.

2. Prove that any two degenerate hyperbolas are affinely equivalent. What can be said about degenerate ellipses and parabolas?

3. Using triangle OAB (Fig. IV, 3) and the idea of affine equivalence, prove the affine theorem: *the segment joining the midpoints of two sides of a triangle is parallel to the third side, and half as long.*

4. Using $y = 1/x$ and affine equivalence, prove the affine theorem: *the midpoints of all parallel chords of a hyperbola are collinear, and lie on a diameter of the hyperbola.* (A diameter of a hyperbola is any line through the center of the hyperbola.)

5. Using $y = x^2$ and affine equivalence, prove the affine theorem: *the midpoints of all parallel chords of a parabola are collinear, and lie on a diameter of the parabola.* (By a diameter of a parabola is meant its axis or any line parallel to the axis.)

6. Make use of a square and the fact that all parallelograms are affinely equivalent to prove the affine theorem: *the diagonals of a parallelogram bisect each other.*

7. Prove that three collinear points A, B, C are affinely equivalent to three collinear points A', B', C', respectively, if and only if the ratio in which C divides $\overline{AB}$ equals the ratio in which C' divides $\overline{A'B'}$.

8. Are all trapezoids affinely equivalent? Justify your answer.

9. What meaning would you attach to the term "metrically equivalent"?

10. The *normal distribution curves* of statistics,

$$y = \frac{1}{\sigma\sqrt{2\pi}} e^{-(x-m)^2/2\sigma^2},$$

where, for each curve, m, σ are constants denoting the mean and standard deviation, respectively, are affinely equivalent. Prove this by showing that the particular curve $y = \dfrac{1}{\sqrt{2\pi}} e^{-x^2/2}$, corresponding to $m = 0$, $\sigma = 1$, goes into the general one given above under the transformation $x' = \sigma x + m, \quad y' = y/\sigma$.

11. Assuming that an affine transformation multiplies *all* areas by the value of its determinant, prove that the area under the particular normal curve of Ex. 10 from $x = 0$ to $x = x_1$ equals the area under the general normal curve from $x = m$ to $x = \sigma x_1 + m$.

10. Affine Properties and 1-1 Transformations. As we have seen, affine properties are properties which are preserved by all affine transformations but not by all projective transformations. When we study the latter in Chapter VI we shall find that some of them are not 1-1. It is therefore of interest to ask whether there are any 1-1 transformations besides the affine transformations which preserve affine properties.* We answer this question in the theorems which follow, and in the exercises.

10.1 *Affine transformations are the only* 1-1 *mappings of the plane onto itself in which the image of a line is always a line.*

Proof. Let T be a 1-1 mapping of the plane onto itself which carries lines into lines. T must then preserve parallelism. For if g, h are parallel lines and their images met in a point, the latter would have to be the image of a point on both g and h, which is impossible. Let A', O', B' be the images of the noncollinear points $A(1, 0)$, $O(0, 0)$, $B(0, 1)$. Then A', O', B' are also noncollinear. For if B' were on line $A'O'$, it would have to be the image of a point on line AO, whereas B is not on this line.

Now let S be the affine transformation which, by 2.4, maps A', O', B' on A, O, B, respectively. Then TS is a 1-1 mapping of the plane onto itself in which A, O, B are fixed points, the image of a line is a line, and parallelism is preserved. TS, in fact, leaves every point of the plane fixed, as we now show.

TS leaves each of the lines $x = 0$, $y = 0$, $x + y = 1$ fixed since it leaves A, O, B fixed (Fig. IV, 4). Then the line $x = 1$, which passes through $A(1, 0)$ and is parallel to the line $x = 0$, must go into the line through A which is parallel to the line $x = 0$; that is, the line $x = 1$ is fixed. Similarly, the line $y = 1$ is fixed. Hence the point $(1, 1)$ is fixed.

Next, consider the line $x + y = 2$. It passes through $(1, 1)$ and is parallel to the line $x + y = 1$, and hence must be fixed. Each of its intersections with the axes, $(2, 0)$ and $(0, 2)$, is therefore fixed. Then, since the lines $x = 2$ and $y = 1$ are fixed, so is the point $(2, 1)$. The line $x + y = 3$, which passes through $(2, 1)$ and is parallel to $x + y = 2$, is therefore fixed, and hence so is each of its points of intersection with the x-axis, $(3, 0)$ and $(0, 3)$. Clearly, any point whose coordinates are nonnegative integers can be shown in this way to be fixed, and a slight change in the method takes care of all other points with integral coordinates.

The remaining points with rational coordinates can now be handled, points on the axes being considered first. A single example, say $(\frac{3}{5}, 0)$, suffices to illustrate the procedure. The line g joining this point to $B(0, 1)$ has the slope $-\frac{5}{3}$. Then g is parallel to the line whose x and y intercepts are -3 and -5, respectively. Since the latter line and B are fixed, so is g, and hence also $(\frac{3}{5}, 0)$.

We omit the argument† showing that each of the remaining points of the

---

* Nonaffine 1-1 transformations of the plane into itself are very numerous. A simple example is $x' = x^3$, $y' = y^3$.

† See O. Veblen and J. H. C. Whitehead, *The Foundations of Differential Geometry*, pp. 14. 15.

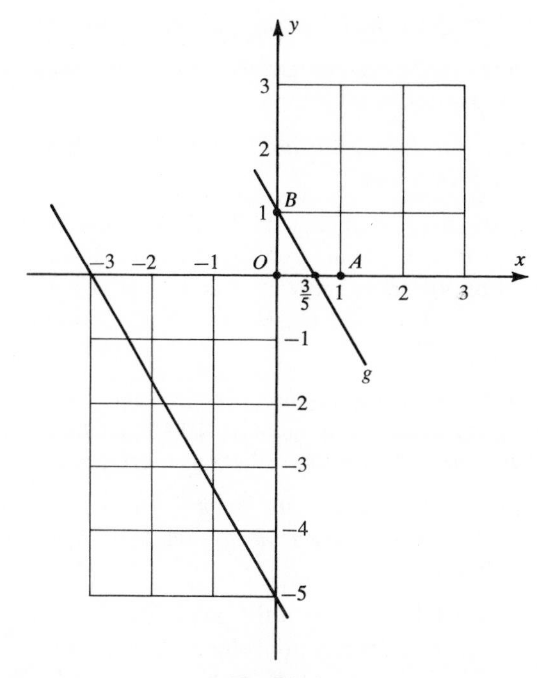

Fig. IV, 4

plane is fixed. Assuming this has been done, we then know that TS is the identity. From $TS = I$ we obtain $T = S^{-1}$. Thus, T is the inverse of an affine transformation. It is therefore affine.

According to 10.1, a 1-1 transformation of the plane into itself is necessarily affine if it sends complete lines into complete lines. The following theorem shows that it is affine if it is merely known to preserve collinearity.

10.2 *Affine transformations are the only* 1-1 *mappings of the plane onto itself which preserve collinearity.*

Proof. Let T be a 1-1 mapping of the plane onto itself which preserves collinearity. Let g be any line, A and B distinct points on it, A' and B' their images, and g' the line $A'B'$. If C is any other point of g, its image, being collinear with A' and B', must lie on g'. Thus all points of g go into points of g'. Conversely, let P' be any point of g' other than A' or B', and let it be the image of P. Assume P is not on g. Then A, B, P are the vertices of a triangle. Since A, B, P go into points of g', so must all points of the lines AB, AP, BP. Any point of the plane not on one of these lines is collinear with points lying, respectively, on two of these lines, and hence goes into a point of g'. Thus, T sends all points of the plane into points of g'. This contradicts our hypothesis that T maps the plane on itself. It follows that P is on g, that the image of g is g', and hence that T is affine by 10.1.

We leave the proof of the following theorem as an exercise.

10.3 *Affine transformations are the only* 1-1 *mappings of the plane onto itself which preserve noncollinearity.*

10.4 *Affine transformations are the only* 1-1 *mappings of the plane onto itself which preserve betweenness.*

Proof. Let T be a 1-1 mapping of the plane onto itself which preserves betweenness. Let A, B, C be distinct collinear points. Then one of them is between the others. Suppose B is between A and C. If A', B', C' are the image points, then B' is between A' and C'. Hence A', B', C' are collinear, and T is affine by 10.2.

Since segments, rays, angles, and triangles are defined in terms of the between-relation, propositions analogous to 10.4 must hold for them too. We prove the following one and leave the others as exercises.

10.5 *Affine transformations are the only* 1-1 *mappings of the plane onto itself in which the image of an angle is always an angle, vertex corresponding to vertex, and each side to a side.*

Proof. Let T be a 1-1 mapping of the plane onto itself which possesses the stated property. Let A, B, C be any noncollinear points, and A', B', C' be their images (Fig. IV, 5). The image of $\angle BAC$ is then an angle (1) whose vertex is the image of A, and hence is A', (2) one of whose sides is the image of side AB, and hence contains B', and (3) the other of whose sides is the image of side AC, and hence contains C'. Thus, A', B', C' are noncollinear. It follows that T preserves noncollinearity and hence is affine by 10.3.

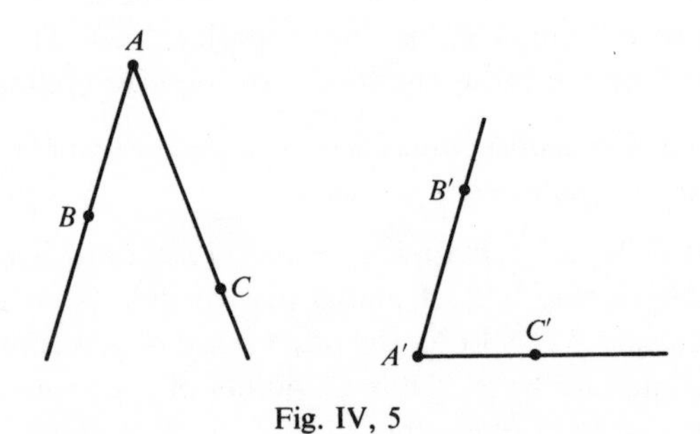

Fig. IV, 5

EXERCISES

1. Show that affine transformations are the only 1-1 mappings of the plane onto itself in which (a) the image of a segment is always a segment (endpoints corresponding to endpoints); (b) ratio of division is preserved; (c) the image of a triangle is always a

triangle (vertices corresponding to vertices, and sides to sides); (d) all distances in a common direction are multiplied by the same constant.

2. Prove 10.3.

11. Axioms for Affine Geometry. Throughout this chapter we have been regarding affine geometry simply as a part of Euclidean geometry. Since we were working within the framework of Euclidean geometry, it was natural, when making affine definitions or proving affine theorems, to use any methods that seemed most convenient. Many of these methods were metric. For example, we often used a rectangular coordinate system, which involves the metric concept of a right angle, and we used the general distance formula, which presumes such a coordinate system. Even so basic an affine concept as betweenness was understood metrically, in terms of distance, for the statement "B is between A and C" was always taken to mean that A, B, C are distinct points such that $AB + BC = AC$.

This dependence of affine geometry on concepts that are foreign to it was characteristic of the subject as it originated historically. In time, however, one came to realize that affine geometry could be developed without such extraneous ideas if it were based on axioms of its own, involving only affine concepts. Choosing such axioms is now comparatively easy, and can be done by excluding certain axioms from a suitable set of axioms for Euclidean geometry and adding a few others. The sets of axioms given by Hilbert and Veblen, and referred to in §7, are useful for this purpose. Consider Hilbert's axioms,* for example. Since we are concerned only with plane geometry, whereas Hilbert's axioms are for space geometry as well, we first of all exclude his so-called space axioms. One such axiom, for example, states that if two planes have a common point, then they have at least one other common point. Next, we exclude all of his axioms that deal with congruence. After combining some of the remaining axioms, making word changes in others, and adding one additional axiom, we obtain the following ten axioms:

Axioms for Affine Plane Geometry

1. *Any two points are on one and only one line.*
2. *On any given line there are at least two points.*
3. *There are at least three points which are not on the same line.*
4. *Of any three points on a line, there is not more than one which is between the other two.*†

---

* In this discussion we use a later version of the axioms than that mentioned in §7. See D. Hilbert, *Grundlagen der Geometrie*, Seventh Edition, 1930.

† It can be proved, as a theorem, that there is always exactly one point which is between the other two.

5. *If a point B is between a point A and a point C, then A, B, C are distinct points on a line and B is also between C and A.*
6. *If A and B are distinct points on a line, then there exists at least one other point C on the line such that B is between A and C.**
7. *Let A, B, C be three points not on the same line and let g be a line not containing A, B, or C. Then if g contains a point between A and B, it will also contain either a point between A and C, or a point between B and C.*
8. *Through any point A not on a line g there passes one, and only one, line not meeting g.* (This line is called *the parallel* to g through A. Also, any two lines which do not meet are called *parallel.*)
9. *If each of three pairs of distinct points A and A', B and B', C and C' is collinear with another point O, and lines AB, $A'B'$ are parallel, and also lines BC, $B'C'$, then lines AC, $A'C'$ are parallel* (Fig. IV, 6).
10. *If the points of a line are divided into two sets such that no point of either set is between points of the other, then one set has a point which is between every other point of that set and each point of the second set.*

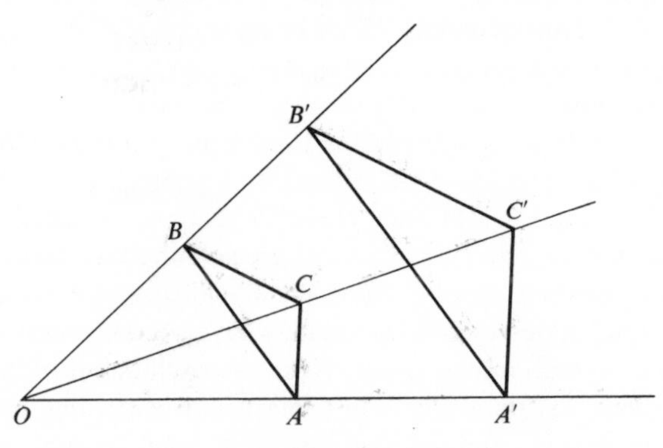

Fig. IV, 6

To illustrate Axiom 10 let us suppose that the points of the x-axis are divided into sets α and β such that α contains all points with positive abscissas (and no others) and β contains the remaining points of the axis. Then clearly no point of either set is between points of the other. This is therefore a division of the points of the x-axis into two sets satisfying the hypothesis of Axiom 10. The conclusion of the axiom is also satisfied, for β has a point, the origin, which is between every other point of β and each point of α. Axiom 10 is called an

* That there exists at least one point on the line between A and B can be proved as a theorem.

axiom of continuity. Without it or its equivalent one could not prove that the points of a line are in 1-1 correspondence with the real numbers.

Examination of these axioms shows that they are concerned only with two kinds of geometrical objects, *points* and *lines*, and with two relations among these objects, one expressed by the word *between*, the other by the word *on*. These four terms are basic terms of affine geometry as it would be developed from the above axioms. So basic are they, in fact, that they are left undefined,* and are called **undefined terms**. Every other geometrical term used in affine geometry is defined by means of them, directly or indirectly. Thus, *segment* $\overline{AB}$ means the set of points consisting of A, B, and all points between A and B, *triangle* ABC means the set of points consisting of the segments $\overline{AB}$, $\overline{AC}$, $\overline{BC}$, and so forth. Some terms, such as *containing*, *meeting*, and *passing through*, which occur in Axioms 7 and 8, are only linguistic conveniences and can easily be rephrased in terms of *on*. Thus, instead of saying "g contains A" we can say "A is on g." On the other hand, the definitions of *ratio of distances* and *ratio of areas* in terms of *point*, *line*, *on*, and *between* are far from obvious and can be given only after lengthy discussions. That such definitions can be given at all, in terms of concepts having no apparent connection with measurement, is one of the most striking aspects of the axiomatic development of affine geometry.

The relation *between*, it should be noted, deals only with the order of points on a line, and is therefore called an *order relation*. The relation *on*, viewed intuitively, is concerned only with whether or not a point and a line meet, and is therefore called an *incidence relation*.

The totality of points and lines which satisfy the above axioms are said to constitute the *affine plane*. The effect of adding congruence axioms to these axioms, so as to obtain the Euclidean plane, is not to increase the number of such points and lines, but only to make for more extensive relations among them, notably relations involving measurement. In the affine plane, for example, two segments can be compared in length only if they are on the same line or on parallel lines; in the Euclidean plane no such restriction is necessary.

The points and lines of the affine plane, then, are just as numerous as those of the Euclidean plane. To be precise, there is a 1-1 correspondence between the points of the two planes, and a 1-1 correspondence between the lines of the two planes. It follows that there is a 1-1 correspondence between the point of the affine plane and the set of all pairs of real numbers (x, y), and a 1-1 correspondence between the lines of the affine plane and the set of all first-degree equations $Ax + By + C = 0$ (two such equations being regarded as distinct only if their coefficients are not proportional). As this strongly suggests, there is an *affine analytic geometry*, and it has much in common with the familiar (Euclidean) analytic geometry.

---

* To define every term used in affine geometry, or in any other mathematical system, would involve circular reasoning and must be avoided. The undefined terms are therefore analogous to the unproved propositions, that is, the **axioms**.

EXERCISES

The following exercises are concerned only with affine plane geometry. Proofs are to be based on Axioms 1–10.

1. Using *on*, restate Axioms 7, 8 so as to avoid terms such as *contains*, *passes through*, *meets*, etc.

2. (a) Define parallel lines and intersecting lines in terms of *on*. (b) Define segment $\overline{AB}$, triangle ABC, and ray AB (with endpoint A) in terms of *between*.

3. Prove that there are at least three points on a line.

4. If g is any line, prove that there are at least four points not on g.

5. Prove that two lines are parallel to each other if they are parallel to the same line.

6. Prove that there are at least seven lines.

7. Assuming that there are infinitely many points on any line, prove that there are infinitely many lines through any point.

8. Define a parallelogram and prove that parallelograms exist.

12. Distance in Affine Geometry. In discussing affine geometry of the Euclidean plane (§7), we noted that some of its definitions and theorems involve distance, and always in a certain limited way: distances are compared in value only when they are on the same line or parallel lines. Distance is subject to the same limitations when affine geometry is developed on the basis of its own axioms, and we mentioned this in §11. Changing the terminology slightly, we said that two segments can be compared in size only if they are on the same line or parallel lines, and contrasted this with the situation in Euclidean geometry, where there is no such restriction. We now go into these matters in more detail. Proofs are omitted since they would carry us too far afield.

Taking the Euclidean situation first, let $\overline{AB}$ and $\overline{CD}$ be any two segments. There is an axiom on congruence* according to which there is a unique point E of line CD, situated on the same side of C as D, such that $\overline{CE}$ is congruent to $\overline{AB}$. Because of axioms dealing with the between relation, there are then exactly three possible positions for E: either $D = E$, or (CED), as in Fig. IV, 7a, or (CDE), as in Fig. IV, 7b. In the first case we say that $\overline{AB}$ and $\overline{CD}$ are **congruent**, in the second that $\overline{AB}$ is **smaller** than $\overline{CD}$, and in the third that $\overline{AB}$ is **greater** than $\overline{CD}$. Thus, any two segments can be compared as to size in a purely geometrical way, for, despite the terminology, no numbers have been used. Now, by choosing an arbitrary segment to serve as unit segment, one could go further and show that with each segment in the plane there can be associated a positive

* D. Hilbert, *The Foundations of Geometry*, translated by E. J. Townsend, p. 12.

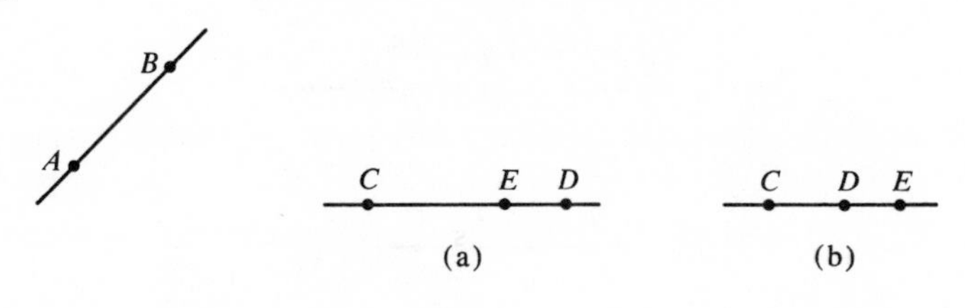

Fig. IV, 7

number, which is called the **length** of the segment, or the **distance** between its endpoints. This is basic in setting up a number scale on a line and a Cartesian coordinate system in the plane.

Turning now to affine geometry, as developed from its own axioms, **let us see how two segments can be compared in size.** This must, of course, be done without the help of axioms of congruence since there are no such axioms among the ten that we gave in §11. In fact, examination of these ten shows that not a single one deals with measurement of any sort.

We note, first, that a segment is defined just as in Euclidean geometry and that a parallelogram is defined as a quadrilateral whose opposite sides lie on parallel lines.

Let $\overline{AB}$, $\overline{CD}$ be two segments. We say that they are **congruent** if and only if (1) they are noncollinear and there is a parallelogram of which they are opposite sides, or (2) they are collinear and there are two parallelograms, one with opposite sides $\overline{AB}$, $\overline{EF}$, the other with opposite sides $\overline{CD}$, $\overline{EF}$ (Fig. IV, 8). It

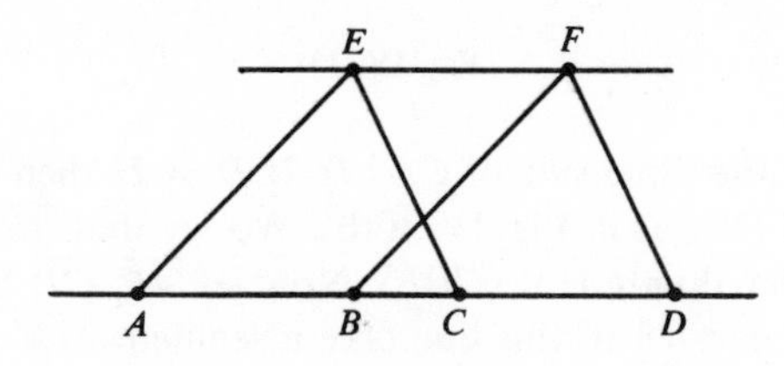

Fig. IV, 8

can be proved that if $\overline{PQ}$, $\overline{RS}$ are congruent, and likewise $\overline{RS}$, $\overline{TU}$, then $\overline{PQ}$, $\overline{TU}$ are congruent. It can also be proved that if $\overline{PQ}$ is any segment (Fig. IV, 9), and R is any point on line PQ or a parallel to line PQ, then there exist unique points S_1 and S_2 on opposite sides of R such that $\overline{PQ}$ is congruent to $\overline{RS_1}$ and to $\overline{RS_2}$.

Now let us see how to compare two segments on the same line or on parallel lines when they are not congruent. Let $\overline{AB}$, $\overline{CD}$ be two segments on parallel lines (Fig. IV, 10). To be precise we assume that B and D are on the same side of line AC. The parallel to line AC through B then meets line CD in a

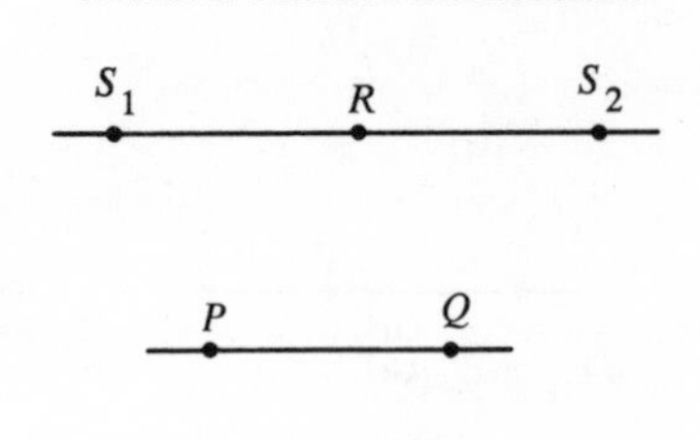

Fig. IV, 9

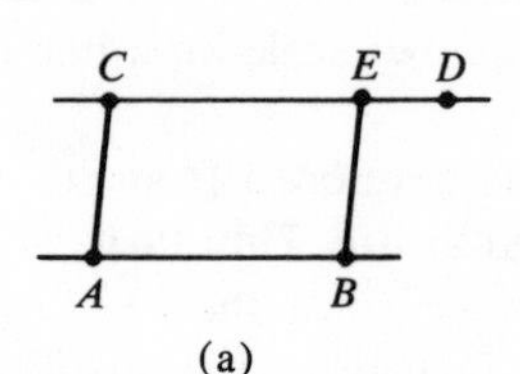

(a)

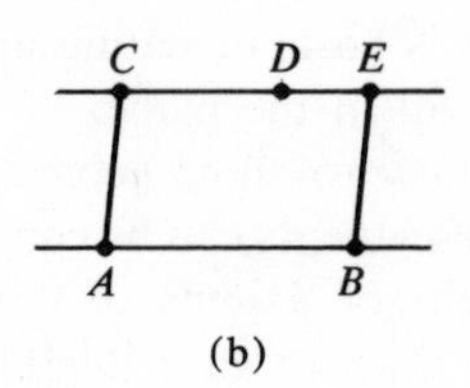

(b)

Fig. IV, 10

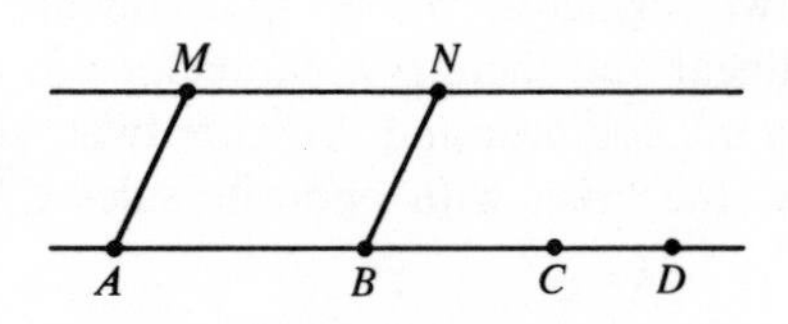

Fig. IV, 11

point E which is on the same side of C as D. If $D \neq E$, then either (CED), as in Fig. IV, 10(a), or (CDE), as in Fig. IV, 10(b). We say that $\overline{AB}$ is **smaller** than $\overline{CD}$ if (CED), and **greater** than $\overline{CD}$ if (CDE). Now let $\overline{AB}$, $\overline{CD}$ be on the same line (Fig. IV, 11). On a parallel to this line take a segment $\overline{MN}$ which is congruent to $\overline{AB}$. If $\overline{MN}$ is not congruent to $\overline{CD}$, then, by the preceding case, either $\overline{MN}$ is smaller than $\overline{CD}$, or it is greater than $\overline{CD}$. In the former situation we say that $\overline{AB}$ is **smaller** than $\overline{CD}$, and in the latter that it is **greater** than $\overline{CD}$. A simple consequence of this definition is that a segment $\overline{AB}$ is always greater than any of its subsegments $\overline{AC}$.

We have now seen that by means of parallelism and betweenness any two segments $\overline{AB}$, $\overline{CD}$ on the same line or on parallel lines can be compared in size purely geometrically. But it is not possible in affine geometry to compare two segments in this way when they are not on the same line or on parallel lines. This can be done in Euclidean geometry, as we saw, by means of congruence axioms. It cannot be done in affine geometry since that geometry lacks such axioms or their equivalent.

By choosing a segment on any given line to serve as unit segment, one could now show, just as in Euclidean geometry, that a positive number can be associated with every segment on that line or any parallel line, a number which, again, is called the **length** of the segment, or the **distance** between its endpoints. Then, if A, B are points on one of these lines and $\overline{AB}$ is congruent to the unit segment, the distance AB is 1; if C is a third point on line AB such that $\overline{BC}$ is congruent to the unit segment, the distance AC is 2; if D is a fourth point on that line such that $\overline{AD}$ is congruent to $\overline{DB}$, then each of the distances AD, DB is $\frac{1}{2}$; and so forth. Each family of parallel lines has such a distance system. Since it is not possible to compare two segments geometrically if they belong to different families, there is no *two-dimensional distance system* as in Euclidean geometry, a system in which every point-pair in the plane has a distance relative to the same unit segment.

Since lengths are numbers, the length of a segment on one line can, of course, be compared *numerically* with the length of a segment on an intersecting line, but such comparison is meaningless inasmuch as each length is relative to its own unit segment. However, it is to be noted that the *ratio* of the lengths of two segments on a line is independent of the choice of unit segment on the line. Such a ratio can sometimes be compared in a geometrically significant way with the corresponding ratio on an intersecting line. The following theorem offers an important example of this.

12.1 *Let A, B, C be distinct points on a line, and A', B', C' be distinct points on an intersecting line. If lines AA', BB', CC' are parallel* (Fig. IV, 12a), *or if $A = A'$ and lines BB', CC' are parallel* (Fig. IV, 12b), *or if $B = B'$ and lines AA', CC' are parallel* (Fig. IV, 12c), then

$$\frac{AB}{BC} = \frac{A'B'}{B'C'}.$$

In this formula, of course, lengths AB, BC are relative to a unit segment on *their* line, and lengths $A'B'$, $B'C'$ are relative to a unit segment on *their* line. It can be verified that the cases illustrated in Fig. IV, 12(a) and 12(b) are deducible from the case shown in Fig. IV, 12(c). By means of this theorem a segment can be subdivided into any number of equal parts just as in Euclidean geometry.

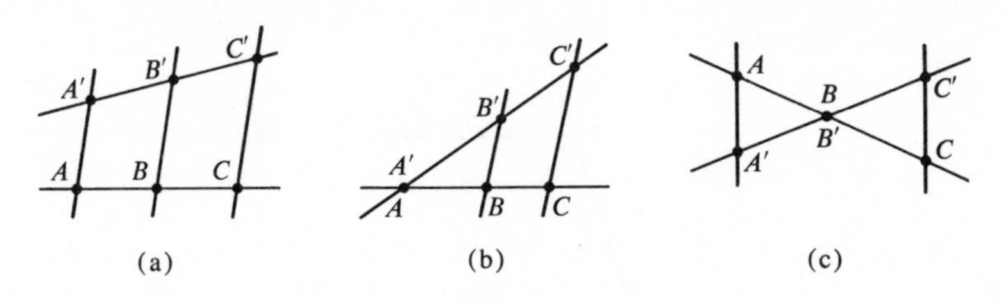

Fig. IV, 12

The fact that there is no two-dimensional distance system in affine geometry does not prevent us from setting up a *two-dimensional coordinate system*. This is done much as in Euclidean geometry. We take as *x- and y-axes* any two lines that meet, call their common point O the *origin of coordinates*, and set up number scales on the axes, each axis having its own unit segment. Since there is no concept of angle-size in affine geometry, it would make no sense to require the axes to be perpendicular.

If P is any point in the affine plane not on an axis, let the parallels through P to the axes meet the x-axis in M and the y-axis in N (Fig. IV, 13). The coordinates (x, y) of P are then defined by the equations

$$x = OM, \qquad y = ON,$$

where OM and ON are *directed distances*. If P is on the x-axis, we define its coordinates to be $x = OP$, $y = 0$, and if it is on the y-axis we define them to be $x = 0$, $y = OP$, where OP is a directed distance.

Let $P_1(x_1, y_1)$, $P_2(x_2, y_2)$ be two points. One easily sees that if they are on the x-axis or a parallel to it, the directed distance P_1P_2 is given by the formula

$$P_1P_2 = x_2 - x_1,$$

and that if they are on the y-axis or a parallel to it, this distance is given by

$$P_1P_2 = y_2 - y_1.$$

Using these formulas and theorem 12.1, and defining *ratio of division* in the usual way, one can prove the following just as in Euclidean geometry:

12.2 *Let $P_1(x_1, y_1)$, $P_2(x_2, y_2)$ be distinct points. If the point $P(x, y)$ divides $\overline{P_1P_2}$ in the ratio $r = P_1P/PP_2$, then*

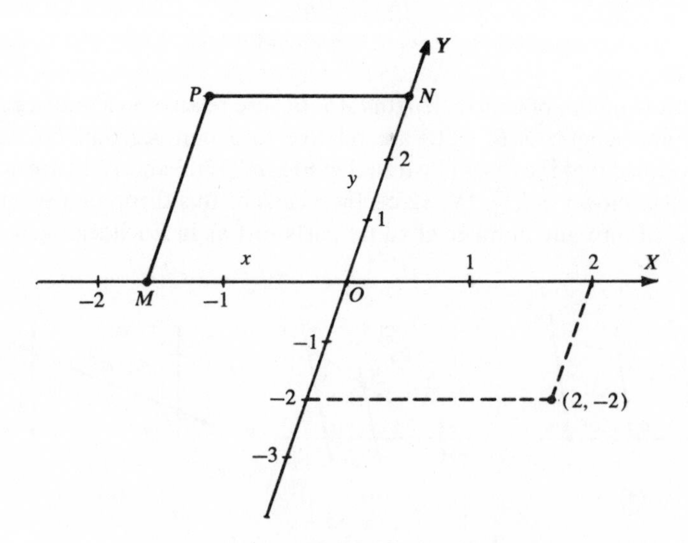

Fig. IV, 13

$$r = \frac{x - x_1}{x_2 - x} = \frac{y - y_1}{y_2 - y} \tag{1}$$

if neither denominator is zero, and

$$x = \frac{x_1 + rx_2}{1 + r}, \qquad y = \frac{y_1 + ry_2}{1 + r}; \tag{2}$$

conversely, the point P which has the coordinates (x, y) given by (2) divides $\overline{P_1P_2}$ in the ratio $r = P_1P/PP_2$.

The axes and lines parallel to them clearly have the equations $x =$ constant and $y =$ constant. Using 12.2 we now show that a line not parallel to an axis likewise has a first-degree equation. Let such a line be given, let P_1, P_2 be any two of its points, and let $P(x, y)$ be another point of the line. Then P divides $\overline{P_1P_2}$ in a certain ratio r, so that, using (1), we obtain

$$\frac{x - x_1}{x_2 - x} = \frac{y - y_1}{y_2 - y},$$

or, after simplifying,

$$(y_2 - y_1)x - (x_2 - x_1)y + x_2y_1 - x_1y_2 = 0. \tag{3}$$

This is an equation which is necessarily satisfied by the coordinates of P. Since the steps that led to the equation can be reversed, it is the equation of the line. Thus, all lines have first-degree equations. The converse is also true, and we leave it as an exercise.

12.3 *All lines have first-degree equations and, conversely, the graphs of all first-degree equations are lines.*

From our earlier discussion in this section it is clear that an individual directed distance on the x-axis has no meaningful relation to an individual directed distance on the y-axis. According to 12.3, however, the *set* of directed distances OM on the x-axis and the *set* of directed distances ON on the y-axis do have significant relations. An example will make this clear. Take any directed distance OM on the x-axis. Then take the directed distance ON on the y-axis such that $ON = 3 \cdot OM + 2$. Let P be the point of intersection of the parallel to the x-axis through N and the parallel to the y-axis through M. If OM is permitted to vary, and the corresponding value of ON is always determined from the preceding equation, then the locus of P is a line according to 12.3, for this equation is of the first degree in OM and ON. In more familiar notation the equation, of course, is $y = 3x + 2$.

Two final remarks. (1) In the coordinate systems used for drawing most practical graphs the unit on one scale bears no relation to the unit on the other scale, just as in an affine coordinate system. One scale, for example, might represent dollars, and the other years. (2) In doing graph work with an affine coordinate system, we draw parallel lines in the conventional way since all that is wanted are physical (pencil, ink, chalk, etc.) lines that would not meet even if greatly extended.

EXERCISES

1. Using 12.1, show how to bisect and trisect a segment.

2. Prove 12.2.

3. Prove the converse part of 12.3.

4. Using affine coordinates, prove that the segment joining the midpoints of two sides of a triangle is half as long as the third side and that the lines containing the segment and the third side are parallel.

5. Show that (3) can be written

$$\begin{vmatrix} x & y & 1 \\ x_1 & y_1 & 1 \\ x_2 & y_2 & 1 \end{vmatrix} = 0,$$

which is, then, the *equation of the line through two points* (x_1, y_1) and (x_2, y_2).

6. If $Ax + By + C = 0$ is a line not parallel to an axis, show that the ratio $(y_2 - y_1)/(x_2 - x_1)$ has the same value for all pairs of points (x_1, y_1), (x_2, y_2) on the line, and interpret the ratio geometrically.

7. Give the equation of the line with intercepts a and b on the x- and y-axes, respectively.

8. Using axes that do not look perpendicular, unit segments on them that do not look equal, and at least six points in each case, draw the graphs of the following:
(a) $2x - 3y + 12 = 0$; (b) $y = x^2$; (c) $x^2 + y^2 = 25$.

9. Which of the following statements are correct?
(a) If two segments have equal lengths, they are congruent.
(b) The distance between the points (1, 0) and (0, 1) is $\sqrt{2}$.
(c) If two segments are congruent, their lengths are equal.
(d) The line through (1, 0), (0, 1) is parallel to the line through (2, 0), (0, 2).
(e) The length of the perimeter of the quadrilateral with consecutive vertices (0, 0), (1, 0), (1, 1), (0, 1) is 4.
(f) If $b_1 \neq b_2$, the equations $y = mx + b_1$ and $y = mx + b_2$ represent parallel lines.
(g) The segments with endpoints (0, 0), (1, 1) and (1, 1), (2, 2) are congruent.

CHAPTER V

Projections, a Transition

1. Introduction. Having detoured slightly to indicate how affine geometry can be developed as an independent system, let us now resume our study of transformations and geometries within the Euclidean framework. As we have mentioned, the affine properties of geometric figures are only the first of three important types of nonmetric properties that we intend to explore, the others being the projective properties and the topological properties. In this chapter and the following one projective properties are studied. We first consider certain simple transformations known as *projections* and show how affine transformations can be viewed in terms of them. This new way of thinking of affine transformations leads us to the concept of a projective transformation of the plane into itself, and hence to the projective geometry of that plane.

2. Parallel Projection of a Line. Let us review the familiar idea of projecting the points of a line g onto a coordinate axis. If A is any point of g, and A' is the foot of the perpendicular from A to the x-axis, we call A' the ***orthogonal projection*** of A on the x-axis (Fig. V, 1). If g is not perpendicular to

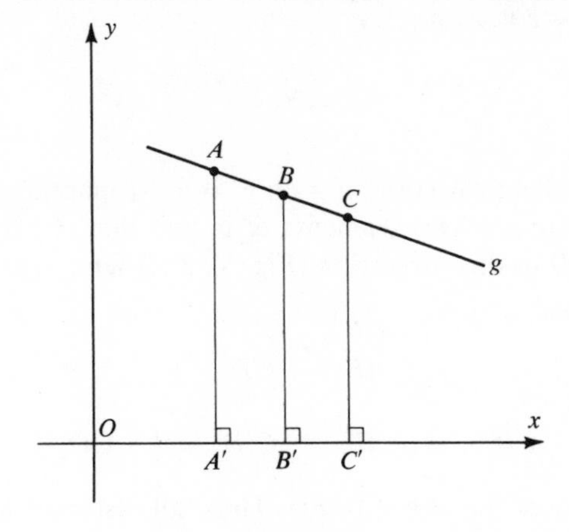

Fig. V, 1

the x-axis, this process of projecting points establishes a 1-1 transformation of the points of g into those of the x-axis which is called an *orthogonal projection.* (The double use of this term will cause no confusion.) Since the lines joining pairs of corresponding points are parallel to the y-axis, the transformation is also called a *parallel projection* in the direction of the y-axis. Viewed in this way the transformation is just one of the many parallel projections of the points of g into the points of the x-axis which result from using other directions than that of the y-axis. Fig. V, 2 shows one of these other parallel projections.

We now generalize the above. Let g, g' be distinct coplanar lines (Fig. V, 3). Let a family of parallel lines coplanar with them, but containing neither, be specified. If P is any point of g, the line of the family passing through P will meet g' in a point P'. The transformation of the points of g into those of g' in which this point P' is the image of P is called a **parallel projection of g on g'.** In particular, it is an *orthogonal projection of* g *on* g' if the lines of the family are perpendicular to g'.

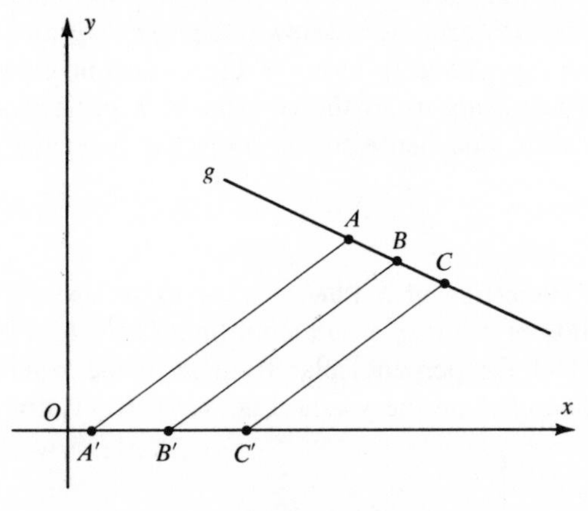

Fig. V, 2

Clearly, a parallel projection of g on g' is a 1-1 mapping of g on g'. Suppose that $\overline{AB}$, $\overline{CD}$ are any two segments of g, and that A', B', C', D' are the images of A, B, C, D in this projection (Fig. V, 3). Then

$$\frac{A'B'}{AB} = \frac{C'D'}{CD},$$

or

$$A'B' = k \cdot AB, \qquad C'D' = k \cdot CD,$$

where k is the value of the above ratios. Thus, all distances are multiplied by the factor k. Ratio of division is therefore preserved, and hence betweenness, too. Distance is not preserved unless $k = 1$.

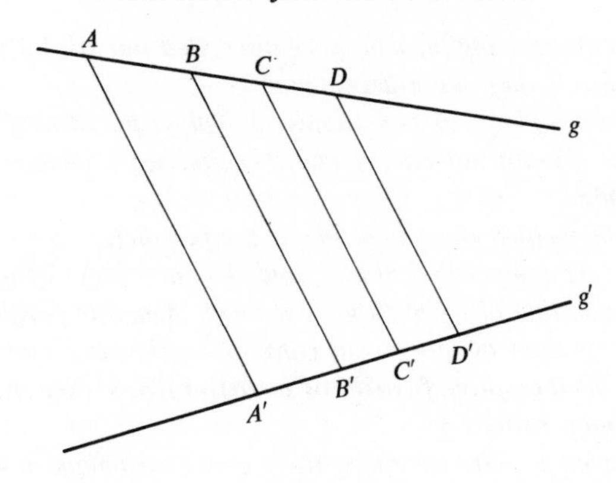

Fig. V, 3

2.1 *A parallel projection of a line g on a line g′ maps g onto g′ in a* 1-1 *way, multiplies all distances by the same factor* (*which may or may not be* 1), *and thus preserves ratio of division and betweenness.*

EXERCISES

1. Find the images of (1, 0), (2, 0), (3, 0) in the projection of the x-axis on the y-axis parallel to lines with the following slopes:

(a) 1; (b) -1; (c) 2.

2. Find the images of $(-1, 2)$, $(1, 2)$, $(3, 2)$ in the parallel projection of the line $y = 2$ on the x-axis if the projection is (a) orthogonal, (b) parallel to lines with inclination 30°.

3. If g and g' meet, show that there are just two parallel projections of g on g' which preserve distance. What can be said if g and g' are parallel?

4. In the orthogonal projection of a line on a coplanar line a segment goes into another which is $\cos\theta$ times as long, θ being the acute angle between the lines. Verify this by finding the images of (0, 2) and (3, 2) in the orthogonal projections of the line $y = 2$ on (a) the line $y = x$, (b) the line $y = \frac{1}{2}x$.

3. Parallel Projection of a Plane. Although we did not do this in any previous chapter, it is now necessary, in continuing the study of projections, to consider mappings of one plane onto a *different* plane. In doing this we make use of the following elementary facts of space geometry:

1. *A line in space is determined by any two points.*
2. *A plane in space is determined by three noncollinear points, or by a line and a point not on it.*

3. *Two nonintersecting lines in space are called* **parallel** *if they are coplanar, and* **skew** *if they are not coplanar.*
4. *A line and a plane, or two planes, are called* **parallel** *if they do not meet.*
5. *Through a point not on a plane there passes a unique plane parallel to that plane.*
6. *Two planes either meet in a line or are parallel.*
7. *If α, α' are nonparallel planes, and A is any point of α not on α', there is a unique line of α which goes through A and is parallel to α'.*
8. *The set of lines in space consisting of a given line and all lines parallel to it is called a* ***space family of parallel lines***. (We shall use the briefer term ***space family***.)
9. *Through each point of space there passes a unique line belonging to a given space family.*
10. *Those lines of a space family which lie in the same plane are called a* ***plane family of parallel lines.***
11. *The lines of a space family which meet a line not in the family constitute a plane family of parallel lines and form a plane.*
12. *A space family is said to be* **parallel to a plane** *only if it contains lines parallel to the plane.*
13. *There are infinitely many space families that are parallel to a given plane.*

Now consider any distinct planes α, α' (Fig. V, 4(a), (b)), and a space family not parallel to either. If A is any point of α, a unique line of the family goes through A, and it meets α' in a unique point A'. The transformation of the points

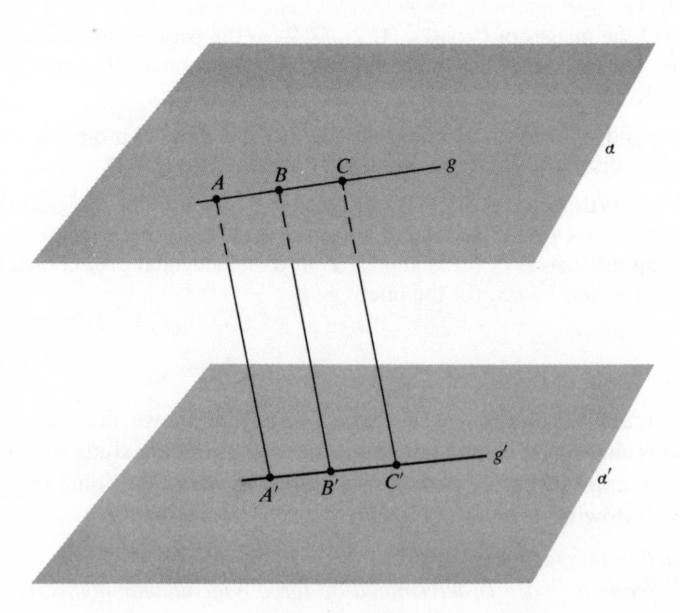

Fig. V, 4(a)

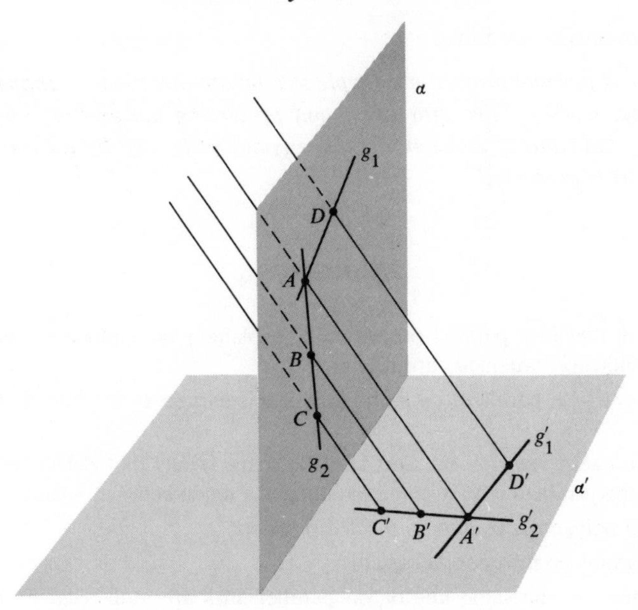

Fig. V, 4(b)

of α into those of α' in which this point A' is the image of A is called a **parallel projection of α on α'**. There is clearly one such transformation of α into α' for each space family of the specified kind, and it is always 1-1.

Let T be a parallel projection of α on α'. If g is any line in α, the lines of the space family which go through the points of g form a plane family of parallel lines. The plane of this family meets α' in a line g' consisting of the images of the points of g. Thus g' is the image of g, and T effects a parallel projection of g on the coplanar line g'. By 2.1, then, all distances on g go into distances on g' which are some positive constant k times as great. Since g is any line of α, it follows that T preserves betweenness and ratio of division. Also, since T is 1-1 and sends lines into lines it must preserve their parallelism or concurrence.

Let us look more closely at the effect of T on distance. If α, α' are parallel (Fig. V, 4a), so are g, g', and hence $k = 1$; in this case, T preserves distance. If α, α' meet, the value of k will usually be different for different positions of g. Fig. V, 4(b) shows two positions, g_1 and g_2; k is less for g_2 than for g_1. When α, α' meet, then, T does not usually multiply all distances by the same constant. When it does multiply all distances by the same constant, however, the constant must be 1. To show this, let A be any point of α not on α', and let α'' be the plane through A parallel to α. Let g be the line of intersection of α, α''. Then g and its image g' are coplanar but do not meet. They are therefore parallel and for them $k = 1$. Hence if T multiplies all distances by the same constant, the latter must be 1.

To summarize, we have:

3.1 *A parallel projection of a plane α on another plane α' maps α onto α' in a* 1-1 *way, sending lines into lines, and preserving parallelism, concurrence, betweenness, and ratio of division. Distance is not multiplied by a constant factor except when it is preserved.*

EXERCISES

1. Prove that in a parallel projection of a plane α on a plane α', as discussed above, the following statements are true:

(a) if α, α' meet, g' is parallel to g if and only if g is parallel to the line of intersection of α, α';

(b) distance is preserved if α, α' meet and the space family that determines the projection is perpendicular to a plane bisecting the angle between α and α';

(c) an n-sided polygon goes into an n-sided polygon;

(d) parallelograms go into parallelograms;

(e) all distances on the same line or on parallel lines are multiplied by a constant factor.

4. Parallel Projections and Affine Transformations. Suppose that we start with a plane α, project it on another plane α_1, then project α_1 on a plane α_2, then project α_2 on α_3, and so forth, continuing in this way until we have achieved n parallel projections. Let us suppose, also, that $\alpha_n = \alpha$, that is, the last projection maps α_{n-1} on α. If we denote these projections by $T_1, T_2, \ldots, T_n$, respectively, then $T_1(\alpha) = \alpha_1, T_2(\alpha_1) = \alpha_2, \ldots, T_n(\alpha_{n-1}) = \alpha$ and $T_1T_2T_3\ldots T_n$, the resultant of the n projections, is a mapping of α onto itself. Symbolizing this resultant by T, we can write $T(\alpha) = \alpha$. Since each of the projections is 1-1 and preserves collinearity, the same is true of T. According to IV, 10.2, then, T must be affine. Thus we have:

4.1 *A mapping of a plane onto itself which is the resultant of parallel projections is an affine transformation.*

The converse of this is also true. For let T be an affine transformation of α, and let A', B', C' be the images of any noncollinear points A, B, C in α. If α_1 is a plane that intersects α along line AB, let T_1 be any parallel projection of α on α_1. Then T_1 leaves A and B fixed and sends C into some point C_1 which is in α_1 but not in α. Thus, $T_1(A, B, C) = A, B, C_1$. Now project α_1 back on α so that C_1 goes into C'. Denoting this parallel projection by T_2, we have $T_2(A, B, C_1) = A, B, C'$. The resultant T_1T_2 thus maps α on itself so that A, B, C go into the noncollinear points A, B, C', respectively. We now repeat the foregoing procedure, working with A, B, C' instead of A, B, C. Let α_2 be a plane intersecting α along line AC', and T_3 a parallel projection of α on α_2. Then $T_3(A, B, C') = A, B_2, C'$, where B_2 is some point in α_2 but not in α. Project α_2 on α so that B_2 goes into B'. If T_4 is this parallel projection, then $T_4(A, B_2, C') = A, B', C'$. Thus T_3T_4 maps α on itself so that A, B, C' go into the noncollinear points A, B', C'. Finally, working with A, B', C' instead of A, B, C', we can find two parallel projections T_5, T_6 whose resultant maps α on itself so that A, B', C' go into A', B', C'. The transformation $T_1T_2T_3T_4T_5T_6$ then maps α on itself so that A, B, C go into A', B', C'. This transformation is affine by 4.1 and identical with T by IV, 2.4. We have thus proved:

4.2 *Every affine transformation (of a plane into itself) is the resultant of parallel projections.*

Thus, Theorems 4.1 and 4.2 give us a new way of viewing the affine transformations dealt with in Chapter IV, a way which involves the use of three-dimensional space. The views obtained in Chapter IV, in terms of primitive transformations, of course, do not require that we leave the plane (see IV, §5).

Motions and similarities, being affine transformations of special kinds, can therefore be viewed in terms of projections, as well as in the various ways with which we became acquainted in Chapters II and III. This will be illustrated in the exercises. As the case of a similarity then shows, the resultant of parallel projections can be a transformation in which all distances are multiplied by the same constant different from 1, despite the fact that no parallel projection has this property.

According to the proof of 4.2, not more than six parallel projections are required to generate any affine transformation. In the case of affine transformations of special type, however, fewer than six projections may suffice, as will be seen in the exercises.

EXERCISES

1. Show that a reflection can be generated by (that is, is the resultant of) two parallel projections.

2. Deduce from Ex. 1 that four parallel projections, at most, are required to generate a rotation or a translation. (*Hint*. See II, 13.3.)

3. Show that a translation can be generated by two parallel projections.

4. Show that the affine transformation which sends (0, 0), (1, 0), (0, 2) into (0, 0), (1, 0), (4, 3), and hence which is not a similarity, is the resultant of two parallel projections.

5. Give an example of an affine transformation, not a similarity, which is the resultant of four parallel projections. (*Hint*. Build on the transformation of Ex. 4.)

6. Can an affine transformation of a plane into itself be generated by an odd number of parallel projections? Explain.

5. Central Projections. As we have seen, parallel projections are those transformations in which the lines joining points to their images are all parallel. Next, we shall consider the type of transformation in which the lines joining points to their images are all concurrent.

Let g, g' be two coplanar lines, and let O be a point which is in their plane but not on g or g' (Fig. V, 5). If P is any point of g, and line OP meets g' in P', we call the transformation in which this point P' is the image of P the **central projection of g on g' with center O** (the word "central" is often omitted). In this section we consider only the case in which g, g' are parallel.

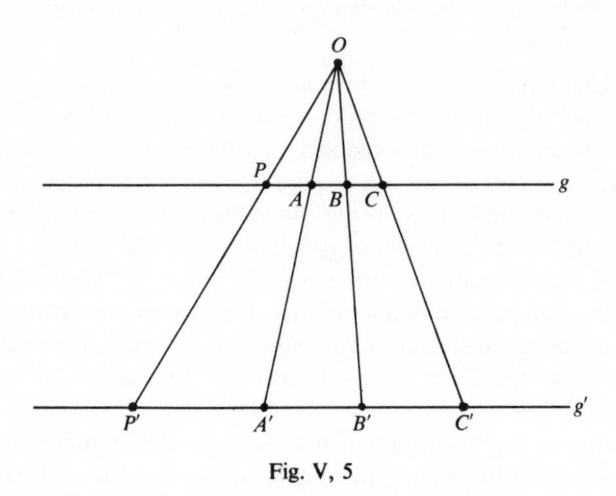

Fig. V, 5

Then we clearly have a 1-1 mapping of g onto g'. Also, if A, B, C go into A', B', C', we know from similar triangles that

$$\frac{A'B'}{AB} = \frac{A'C'}{AC} = \frac{B'C'}{BC}.$$

Hence, denoting these equal ratios by k, we have

$$A'B' = k \cdot AB, \quad A'C' = k \cdot AC, \quad B'C' = k \cdot BC.$$

Thus all distances are multiplied by the same constant k. It is easily verified that k is not 1 except when O is midway between g and g'. In Fig. V, 5, for example, where this condition is not met, $k \neq 1$ since lines AA', BB', CC' are not parallel. To summarize, we have:

5.1 *A central projection of a line g on a parallel line g' maps g onto g' in a* 1-1 *way and multiplies all distances by the same factor k (generally not* 1), *thus preserving ratio of division and betweenness.*

It may be noted that this projection differs in its basic properties from a parallel projection of a line on an intersecting line only in that it has no fixed point.

Now let α, α' be distinct planes, and O be a point on neither (Fig. V, 6). If A is any point of α, and line OA meets α' in a point A', we call the transformation in which this point A' is the image of A the **central projection of α on α' with center O** (the word "central" is often omitted).

In this section we consider only the case in which α, α' are parallel. Then the transformation, which we denote by T, clearly maps α onto α' in a 1-1 way. If g is any line in α, the lines joining O to the points of g form a plane, and this plane meets α' in a line g'. Hence $T(g) = g'$. Thus T sends lines into lines and, being 1-1, preserves parallelism and concurrence.

Now g, g' are coplanar, but do not meet, for one is in α, and the other in α'. Thus, g and g' are parallel, and T achieves a projection of g on g' with center O. By 5.1, then, all distances on g are multiplied by a constant k. The value of k is the same for all lines g, as is seen from Fig. V, 6. In this figure, AB and $A'B'$ are corresponding distances on g and g', line OCC' is perpendicular to α and α', triangles OAB, $OA'B'$ are similar, and so are triangles OAC, $OA'C'$. Hence

$$k = \frac{A'B'}{AB} = \frac{OA'}{OA} = \frac{OC'}{OC}.$$

Thus, k equals the ratio of the perpendicular distances from O to α' and α. **It is therefore independent of g.**

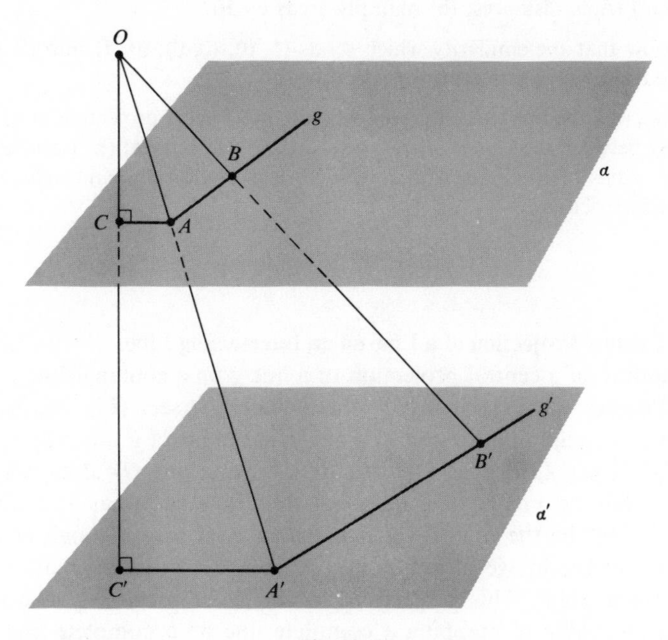

Fig. V, 6

From what has already been shown, it follows that ratio of division and betweenness are also preserved. In summary, then, we have:

5.2 *A central projection of a plane α on a parallel plane α' maps α onto α' in a* 1-1 *way, sends lines into lines, and multiplies all distances by the same constant k, thus preserving parallelism, concurrence, ratio of division, and betweenness.*

It is to be noted that when $k \neq 1$, the central projections of this theorem have a distance property possessed by no parallel projection of one plane on another. On the other hand, a mapping of a plane onto itself which is the resultant of such central projections is necessarily a similarity, and hence it can also be generated by parallel projections (see 4.2).

EXERCISES

1. Verify that it is possible for a central projection of a line on a parallel line, or of a plane on a parallel plane, to preserve distance.

2. Prove that a central projection of a plane on a parallel plane preserves (a) parallelism; (b) concurrence; (c) ratio of division; (d) betweenness; (e) angle-size.

3. Find the images of (1, 2) and (3, 2) in the projection of the line $y = 2$ on the x-axis from the center (0, 4).

4. Where should the center be if the central projection of a plane on a parallel plane is to (a) triple distances; (b) multiply areas by 10?

5. Show that the similarity which sends (1, 0), (0, 0), (0, 1) into (2, 0), (0, 0), (0, 2) is the resultant of two central projections.

6. Describe the images of the following figures for the projection in which plane α is midway between the center and plane α': (a) right triangle; (b) isosceles triangle; (c) circle of radius r; (d) ellipse with semimajor axis a and semiminor axis b; (e) conic section of eccentricity e.

6. Central Projection of a Line on an Intersecting Line. We now return to the definition of a central projection of a line g on a coplanar line g' given in §5, and consider the case in which these lines intersect (Fig. V, 7). Let the parallel to g' through O meet g in V. Then every point of g other than V has an image on g'. Thus A, B, C go into A', B', C'. Since line OV does not meet g', the point V has no image. The mapping thus fails as far as V is concerned. Similarly, let W' be the point in which the parallel to g through O meets g'. Then W' is not the image of any point, whereas every other point of g' is the image of some point. This projection, unlike those previously considered, is seen to be incapable of mapping a complete line on a complete line. It does,

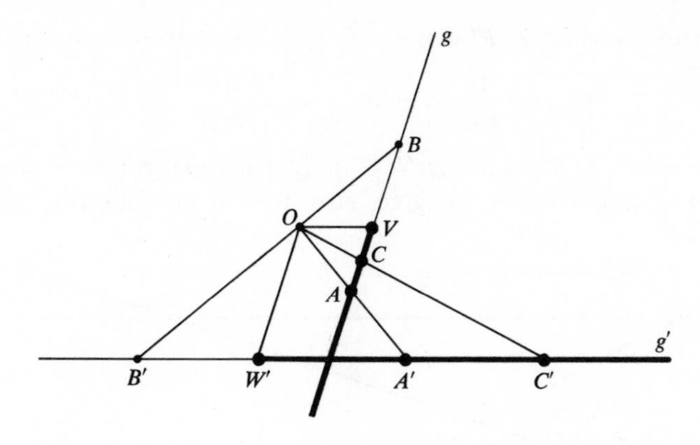

Fig. V, 7

however, establish a 1-1 correspondence between the points of g, excluding V, and the points of g', excluding W'.

If, in Fig. V, 7, we take A closer and closer to V, we find that A' is more and more distant on g'. But when A coincides with V, A ceases to have an image. The latter has vanished, as it were. For this reason, in this projection, V is called the *vanishing point on* g, and W' is called the *vanishing point on* g'. It is in terms of the vanishing points that one obtains a firm over-all grasp of the mapping. Thus the part of g above V goes into the part of g' to the left of W', and the part below V goes into the part to the right of W'. This correspondence is indicated by weighted lines in Fig. V, 7. Points far from V, for example, go into points near W'.

The projection has other interesting features. Betweenness is not always preserved. For example, C' is not between A' and B', although C is between A and B. In fact, no point between A and B goes into a point between A' and B', and, in particular, the midpoint of $\overline{AB}$ does not go into the midpoint of $\overline{A'B'}$. Segments, therefore, do not always go into segments. $\overline{AV}$, for example, goes into a ray, namely, the part of g' which is to the right of A'. Similarly, $\overline{BV}$ goes into the part of g' to the left of B'. Distances are therefore subject to great distortions. Thus, two points very close to V, one in $\overline{AV}$ and the other in $\overline{BV}$, go into points very distant from each other, one being far to the right on g', the other far to the left. By contrast, $A'C'$ is not more than two or three times AC. Therefore, distances cannot all be multiplied by the same constant. Because of this, and what was noted above concerning betweenness, it is clear that ratio of division is not invariant.

We now determine how ratio of division changes. Let A, B, P be distinct points of g different from V, and let their images be A', B', P'. In our discussion A, B are on the same side of V (Fig. V, 8a). The case in which this is not so is left as an exercise. Also, we have indicated positive directions on g, g'.

The ratios in which P, P' divide $\overline{AB}$, $\overline{A'B'}$, respectively, are

$$r = \frac{AP}{PB} \quad \text{and} \quad r' = \frac{A'P'}{P'B'}.$$

Let us first assume (APB), as in Fig. V, 8(a). Then each length in the above ratios is positive. For simplicity let us agree, also, that in the following analysis all

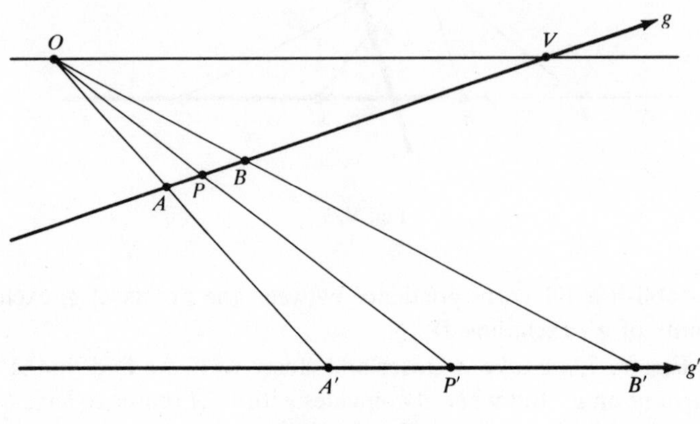

Fig. V, 8(a)

angles and other measures are undirected, and hence positive. By the Law of Sines, we have, from triangle AOP,

$$AP = AO \cdot \frac{\sin \angle AOP}{\sin \angle OPA}, \tag{1}$$

and from triangle BOP,

$$PB = BO \cdot \frac{\sin \angle BOP}{\sin \angle OPB} = BO \cdot \frac{\sin \angle BOP}{\sin \angle OPA}, \tag{2}$$

since $\angle OPB$ and $\angle OPA$ are supplementary and have equal sines. Dividing (1) by (2) gives

$$\frac{AP}{PB} = \frac{AO \cdot \sin \angle AOP}{BO \cdot \sin \angle BOP}. \tag{3}$$

In the same way, using the image triangles we get

$$\frac{A'P'}{P'B'} = \frac{A'O \cdot \sin \angle A'OP'}{B'O \cdot \sin \angle B'OP'}. \tag{4}$$

Dividing (4) by (3) and noting that $\angle AOP = \angle A'OP'$, $\angle BOP = \angle B'OP'$, we obtain

$$\frac{r'}{r} = \frac{A'O \cdot BO}{AO \cdot B'O},$$

$$r' = \frac{A'O \cdot BO}{AO \cdot B'O} r. \tag{5}$$

With slight changes the above analysis yields this same result when P is not between A, B. Suppose, for example, that (ABP), as in Fig. V, 8(b). Since PB is then negative, the right-hand side of (2) must be preceded by a negative sign. Likewise for the equation containing $A'P'$ which is analogous to (1). Also, $\angle AOP$ and $\angle A'OP'$ are now supplementary, as are $\angle BOP$ and $\angle B'OP'$.

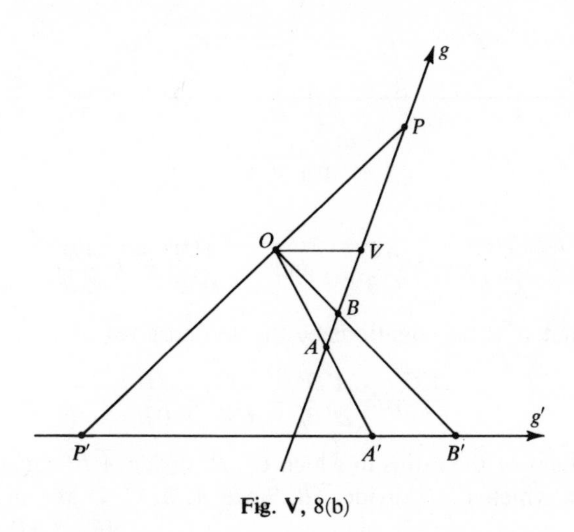

Fig. V, 8(b)

From (5) we see, first, that $r' \neq r$, and hence that ratio of division is not preserved. For if the coefficient of r were 1, then $A'O/B'O$ would equal AO/BO, and this would imply that triangles AOB, $A'OB'$ are similar, which is impossible since g and g' are not parallel.

Next we note that the coefficient of r does not involve P or P', and hence has the same value regardless of their positions. Denoting this value by k, and replacing r by AP/PB and r' by $A'P'/P'B'$, we can then write

$$\frac{A'P'}{P'B'} = k\,\frac{AP}{PB}. \tag{6}$$

In other words, if A, B are kept fixed, and P is allowed to vary, the ratio in which P' divides $\overline{A'B'}$ is always the same multiple k of the ratio in which P divides $\overline{AB}$. The sign of k is positive or negative according as A, B are on the same side or on opposite sides of V. In Figures V, 8(a), 8(b), for example, $k > 0$, with AO, $A'O$, BO, $B'O$ all positive by agreement. If A, B were above V, and these distances were regarded as directed, AO, $A'O$ would have opposite signs, and so would BO, $B'O$, and hence k would be positive again.

Equation (6) leads immediately to the discovery of a quantity which is invariant for the projection. For let C, D be two positions of P, and C', D' the corresponding positions of P' (Fig. V, 9). Then, by (6),

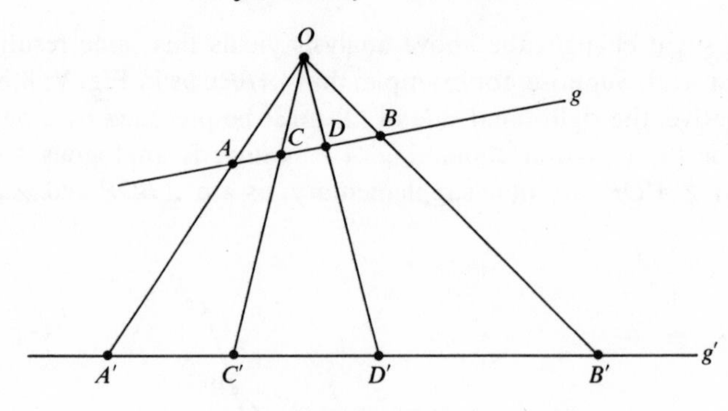

Fig. V, 9

$$\frac{A'C'}{C'B'} = k\frac{AC}{CB} \quad \text{and} \quad \frac{A'D'}{D'B'} = k\frac{AD}{DB}.$$

Dividing the first of these equations by the second gives

$$\frac{A'C'}{C'B'} \div \frac{A'D'}{D'B'} = \frac{AC}{CB} \div \frac{AD}{DB}. \tag{7}$$

Thus, the quotient of the ratios in which C', D' divide $\overline{A'B'}$ equals the quotient of the ratios in which C, D divide $\overline{AB}$. Since A, B, C, D are any four distinct points of g which have images, we see that the quantity $\dfrac{AC}{CB} \div \dfrac{AD}{DB}$ is an invariant of the projection.

6.1 Definition. *If A, B, C, D are distinct collinear points, the quantity*

$$\frac{AC}{CB} \div \frac{AD}{DB}$$

is called the cross-ratio in which C, D divide the point-pair A, B (or the segment $\overline{AB}$), and is represented by the symbol (AB, CD), which is read "the cross-ratio of A, B, C, D."

Equation (7), which expresses the invariance of cross-ratio in the projection of g on g', can now be written as

$$(A'B', C'D') = (AB, CD). \tag{8}$$

To summarize the present section, then, we have:

6.2 *A central projection of a line g on an intersecting line g' maps g onto g' in a* 1-1 *way except for one point on each line, and preserves cross-ratio for points that have images, but does not preserve ratio of division.*

It is important to note that all of the projections considered earlier likewise preserve cross-ratio, for they leave ratio of division unchanged, and hence also the quotient of two such ratios.

EXERCISES

1. Derive (5) when A, B are above V in Fig. V, 8(b).

2. Derive (5) when A, B are on opposite sides of V in Fig. V, 8(b).

3. Show that the projection considered in §6 never preserves ratio of division.

4. Find the images A', B', C', D' of $A(0, 1)$, $B(0, 2)$, $C(0, 3)$, $D(0, 4)$ in the projection of the y-axis on the x-axis from the center $(-1, 5)$.

5. Find (AB, CD) and $(A'B', C'D')$ for the points of Ex. 4, and verify that they are equal.

6. If A, B, C, D are distinct collinear points, and O is any point not on their line, then

$$|(AB, CD)| = \frac{\sin \angle AOC \cdot \sin \angle BOD}{\sin \angle BOC \cdot \sin \angle AOD},$$

where the angles are positive and less than π. Prove this when the points are situated: (a) as in Fig, V, 9; (b) as in Fig. V, 9, but with B and D interchanged.

7. Central Projection of a Plane on an Intersecting Plane. Returning to the definition of a central projection of a plane α on another plane α' given in §5, we now consider the case in which α, α' meet (Fig. V, 10).

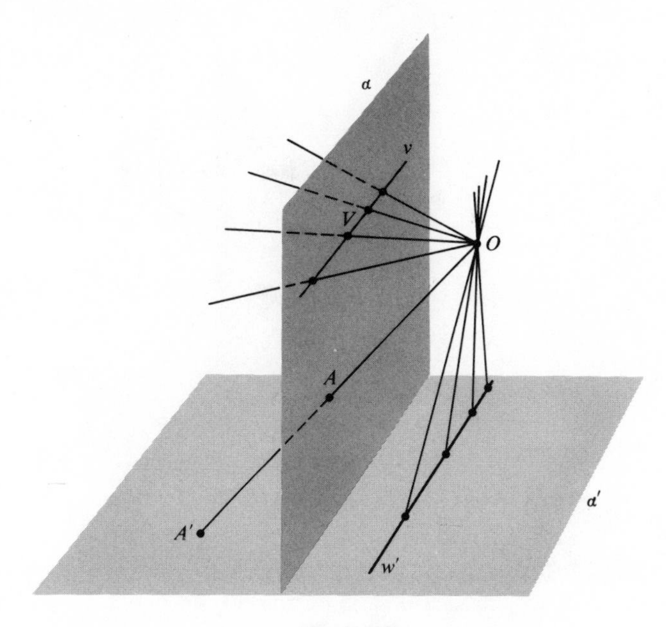

Fig. V, 10

A point of α can fail to have an image only if the line joining it to O is parallel to α'. All such points lie at the intersection of α with the plane through O parallel to α', and therefore constitute a line v. We call v the *vanishing line in* α, and its points the *vanishing points in* α. Similarly, there are *vanishing points in* α', points without any originals in α. They constitute a line w', the *vanishing line in* α', which is the intersection of α' with the plane through O parallel to α. Clearly, each point of α not on v has a unique image, and each point of α' not on w' has a unique original. Thus, in Fig. V, 10, A' is the image of A. We see that, except for the points of v and w', the projection maps α in a 1-1 way on α'.

Let us denote this projection by T and study it further. If g is any non-vanishing line in α, the plane determined by g and O meets α' in a line g'. Since g, g', and O are coplanar, and O is on neither line, T effects a central projection of g on g' with center O. If g is parallel to v (Fig. V, 11(a)), then g, g' are parallel (Ex. 1(c)) and the complete line g goes into the complete line g' by 5.1. If g meets v (Fig. V, 11(b)), then g, g' intersect (Ex. 1(c)) and g is not mapped completely on g' according to 6.2. Hence we cannot say that T always sends complete lines into complete lines. But T does pair the lines of α with those of α', excluding v and w', and its preserves collinearity in the sense that if A, B, C are collinear and have images, the latter are also collinear. Also, in view of 6.2 and the last paragraph of §6, T preserves cross-ratio but not ratio of division. Of course, then T cannot always preserve betweenness. We summarize these results as follows:

7.1 *A central projection of a plane on an intersecting plane establishes a* 1-1 *correspondence between the points of the planes, except for the points of one*

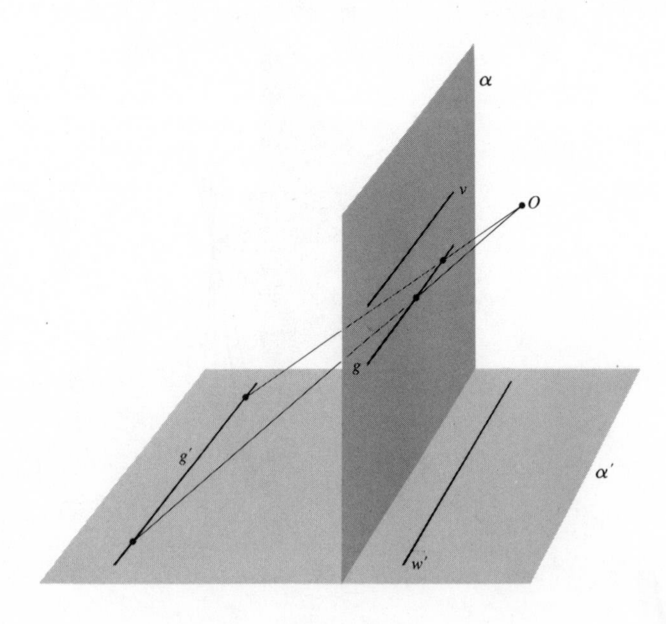

Fig. V, 11(a)

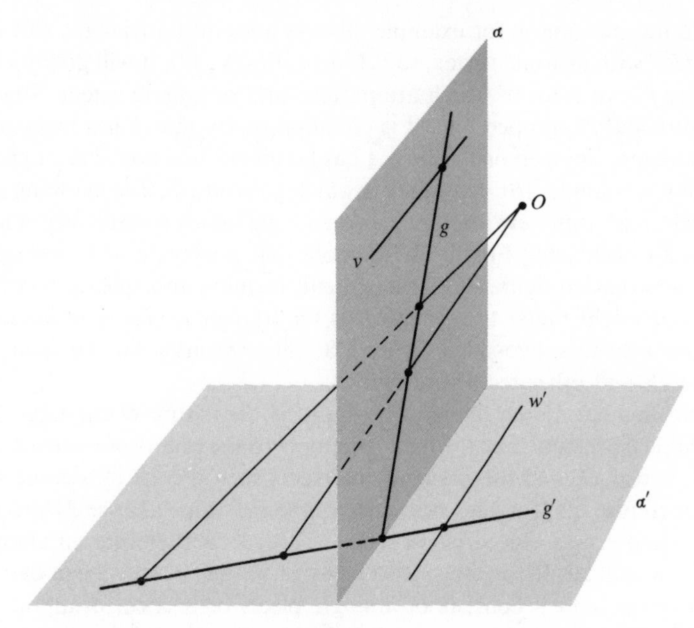

Fig. V, 11(b)

line in each plane, and, except for those two lines, establishes a 1-1 *correspondence between the lines of the planes. Collinearity and cross-ratio are preserved, but not betweenness or ratio of division, and complete lines do not always go into complete lines.*

We shall nevertheless say, in the interest of verbal simplicity, that line g goes into line g' even when there is an exceptional point on each line.

The vanishing points of the projection are the source of some of its most striking properties. Intersecting lines, for example, may go into parallel ones. To show this let g, h be lines of α that meet in a vanishing point V. The image lines g', h' are then parallel. For if they met, it would have to be in α'. Also, since g' is in plane O, g and h' is in plane O, h, if they met, it would have to be in a point of line OV, which is common to these planes. But line OV is parallel to α'. Hence g', h' cannot meet.

Conversely, parallel lines may go into intersecting lines. To show this we make use of T^{-1}, which is the projection of α' on α from center O. This projection clearly has the same vanishing points and lines as T. By the preceding paragraph, then, T^{-1} sends two lines m', n' which meet in a vanishing point of α' into two parallel lines m, n of α. It follows that T sends m, n into m', n', which is what was to be shown.

Since T preserves collinearity, the image of a rectilinear figure f (not situated entirely on v) is always a rectilinear figure. The extent to which these figures resemble each other, however, depends on the location of f relative to v.

A triangle not meeting v, for example, always goes into a triangle. But if f is a triangle ABC with just one vertex, say A, on v (Fig. V, 12), it will go into a three-sided figure f' two sides of which are parallel and of infinite extent. Since f is a closed figure and f' an open one, it is common to say that T has transformed a closed figure into an open one. (Since A has no image, however, one might prefer to say that it is triangle *ABC excluding A* which goes into f'. The mapping appears less radical under this view inasmuch as a triangle minus a vertex is not a closed figure but an open one.) Finally, let us note that a triangle with one side on v goes into a two-sided figure of infinite extent, forming an angle, in fact (Fig. V, 13). (Or, one might prefer to say that it is the triangle *exclusive of the side on v* which goes into this two-sided figure.) Similar remarks can be made about quadrilaterals and other rectilinear figures.

More familiar to the reader, no doubt, is the nature of the figure c' into which a circle c projects. For the lines joining O to the points of c form a circular cone with vertex O, and the plane α' intersects such a cone in a conic section, which is therefore c'. If c does not meet v, then c' runs completely around the cone, and, like c, is a closed curve (Fig. V, 14). It is therefore an ellipse, and generally not a circle. If c is tangent to v, say at V, then c' is an open curve since V has no image, and it consists of a single piece, or branch, of infinite extent (Fig. V, 15). The plane α' being parallel to line OV, which is an element of the cone, c' is a parabola. According as a variable point P of c approaches V from one side or the other, its image P' recedes indefinitely far along c' in one direction or the other. Finally, let c meet v in two distinct points V_1, V_2 (Fig. V, 16).

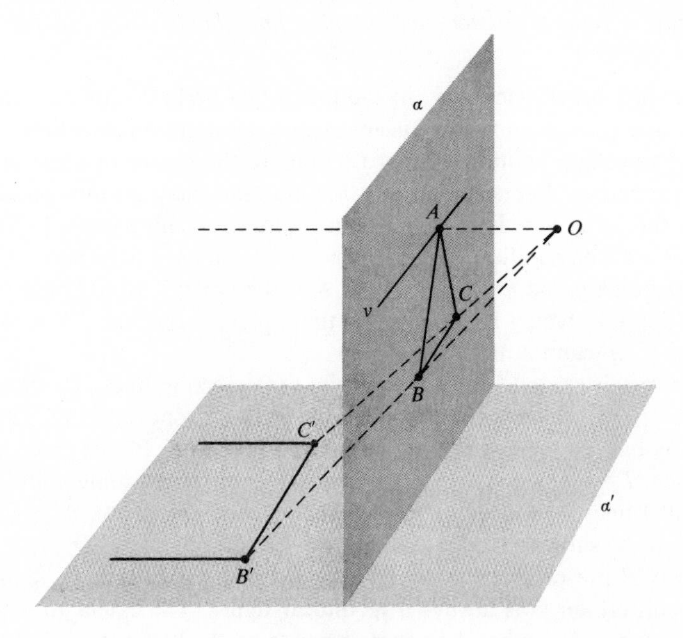

Fig. V, 12

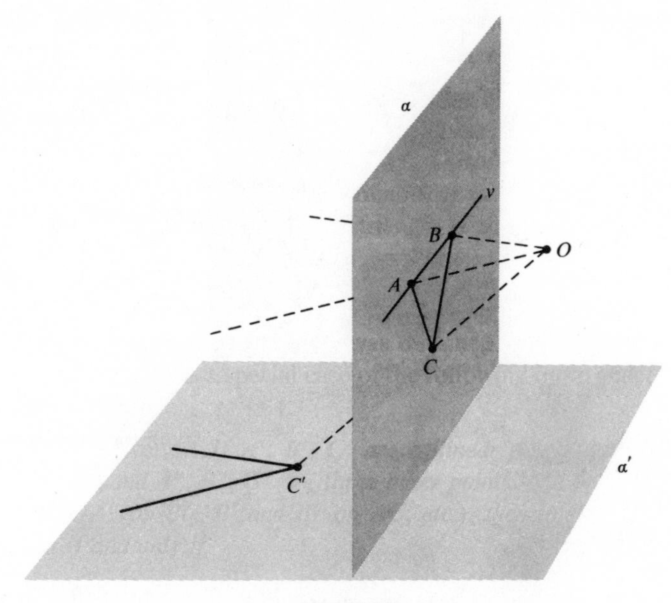

Fig. V, 13

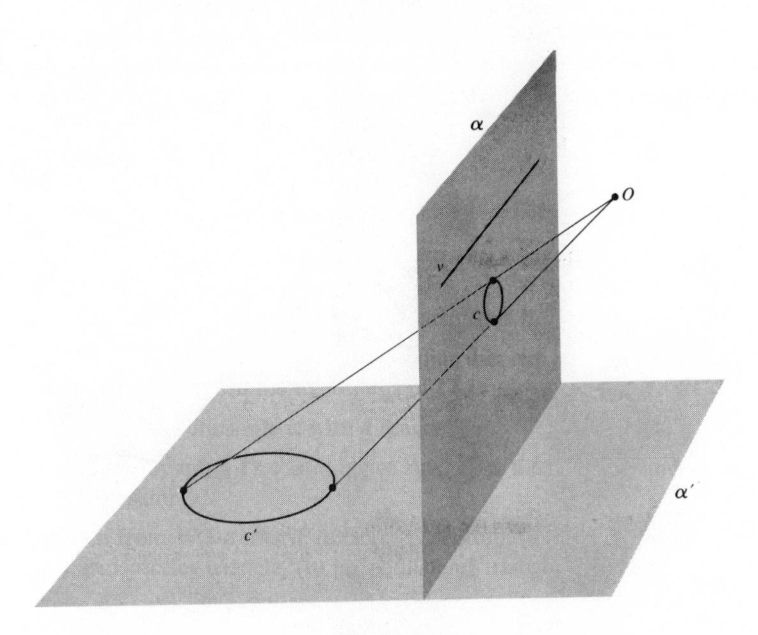

Fig. V, 14

Then c' consists of two pieces or branches, c_1' and c_2', each open and of infinite extent, c_1' corresponding to c_1, the upper part of c, and c_2' corresponding to c_2, the lower part of c. As a variable point on c_1 approaches V_1 or V_2, its image

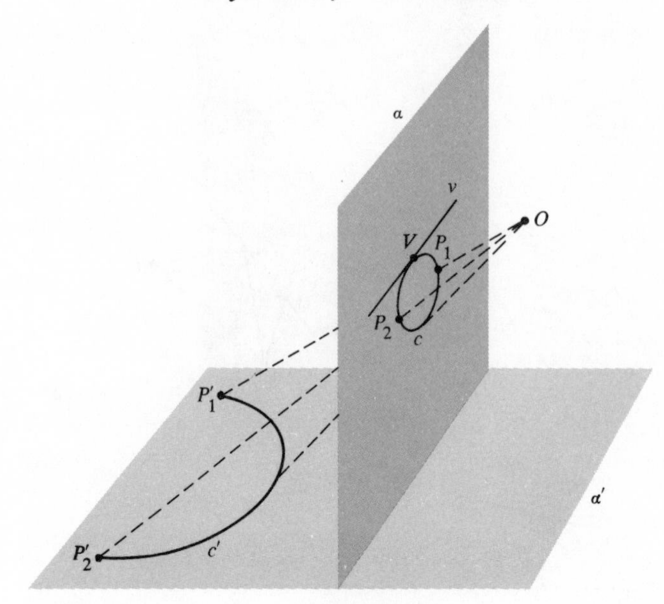

Fig. V, 15

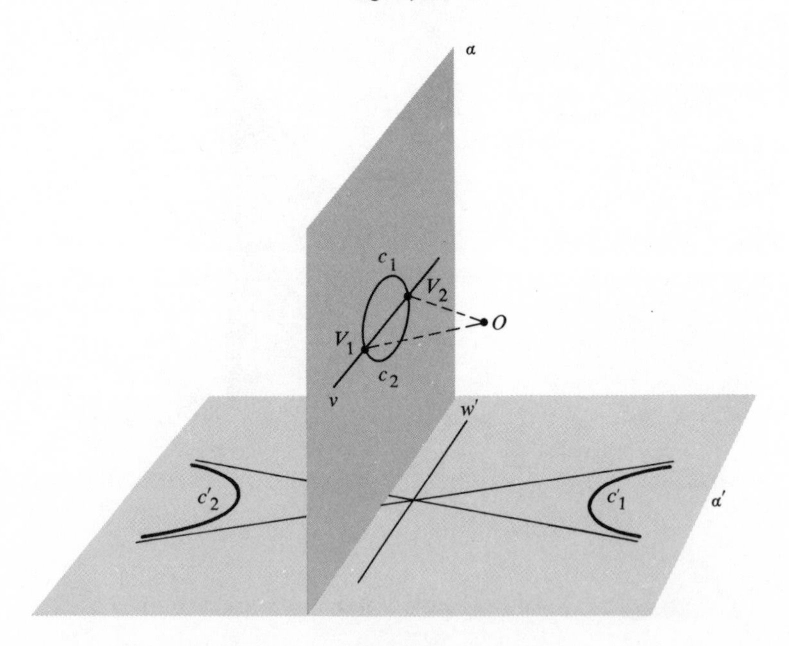

Fig. V, 16

moves off indefinitely far on c_1'; as a variable point on c_2 approaches V_1 or V_2, **its** image moves off indefinitely far on c_2'. The points of c_1' are on the **nappe of the** cone which is to the right of O, and the points of c_2' are on the nappe to the left of O. In other words, α' intersects both nappes. Hence the curve of intersection c' is a hyperbola. We took c so that V_1, V_2 are the ends of a diameter. The tangents at V_1, V_2 being parallel, their images, the asymptotes of c', meet on w'.

Thus, depending on its position relative to v, a circle projects into an ellipse, parabola, or hyperbola. (Or one may prefer the alternative statement that it is the circle *excluding one point* which projects into the parabola, and the circle *excluding two points* which projects into the hyperbola.) This result can be generalized: every nondegenerate conic (perhaps excluding one or two points) projects into a nondegenerate conic. For the lines joining O to the points of a nondegenerate conic c in α always form a type of cone, called a *quadric cone*, which is intersected by α' in a nondegenerate conic c'. Moreover, c' may or may not be the same kind of conic as c. If, for example, c is an ellipse, c' will be an ellipse, parabola, or hyperbola according as c does not meet v, is tangent to v, or crosses v. Clearly, a degenerate conic (two lines, which may coincide, or a point) not contained wholly in v always projects into a degenerate conic.

7.2 *A central projection of a plane on an intersecting plane always sends rectilinear figures into rectilinear figures, and conic sections into conic sections. Images can differ greatly from the originals, figures of finite extent sometimes going into figures of infinite extent, parallel lines sometimes going into intersecting lines, and vice versa, and conics of one type (that is, parabolic, elliptic, or hyperbolic) sometimes going into conics of another type.*

EXERCISES

The exercises refer only to the projection in the preceding discussion.

1. Prove the following for the projection of α on α': (a) v and w' are parallel; (b) g' meets w' if and only if g meets v; (c) g, g' intersect or are parallel according as g, v intersect or are parallel.

2. In the projection of α on α' show that: (a) g', h' meet in a nonvanishing point if g, h do; (b) g', h' are parallel to w' if g, h are parallel to v; (c) g', h' meet on w' if g, h meet v and are parallel.

3. Determine the image of a triangle which is met by v in two points, neither of which is a vertex.

4. Determine the image of a quadrilateral meeting v in (a) no point; (b) just one point; (c) two adjacent vertices; (d) two opposite vertices; (e) two points, just one of which is a vertex; (f) two points, neither of which is a vertex.

5. Show that a parabola in α can project into a circle (more precisely, a circle excluding one point) in α'. (*Hint*. Consider the projection of α' on α.)

6. Show that a hyperbola in α can project into a circle (more precisely, a circle excluding two points) in α'.

7. (a) If no side of a triangle is on v, show that the image of the triangle, although

not necessarily a triangle, is contained in three lines. (b) Solve (a) after replacing "triangle" by "quadrilateral" and "three" by "four."

8. Projective Properties. Having considered separately the different kinds of parallel and central projections of one plane on another, let us now consider them collectively and note their common properties. These common properties are called **projective properties** and they can be determined from the theorems on the projections of a plane already proved and from the last paragraph in §6. We describe them in the following statement:

8.1 *Every projection of one plane on another preserves collinearity, sends each rectilinear figure and conic section, respectively, into another, not necessarily of the same type, and preserves cross-ratio. In so doing, the projection establishes between the planes* 1-1 *correspondences that relate their points (except perhaps for the points of one line in each plane), their lines (except perhaps for these two lines), and their conic sections, respectively.*

Thus, collinearity, cross-ratio, the property of being a rectilinear figure, and the property of being a conic section are projective properties. Additional ones will be encountered later in the chapter (9.7, 10.1, 13.2).

9. The Values of Cross-Ratios. Our work with cross-ratio thus far has been limited to defining this quantity and showing it to be invariant for projections. Therefore we only expressed it in the form which emphasizes its meaning:

$$(1) \qquad (AB, CD) = \frac{AC}{CB} \div \frac{AD}{DB}.$$

Now, in proceeding further, we shall often write it in the alternative form

$$(2) \qquad (AB, CD) = \frac{AC \cdot DB}{CB \cdot AD}.$$

Also we shall call A, B, in that order, the *first pair of points* in (AB, CD), and C, D, in that order, the *second pair of points*.

Let us suppose that A, B, C, D, with abscissas x_1, x_2, x_3, x_4, are distinct points on a line not perpendicular to the x-axis, and that their orthogonal projections on this axis are P, Q, R, S (Fig. V, 17). Then, from elementary geometry, $AC/CB = PR/RQ$ and $AD/DB = PS/SQ$, so that

$$(AB, CD) = \frac{AC \cdot DB}{CB \cdot AD} = \frac{PR \cdot SQ}{RQ \cdot PS}$$

$$= \frac{(x_3 - x_1)(x_2 - x_4)}{(x_2 - x_3)(x_4 - x_1)} = \frac{(x_3 - x_1)(x_4 - x_2)}{(x_3 - x_2)(x_4 - x_1)}.$$

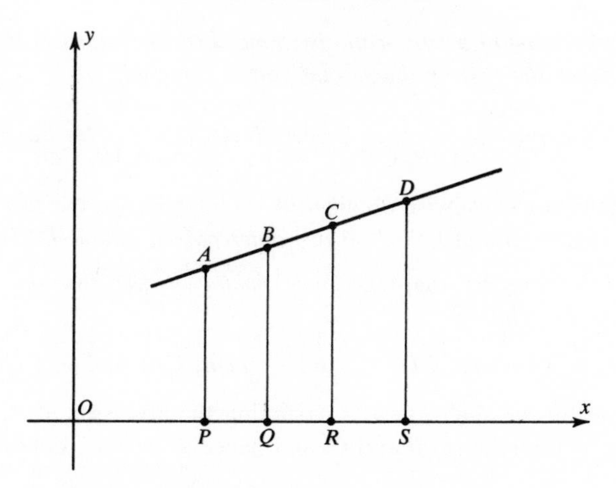

Fig. V, 17

The denominator in this result is not zero since no two abscissas are equal. Similarly, if A, B, C, D, with ordinates y_1, y_2, y_3, y_4, are distinct points on a line not perpendicular to the y-axis, we can obtain for (AB, CD) an analogous expression in these ordinates. Our result is therefore the following:

9.1 *Let $A(x_1, y_1)$, $B(x_2, y_2)$, $C(x_3, y_3)$, $D(x_4, y_4)$ be distinct collinear points. If their line is not perpendicular to an axis, then*

$$(AB, CD) = \frac{(x_3 - x_1)(x_4 - x_2)}{(x_3 - x_2)(x_4 - x_1)} = \frac{(y_3 - y_1)(y_4 - y_2)}{(y_3 - y_2)(y_4 - y_1)};$$

if it is perpendicular to the y-axis, (AB, CD) is given only by the first of these expressions, and if it is perpendicular to the x-axis only by the second.

The value of a cross-ratio depends, of course, not only on the points involved, but also on their order. Thus, (AB, DC) is the quotient of the ratios in which D, C, in that order, divide A, B, in that order, and hence

$$(AB, DC) = \frac{AD}{DB} \div \frac{AC}{CB}.$$

Comparing this with (1) we see that (AB, DC) is the reciprocal of (AB, CD). Similarly, (BA, CD) is the quotient of the ratios in which C, D, in that order, divide B, A, in that order. Hence

$$(BA, CD) = \frac{BC}{CA} \div \frac{BD}{DA} = \frac{BC \cdot DA}{CA \cdot BD} = \frac{CB \cdot AD}{AC \cdot DB}.$$

On referring to (2) we see that (BA, CD) is the reciprocal of (AB, CD). Thus, we have the theorem:

9.2 *The value of a cross-ratio is changed to its reciprocal if the order of the points in either the first or the second pair is reversed*:

$$(BA, CD) = \frac{1}{(AB, CD)}, \qquad (AB, DC) = \frac{1}{(AB, CD)}.$$

A simple result is likewise obtained if we interchange the *two inner points* of (AB, CD), that is, B and C, or the *two outer points*, A and D. This result is:

9.3 *Interchanging the two inner points or the two outer points of* (AB, CD) *gives the relations*:

$$(AC, BD) = 1 - (AB, CD) \quad \text{and} \quad (DB, CA) = 1 - (AB, CD).$$

To prove this we shall use 9.1. Assuming the abscissas of A, B, C, D to be distinct, let us interchange B and C, and hence x_2 and x_3, obtaining

$$(AC, BD) = \frac{(x_2 - x_1)(x_4 - x_3)}{(x_2 - x_3)(x_4 - x_1)}.$$

by 9.1. We must then show that this equals $1 - (AB, CD)$, where (AB, CD) is expressed as in 9.1. This presents no difficulty and is left to the student. The rest of 9.3 is proved in the same way.

The following is an immediate consequence of 9.2 and 9.3:

9.4 *The value of a cross-ratio is unchanged by reversing the order of the points in both the first and the second pairs,*

$$(AB, CD) = (BA, DC),$$

or of the inner points and also of the outer points,

$$(AB, CD) = (DC, BA),$$

and hence of the first and second pairs,

$$(AB, CD) = (CD, AB).$$

Although there are 24 distinct permutations, or linear arrangements, of the letters A, B, C, D, and hence 24 cross-ratios that can be formed from these points, we see from 9.4 that the values of these cross-ratios are not all different. In fact, there are at most 6 distinct values, for the 24 cross-ratios can be divided into 6 sets of 4 cross-ratios each, such that the cross-ratios in each set are equal by 9.4. Thus, for one of these sets we have

$$(AB, CD) = (BA, DC) = (DC, BA) = (CD, AB).$$

We still do not know the actual values that cross-ratios may have. To determine this, let us consider (AB, CD) and denote AC/CB, AD/DB by r_1, r_2, respectively. Then

$$(AB, CD) = \frac{AC}{CB} \div \frac{AD}{DB} = \frac{r_1}{r_2}.$$

Being ordinary ratios of divisions, r_1 and r_2 can each have any value except 0 and -1. Also, $r_1 \neq r_2$ since C and D are distinct. Hence we have:

9.5 *Cross-ratios can have all values except* 0 *and* 1.*

It is to be noted that if $(AB, CD) < 0$, then r_1, r_2 have opposite signs, with the result that just one of the points C, D is between A and B. This implies that just one of the points A, B is between C and D, as a diagram will show. In other words, in traversing the line of these points we can get from A to B only by encountering C or D, and from C to D only by encountering A or B. We describe this situation by saying that the pairs of points A, B and C, D *separate* each other. Stated more completely, if A, B, C, D are distinct collinear points, and just one point in each of the pairs A, B and C, D is between the points of the other pair, $(AB, CD) < 0$ and the pairs are said to **separate** each other. When $(AB, CD) > 0$ it will be seen that the pairs A, B and C, D do not separate each other. The converse statement in each case is also true. Hence we have:

9.6 $(AB, CD) < 0$ *if and only if the pairs of points* A, B *and* C, D *separate each other.*

Since cross-ratio is preserved by all projections, so is the separation or nonseparation of pairs of points, and we have the theorem:

9.7 *The separation or nonseparation of pairs of points is a projective property.*

EXERCISES

1. Find (AB, CD) for each of the following sets of points A, B, C, D, respectively:
(a) $(0, -1)$, $(2, 3)$, $(4, 7)$, $(5, 9)$; (b) $(-5, 0)$, $(-5, 3)$, $(-5, 1)$, $(-5, 4)$;
(c) $(3, 1)$, $(-1, 1)$, $(2, 1)$, $(4, 1)$; (d) $(3, 1)$, $(2, 1)$, $(-1, 1)$, $(4, 1)$.

2. Complete the proof of 9.3.

3. Write the three cross-ratios which are equal to each given cross-ratio:
(a) (AB, DC); (b) (AC, BD); (c) (AC, DB);
(d) (AD, BC); (e) (AD, CB); (f) (AB, CD).

4. If the points in Ex. 1(a) are A, B, C, D, compute the six cross-ratios given in Ex. 3.

5. If R denotes the value of (AB, CD), verify that the values of the other cross-ratios given in Ex. 3 are, respectively,
(a) $1/R$; (b) $1 - R$; (c) $1/(1 - R)$; (d) $(R - 1)/R$; (e) $R/(R - 1)$.

6. Show that the five values given in Ex. 5 are different from R unless R is -1, $\frac{1}{2}$, or 2.

---

* In some books the points A, B, C, D are not required to be distinct, and the values 0 and 1 are then not excluded.

7. If distinct collinear points A, B, C, where C is not the midpoint of $\overline{AB}$, and any number R other than 0 or 1 are given, show that there is a unique point D such that $(AB, CD) = R$. If C is the midpoint of $\overline{AB}$, show that D is still uniquely determined provided that R is not equal to 0, 1, or -1.

8. If A, B, C, D, E are distinct collinear points, show that

$$(AB, DE)(BC, DE) = (AC, DE).$$

9. Show that $0 < (AB, CD) < 1$ if A, B, C, D are in one of the linear orders $ABDC$, $ACDB$, $DCAB$, and that $(AB, CD) > 1$ if they are in one of the linear orders $ABCD$, $ADCB$, $CDAB$ (see VII, §7).

10. Harmonic Division. As in §9, let us again write $(AB, CD) = r_1/r_2$, where $r_1 = AC/CB$ and $r_2 = AD/DB$. If $(AB, CD) = -1$, in which case $r_1 = -r_2$, we say that C, D divide $\overline{AB}$ **harmonically**, or that A, B, C, D, in that order, form a **harmonic set**, and we call D the **harmonic conjugate** of C with respect to A and B. The relative positions of the points are quite simple in harmonic division. One of the points C, D is interior to $\overline{AB}$, dividing it internally, the other is outside this segment, producing a so-called external division, and both ratios have the same absolute value. When, for example, $r_1 = \frac{1}{2}$ and $r_2 = -\frac{1}{2}$, C divides $\overline{AB}$ internally so that AC is half of CB, and D divides $\overline{AB}$ externally so that AD is half of DB (Fig. V, 18). The reason for the term

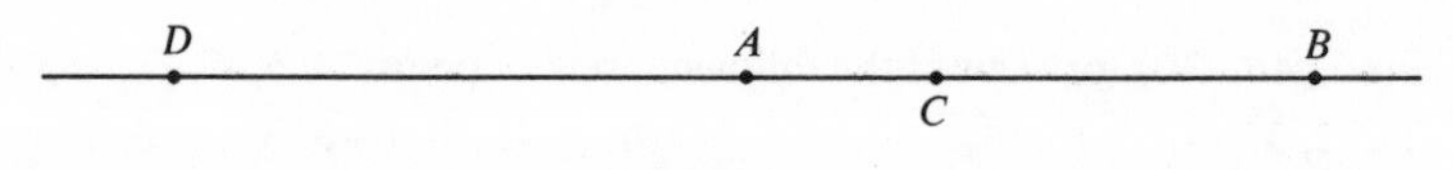

Fig. V, 18

"harmonic" is that the directed distances AC, AB, AD, in that order, are in harmonic progression when $(AB, CD) = -1$. We recall that a series of numbers is called a *harmonic progression* if the series of their reciprocals is an arithmetic progression. In the illustration just given, for example, if we suppose that $AB = 3$, then $AC = 1$, $AD = -3$, and the numbers 1, 3, -3 are a harmonic progression since their reciprocals 1, $\frac{1}{3}$, $-\frac{1}{3}$ are an arithmetic progression (with the common difference $-\frac{2}{3}$).

If A, B, C, D form a harmonic set and they project into A', B', C', D', then the latter also form a harmonic set since cross-ratio is preserved.

10.1 *The property of being a harmonic set of points is a projective property.*

Another way of expressing this is to say that *the property of being the harmonic conjugate of a point with respect to two others is projective.*

We have the following theorem on harmonic conjugates:

10.2 *If A, B, C are distinct collinear points, and C is not the midpoint of $\overline{AB}$, there is a unique point D which is the harmonic conjugate of C with respect to A and B.*

To prove this we note that under the given hypothesis r_1 is presumed known and different from 1 (and, of course, from 0 and -1). Hence, since r_2 is to equal $-r_1$, D must be found so that the ratio r_2 in which it divides A, B is the known number $-r_1$. This can always be done, and uniquely (see IV, §3, Ex. 7).

If $(AB, CD) = -1$, then $(AB, DC) = -1$ by 9.2, and $(CD, AB) = -1$ by 9.4. Thus, we have:

10.3 *If D is the harmonic conjugate of C with respect to A and B, then C is the harmonic conjugate of D with respect to A and B. If C, D divide A, B harmonically, then A, B divide C, D harmonically.*

We saw in §9 that when $(AB, CD) < 0$, the two pairs of points A, B and C, D separate each other. In the special case when $(AB, CD) = -1$, we therefore say that the two pairs of points *separate each other harmonically*.

EXERCISES

1. Show that if C, D divide A, B harmonically, then C, D divide B, A harmonically.

2. In each case find the harmonic conjugate of C with respect to A and B:

(a) $A(1, 0)$, $B(2, 0)$, $C(3, 0)$; (b) $A(1, 0)$, $B(4, 0)$, $C(2, 0)$;
(c) $A(2, 3)$, $B(4, 7)$, $C(5, 9)$; (d) $A(2, 3)$, $B(5, 9)$, $C(4, 7)$.

3. Show that if four points, taken in a certain order, form a harmonic set, the twenty-four cross-ratios of the points have only three distinct values instead of the usual six, namely, $-1, \frac{1}{2}, 2$ (see §9, Exs. 3-6).

4. If $(AB, CD) = 2$, which cross-ratio of these points has the value -1? What would the answer be if $(AB, CD) = \frac{1}{2}$?

5. Prove that the directed distances AC, AB, AD, in that order, are in harmonic progression if $(AB, CD) = -1$.

11.* Cross-Ratio of Concurrent Lines.

Analogous to the cross-ratio of points there is also a *cross-ratio of lines*, and it, too, is invariant under projec-

* Except when projections are being considered, all the points and lines of §§11, 12, 13 are assumed to be coplanar.

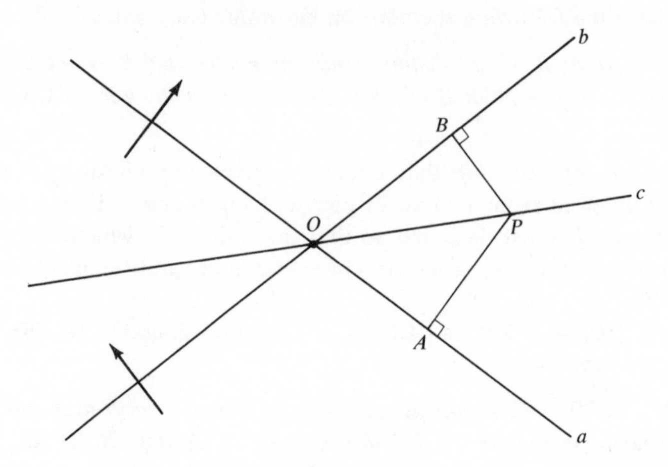

Fig. V, 19

tions. In order to define it we must first define the *ratio of division of two intersecting lines a, b* by another line c through their common point O (Fig. V, 19). Distances from a and b will be taken positive if measured in the directions of the arrows, and negative if in the opposite directions. Thus, in the figure, $AP > 0$ and $BP < 0$. It is actually immaterial which directions are taken positive, but once chosen they are not to be changed. If a, b are the coordinate axes or are parallel to them, however, we should take positive directions in the conventional way to avoid confusion.

From P, which is any point of c other than O, drop perpendiculars to a, b and form the ratio AP/BP of the directed distances from a, b to P. This ratio clearly has the same value regardless of the position of P on c. We call AP/BP the **ratio of division of lines *a*, *b* by line *c***, and denote it by r.

Lines a, b divide the plane into four regions, which we number as in Fig. V, 20. Then $r > 0$ if c traverses regions II and IV, and $r < 0$ if c traverses I and III. Thus $r < 0$ for c as shown in Fig. V, 19. The absolute value of r becomes arbitrarily small as c approaches a, and arbitrarily great as c approaches b. As this suggests, r can have any real value except 0 (since $c \neq a$), and there is exactly one line c which divides a, b in a given ratio. Thus, when $r = 1$, c is the line in regions II, IV which bisects the angles between a and b; when $r = -1$, c is the bisector of regions I, III.

Now, let c, d be distinct lines which divide a, b in the ratios r_1, r_2, respectively. Then $r_1 \neq 0$, $r_2 \neq 0$, and $r_1 \neq r_2$. We call r_1/r_2 a **cross-ratio of lines,** or, more precisely, the **cross-ratio in which lines *c*, *d* divide lines *a*, *b*,** and denote it by (ab, cd), which is read "the cross-ratio of a, b, c, d."

The following is then easily proved:

11.1 *Cross-ratios of lines can have all values except* 0 *and* 1. *If distinct*

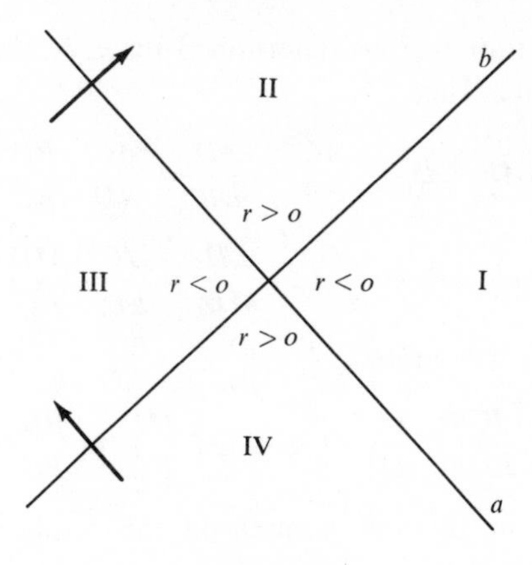

Fig. V, 20

concurrent lines a, b, c and any number not 0 *or* 1 *are given, there is a unique line d such that* (ab, cd) *equals the given number.*

In forming cross-ratios of lines the order, of course, is important. Thus, if $(ab, cd) = r_1/r_2$, then $(ab, dc) = r_2/r_1$, and hence (ab, dc) is the reciprocal of (ab, cd). As this may suggest, all the effects of changing the order are the same as in the case of points. Rather than prove this on the basis of the definition of (ab, cd), we shall deduce it from the following theorem, which expresses a fundamental relation between cross-ratio of lines and cross-ratio of points:

11.2 *If distinct concurrent lines a, b, c, d are met by a transversal in distinct points A, B, C, D, respectively, then* $(ab, cd) = (AB, CD)$.

We give the proof only for the case shown in Fig. V, 21, where c, d lie in

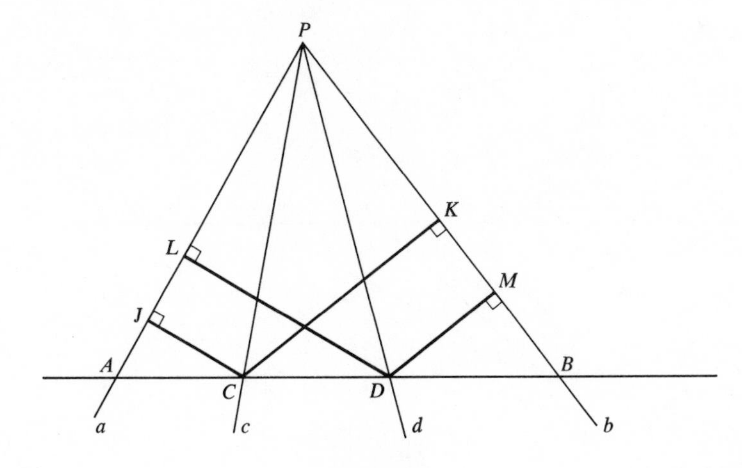

Fig. V, 21

the same two of the four regions determined by a, b, and angles CAP, CPA, DBP, DPB are all acute. Then

$$(AB, CD) = \frac{AC}{CB} \div \frac{AD}{DB} = \frac{AC}{AD} \cdot \frac{BD}{BC}, \tag{1}$$

$$(ab, cd) = \frac{JC}{KC} \div \frac{LD}{MD} = \frac{JC}{LD} \cdot \frac{MD}{KC}. \tag{2}$$

From similar triangles we obtain

$$\frac{JC}{LD} = \frac{AC}{AD} \quad \text{and} \quad \frac{MD}{KC} = \frac{BD}{BC}.$$

On substituting this in (2) and comparing the result with (1) we obtain $(AB, CD) = (ab, cd)$.

In 11.2 we have regarded the lines a, b, c, d as given and fixed, and the points A, B, C, D as dependent on the position of the transversal. If we reverse this point of view, that is, if we consider the points as given and fixed, and the lines as variable, we obtain the following remarkable corollary of 11.2:

11.3 *If four given collinear points are joined to a variable point not collinear with them, the resulting lines have a cross-ratio whose value is the same for all positions of the point.*

Thus, if the given points A, B, C, D are on g (Fig. V, 22), the variable point P can range over the entire plane (excluding the points of g) and (ab, cd) will always have the same value, namely, the fixed number (AB, CD).

Returning now to the question of the order of the lines, we can easily see that 11.2, 9.2, 9.3, and 9.4 together imply the following:

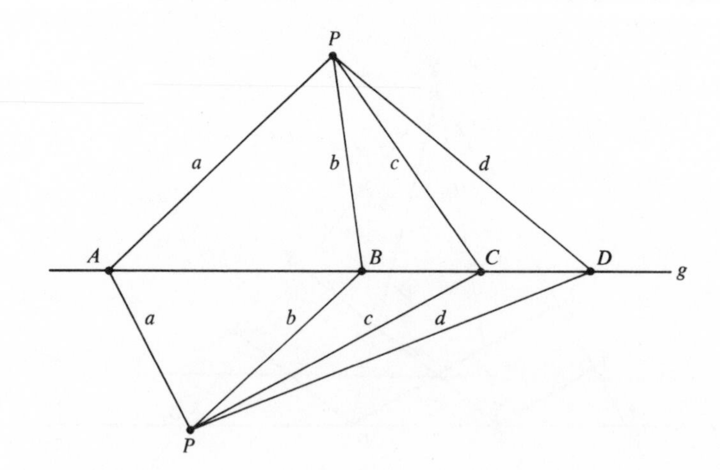

Fig. V, 22

11.4 *Cross-ratio of lines is affected in the same way by a change in order as is cross-ratio of points.*

For example, to see that $(ab, cd) = (cd, ab)$ we need only let lines a, b, c, d be met by a transversal in the distinct points A, B, C, D. Then, by 11.2, $(ab, cd) = (AB, CD)$ and $(cd, ab) = (CD, AB)$. Since $(AB, CD) = (CD, AB)$ by 9.4, it follows that $(ab, cd) = (cd, ab)$.

We mentioned earlier that cross-ratio of lines is invariant under projections. Now we shall prove this for the case in which concurrent lines go into concurrent lines. In §13, after defining the cross-ratio of parallel lines, we shall give the proof for the case in which concurrent lines go into parallel lines, and the case in which the opposite occurs.

11.5 *If, in the projection of one plane on another, four concurrent lines go into four concurrent lines, the cross-ratios of the two sets of lines are equal.*

Proof. Consider any projection of a plane α on a plane α' (Fig. V, 23). Let a, b, c, d be distinct concurrent lines in α and let the distinct concurrent lines a', b', c', d' be their images in α'. Necessarily, a, b, c, d meet in some nonvanishing point, a', b', c', d' meet in the image of that point, and all these lines are nonvanishing. Let g be a nonvanishing line of α which meets a, b, c, d in the distinct, nonvanishing points A, B, C, D. Regardless of the type of projection such a line g can always be chosen. Then, by 11.2,

$$(AB, CD) = (ab, cd).$$

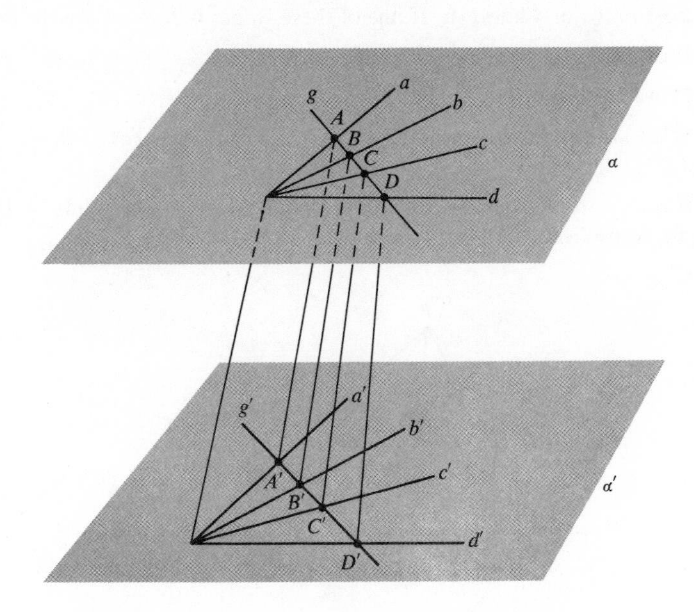

Fig. V, 23

If g' is the image of g, then A, B, C, D go into distinct, nonvanishing points A', B', C', D' of g'. Since cross-ratio of points is preserved by projections,

$$(AB, CD) = (A'B', C'D').$$

Since A', B', C', D' lie, respectively, on a', b', c', d' we have, again by 11.2,

$$(A'B', C'D') = (a'b', c'd').$$

From these three equations we obtain the desired conclusion,

$$(ab, cd) = (a'b', c'd').$$

EXERCISES

1. If a, b in Fig. V, 19 meet at an angle of 75° and c divides this angle by making an angle of 45° with a, find the ratio in which c divides a, b.

2. If lines a, b are perpendicular and r is any given number other than 0, verify that there is a unique line c dividing a, b in the ratio r.

3. Do Ex. 2, taking a, b not perpendicular.

4. Show that $\left|\frac{AP}{BP}\right|$ in Fig. V, 19 equals $\left|\frac{\sin \angle AOP}{\sin \angle BOP}\right|$ and hence becomes arbitrarily small or great according as c approaches a or b.

5. (a) How many distinct values, at most, can be obtained by forming the 24 possible cross-ratios of 4 lines? (b) If one of these values is R, what are the others?

6. Prove 11.1.

7. Prove 11.2 using Fig. V, 24.

8. What are the relative positions of a, b, c, d when $(ab, cd) > 0$, and what are they when $(ab, cd) < 0$?

9. If a, b, c, d are the lines joining the point (0, 5) to the points (0, 0), (1, 0), (2, 0), (3, 0), respectively, find (ab, cd).

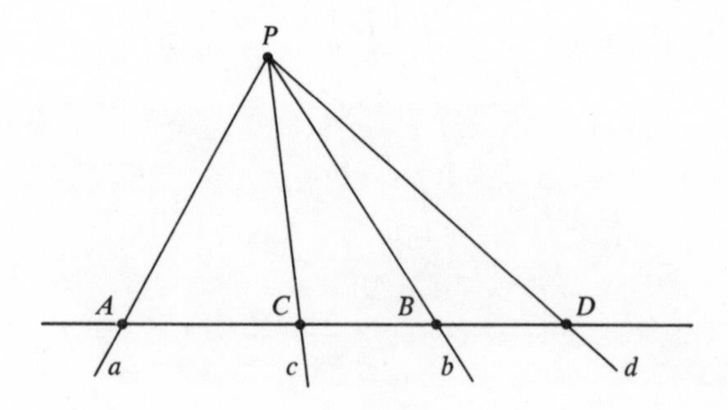

Fig. V, 24

10. Show that the line $y = mx$, where $m \neq 0$, divides the x- and y-axes (in that order) in the ratio m, and hence that the cross-ratio of the lines $x = 0$, $y = 0$, $y = m_3x$, $y = m_4x$ is m_4/m_3.

11. If a, b, c, d are distinct concurrent lines with slopes m_1, m_2, m_3, m_4, respectively, show that $(ab, cd) = \dfrac{(m_3 - m_1)(m_4 - m_2)}{(m_3 - m_2)(m_4 - m_1)}$. (First consider the case in which the lines go through the origin.)

12. Prove theorem 11.4.

13. Deduce from 11.2 and §6, Ex. 6 that

$$|(ab, cd)| = \frac{\sin(a, c) \cdot \sin(b, d)}{\sin(b, c) \cdot \sin(a, d)},$$

where, for example, we use (a, c) to denote the positive angle, less than π, between a and c. Thus, the cross-ratio of four lines depends only on the angles between them.

14. Generalize the result in Ex. 10 by showing that the cross-ratio of the distinct lines $x = 0$, $y = m_2x$, $y = m_3x$, $y = m_4x$, is $(m_4 - m_2)/(m_3 - m_2)$.

15. If a_1, a_2, a_3, a_4 are distinct concurrent lines such that a_1 is vertical and a_2, a_3, a_4 have slopes m_2, m_3, m_4, respectively, show that $(a_1a_2, a_3a_4) = (m_4 - m_2)/(m_3 - m_2)$ (see Ex. 14).

12. Harmonic Division of Concurrent Lines.

Suppose that $(ab, cd) = -1$, where the letters denote concurrent lines. If a, b, c, d are met by a transversal in distinct points A, B, C, D, then $(AB, CD) = -1$ by 11.2, and hence these points form a harmonic set. For this reason, if $(ab, cd) = -1$ we shall say that the lines a, b, c, d, in that order, are a **harmonic set of lines**, or that c, d divide a, b **harmonically**, or that d is the **harmonic conjugate** of c with respect to a, b.

Apart from possessing the transversal relation just noted, the above lines a, b, c, d occupy comparatively simple positions relative to each other inasmuch as the ratios in which c and d divide a, b are negatives of each other. Thus, in the special case when these ratios are 1 and -1, c and d are the bisectors of the angles formed by a, b.

From 11.1, 11.4, 11.5, respectively, we obtain the following three theorems:

12.1 *If a, b, c are concurrent lines, there is a unique line d which is the harmonic conjugate of c with respect to a, b.*

12.2 *If c, d divide a, b harmonically, then c, d divide b, a harmonically, and d, c divide both a, b and b, a harmonically.*

12.3 *If the concurrent lines a, b, c, d are harmonic and they project into the concurrent lines a', b', c', d', then the latter are also harmonic.*

EXERCISES

1. Prove 12.1.

2. Prove 12.2.

3. Prove 12.3.

4. Show that the lines joining the point (0, 5) to the points (0, 0), (3, 0), (1, 0), (−4, 0), respectively, form a harmonic set in that order.

5. If $(ab, cd) = -1$, where the letters denote lines, show that $(ac, bd) = 2$, $(ac, db) = \frac{1}{2}$, and that none of the twenty-four cross-ratios of these lines has a value other than -1, 2, or $\frac{1}{2}$.

6. Show that the axes are divided harmonically by every pair of lines through the origin whose angles of inclination are supplementary. (See §11, Ex. 10.)

7. Using Ex. 6 determine whether the lines $x = 0$, $y = 0$, $x + 2y = 0$, $x - 2y = 0$ form a harmonic set in some order. If they do, write down all of the orders in which they are harmonic.

8. Find the harmonic conjugate of the x-axis with respect to the lines $y = 3x$, $y = -3x$ (see Ex. 6).

9. Show that the lines $2x - 3y = 0$, $x + y = 0$ divide the lines $7x - 3y = 0$, $x - 9y = 0$ harmonically, and vice versa (see §11, Ex. 11).

10. Find the harmonic conjugate of the line $y = 4x$ with respect to the lines $y = x$ and $y = 3x$ (see §11, Ex. 11).

13. Cross-Ratio of Parallel Lines. Parallel lines, we know, can project into concurrent lines, and vice versa. This suggests that it may be possible to define cross-ratio of four parallel lines in such a way that it, too, is preserved by projections. To see how this can be done, let a, b, c, d be distinct

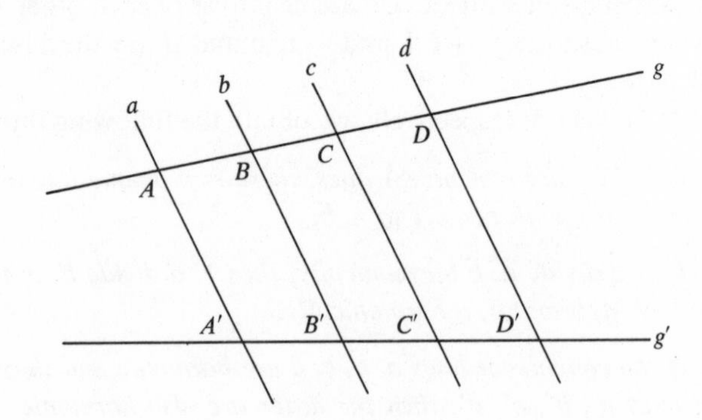

Fig. V, 25

parallel lines, and g, g' two transversals meeting them in A, B, C, D and A', B', C', D', respectively (Fig. V, 25). Then, from elementary geometry,

$$\frac{AC}{CB} = \frac{A'C'}{C'B'} \quad \text{and} \quad \frac{AD}{DB} = \frac{A'D'}{D'B'},$$

so that $(AB, CD) = (A'B', C'D')$. In other words, the cross-ratio of the points in which a, b, c, d are met by a transversal has a value which is the same for all transversals. This suggests the following definition:

13.1 Definition. *The cross-ratio (ab, cd) of four parallel (coplanar) lines a, b, c, d is defined to be the same number as (AB, CD), where A, B, C, D are the points in which a, b, c, d are met by any transversal.*

Since four parallel lines can always be drawn through four collinear points, cross-ratio of parallel lines has the same range of values as cross-ratio of points, and hence as cross-ratio of concurrent lines. Also, the definitions of harmonic set, harmonic division, and harmonic conjugate are the same for parallel lines as for concurrent lines.

If the transversal mentioned in Definition 13.1 is perpendicular to a, b, c, d, then

$$(ab, cd) = (AB, CD) = \frac{AC}{CB} \div \frac{AD}{DB} = \frac{AC}{BC} \div \frac{AD}{BD}.$$

Thus, (ab, cd) equals the ratio of the distances from a and b to c divided by the ratio of the distances from a and b to d. Our definition of cross-ratio of parallel lines is therefore consistent with our definition of cross-ratio of concurrent lines.

In 11.5 we saw that if four concurrent lines project into four concurrent lines, the cross-ratios of the two sets of lines are equal. The analogous statement for parallel lines can now be proved, and in the same way. This method, moreover, will show that the cross-ratios are equal even when the original lines are parallel and their images concurrent, and vice versa. We leave the details to the student. Since cross-ratio of lines is not defined in any other cases than those we have considered, we conclude that:

13.2 *The cross-ratio of lines is a projective property.*

EXERCISES

1. Draw diagrams to illustrate the following values of (ab, cd), where a, b, c, d are parallel lines: (a) 2; (b) -2; (c) -1; (d) $\frac{1}{3}$.

2. If distinct parallel (coplanar) lines a, b, c, where c is not midway between a and b, and any number R, not 0 or 1, are given, show that there is a unique line d such that $(ab, cd) = R$. If c is midway between a and b, show that d is still uniquely determined provided that R is not 0, 1, or -1.

3. Show that the effects of changing the order in cross-ratio of parallel lines are the same as in cross-ratio of concurrent lines.

4. Define *harmonic set* and *harmonic conjugate* for parallel lines.

5. If a, b, c are distinct (coplanar) parallel lines, is there always a harmonic conjugate of c with respect to a, b? Justify your answer. (See Ex. 2.)

6. Prove 13.2.

14. Cross-Ratio and the Conic Sections. We have already seen that the property of being a nondegenerate conic is projective, that is, nondegenerate conics always project into nondegenerate conics. We now prove a fundamental property of these curves which further emphasizes their projective character.

14.1 *If four points of a nondegenerate conic section are given, and the lines joining them to a fifth point of the conic are drawn, the cross-ratio of these lines, always taken in the same order, has a constant value for all positions of the fifth point on the conic.*

Proof. Consider circles first. Since an angle inscribed in a circle is measured by half its intercepted arc, an arc AB subtends the same angle at one point P of the circle as at any other point Q. That is, $\angle APB = \angle AQB$ (Fig. V, 26).

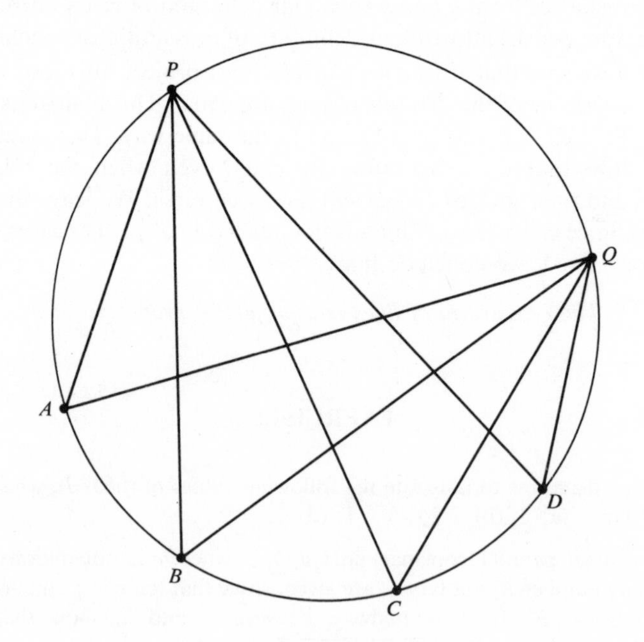

Fig. V, 26

Likewise, $\angle APC = \angle AQC$, $\angle BPD = \angle BQD$, etc. From §11, Ex. 13 we have

$$|(PA\ PB, PC\ PD)| = \frac{\sin APC \cdot \sin BPD}{\sin BPC \cdot \sin APD}$$

and

$$|(QA\ QB, QC\ QD)| = \frac{\sin AQC \cdot \sin BQD}{\sin BQC \cdot \sin AQD},$$

where, for example, PA on the left side means line PA, and APC on the right side means $\angle APC$. Since the two right sides are equal, and $(PA\ PB, PC\ PD)$, $(QA\ QB, QC\ QD)$ clearly have the same sign, we conclude that

$$(1) \qquad (PA\ PB, PC\ PD) = (QA\ QB, QC\ QD),$$

which proves the theorem for circles.

Now consider an arbitrary nondegenerate conic K (Fig. V, 27). We know that in projections of one plane on another a circle always goes into a nondegenerate conic. It is also true, conversely, any given nondegenerate conic can be obtained as a projection of a suitably chosen circle.* Hence K is the projection

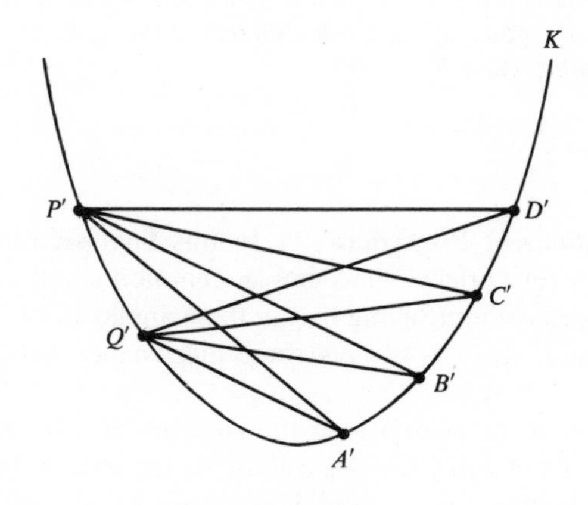

Fig. V, 27

of some circle γ. Let A', B', C', D', P', Q' be any distinct image points of K in this projection, and A, B, C, D, P, Q be their originals on γ. Then equation (1) holds for these points of γ. Since cross-ratio is preserved by the projection, $(P'A'\ P'B', P'C'\ P'D')$ and $(Q'A'\ Q'B', Q'C'\ Q'D')$ are equal, respectively, to the left and right sides of (1), and hence to each other. This completes the proof of 14.1.

If P' approaches A' in Fig. V, 27, lines $P'B'$, $P'C'$, $P'D'$ approach lines $A'B'$, $A'C'$, $A'D'$, respectively, and line $P'A'$ approaches t, the tangent to the

---

* See, for example, George Salmon, *A Treatise on Conic Sections*, p. 330.

conic at A'. Hence $\angle A'P'C'$, $\angle A'P'D'$, $\angle B'P'C'$, and $\angle B'P'D'$ approach the angle between t and $A'C'$, the angle between t and $A'D'$, $\angle B'A'C'$, and $\angle B'A'D'$, respectively. Since $(P'A'\ P'B',\ P'C'\ P'D')$ maintains a constant value as P' approaches A', and $(t\ A'B',\ A'C'\ A'D')$ has a definite value, reference to the formula in §11, Ex. 13 shows that these values are therefore necessarily the same. Thus we have the following corollary to 14.1:

14.2 *A chosen cross-ratio of the tangent to a nondegenerate conic section at any point on it and the lines joining that point to three other points of the conic equals the corresponding cross-ratio of the lines joining the four points to a fifth point on the conic.*

EXERCISES

1. Given the circle $r = 2\cos\theta$, the points A, B, C, D on the circle for which θ is 10°, 20°, 30°, 40°, respectively, and the tangent t at O, compute

(a) $(OA\ OB,\ OC\ OD)$, (b) $(t\ OA,\ OB\ OC)$.

2. From the results of Ex. 1 what can be said about the value of $(PO\ PA,\ PB\ PC)$, where P is an arbitrary point of the circle different from A, B, C, and O? Check by taking P to be the point where $\theta = -50°$.

15. Applications of Projections. In this final section of the chapter we wish to mention very briefly some simple, common situations in which projections occur. It goes without saying that in these applications, as in all applications of mathematics, there is always some discrepancy between theory and reality.

A familiar use of projection is in the showing of a moving picture. The complete film consists of many tiny snapshots. At the instant one of these snapshots f lies parallel to the screen s, the figures on it are transformed by central projection into very much larger ones on the screen, the center of projection being a small source of light L (Fig. V, 28). Since the projection has the properties of a similarity transformation, figures on the screen have the same shape as the corresponding ones on the film. The enlargement is in the ratio $d_s : d_f$, where d_s and d_f can be either the perpendicular distances from L to s and f, or else the oblique distances used in Fig. V, 28.

Projection is also important in photography. Take the simple case in which a snapshot is being taken of a painting. The camera is held so that the sensitized film within it is parallel to the plane of the painting. At the instant the picture is taken, a ray of light emanates from each point of the painting, passes through the tiny aperture of the camera, and strikes the film. The effect is to transform the points of the painting $ABCD$ (Fig. V, 29) into points of the film $A'B'C'D'$

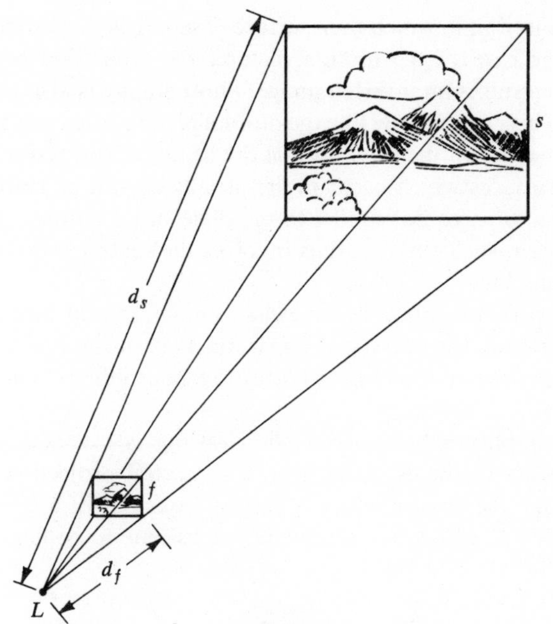

Fig. V, 28

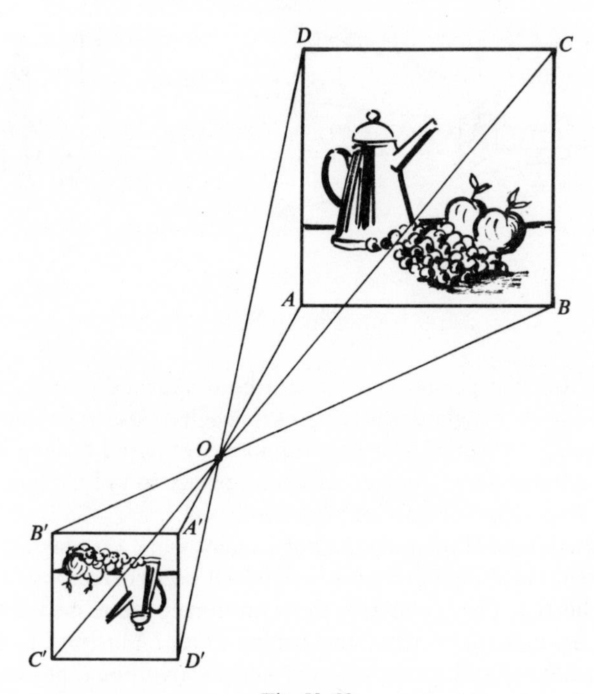

Fig. V, 29

by a central projection in which the aperture O serves as the center of projection. Here, unlike the case of the motion picture, the center lies between the two planes, and the result is an inverted image. Photography is also used for making enlargements. This is essentially like the motion picture situation, the two parallel planes being situated on the same side of the center of projection.

The flat surfaces which occur in life usually cannot be manipulated, as in the case of a painting, to lie parallel to the plate of the camera. Photographing scenes that contain such surfaces thus involves the central projection of a plane on a nonparallel plane.

Central projection plays a somewhat similar role in human vision as it does in photography, for our visual apparatus is basically much like a camera, the plate and aperture of the camera corresponding to the retina and lens of the eye.

Just like a photograph, a realistic drawing is intended to represent a scene as it appears to the eye, and here, too, central projection is a means of achieving it. Take the case in which a person looks out of a closed window W and views a scene L, which, for simplicity, we assume to be flat (Fig. V, 30). If

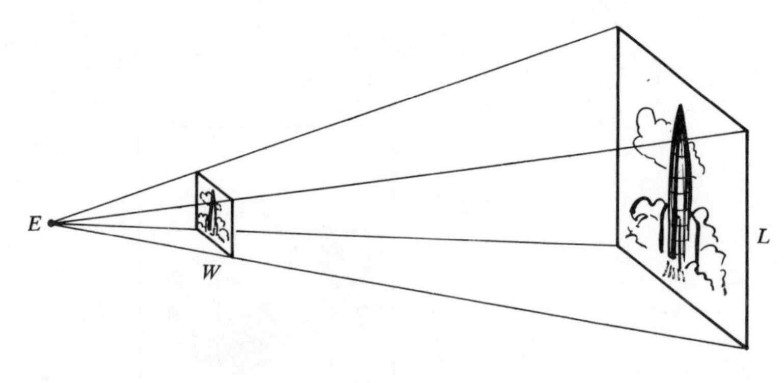

Fig. V, 30

he closes one eye, the person can make a quite realistic drawing of the scene merely by tracing on the glass what he sees through it, just as people often copy a picture by placing a sheet of transparent paper over it and tracing on the paper. The tracing on the glass, of course, achieves a projection of the landscape on the glass, the eye E serving as center of projection.

Since artists and draftsmen do not usually work on glass, their problem of making a realistic drawing, that is, one which uses correct visual perspective, is more complicated. They ordinarily draw on some opaque medium, like canvas or heavy paper, essentially what the person in our illustration traces on his transparent medium, and to do this well usually requires a knowledge of projections, or, at least, a knowledge of the way in which distance and position affect the appearance of objects. An artist knows, for example, that a square in the scene he is depicting need not appear as a square on his canvas, nor as a rectangle, nor even as a parallelogram, but that it should certainly appear as a quadrilateral of some sort. This is consistent with the mathematical fact that the property of being a quadrilateral is projective, whereas the property of being a square or rectangle or parallelogram is not. The artist may not know this fact from any formal study of projective geometry. He may simply know from observation that when square surfaces are viewed from various positions and at various distances, they always appear rectilinear, usually as quadrilaterals, but not necessarily as squares, rectangles, or even as parallelograms.

The idea that accurate drawing could be thought of, and achieved, in terms of projections became prominent in Europe during the Renaissance, and helped the artists of that period to achieve a realism that was hitherto unknown. In fact, the investigation of the new geometrical ideas was carried on to a large extent by the artists themselves. Fig. V, 31, for example, shows an etching by the 16th century painter, Dürer, which illustrates the making of a drawing from life by use of central projection. The designer in this etching is tracing on his transparent medium what he sees through it, using only one eye and keeping it in a fixed position. The interest which artists of the 15th and 16th centuries showed in projections was later shared by professional mathematicians, who developed the subject much further in purely geometrical directions and ultimately created a new branch of geometry, projective geometry.

Fig. V, 31

Projections are of great importance in constructing flat maps of the earth's surface, and this activity, like the making of perspective drawings, came to the fore in Europe during the Renaissance. The construction of such maps involves, at the very simplest, the projection of a sphere onto a plane, but it may also involve projections of a sphere onto a cylinder or a cone. Not having considered such projections previously, nor wishing to study their properties now, we leave the investigation of map-making to the interested student.

CHAPTER VI

Projective Transformations

1. Introduction. In the preceding chapter we saw how the affine transformations of a plane into itself can be generated by means of parallel projections of one plane on another. Now, by using both parallel and central projections, we shall be led to the concept of a still broader type of transformation of a plane into itself, a *projective transformation.* With these new transformations at our disposal, we can resume the study of transformations within a single plane carried on in Chapters II–IV. Having there dealt with motions, similarity transformations, and affine transformations, together with their associated geometries, we shall now consider the even more extensive class of projective transformations and the related system of geometry known as projective geometry.

2. Definition of a Projective Transformation. Projections of one plane on another are *two-dimensional projective transformations* of a particularly simple type: the lines joining points to their images are either concurrent or parallel. More generally, by a **two-dimensional projective transformation** we shall mean any transformation of one plane into another, or into itself, that has the basic projective properties stated in V, 8.1. This means that, apart from preserving collinearity, cross-ratio, the property of being a conic, etc., the transformation should have at most two exceptional lines, one whose points have no images and one whose points have no originals. We shall usually omit the word "two-dimensional."

Clearly, the resultant of any finite sequence of parallel projections of one plane on another is a projective transformation, and in this case every point has an image and an original. Also, the resultant of any finite sequence of projections of one plane on another in which there is *just one* central projection of a plane on an intersecting plane is a projective transformation, and in this case there are always exceptional points, at most two lines of them. A simple example will clarify this. Consider the resultant S of a parallel projection of a plane α on an intersecting plane α', and a central projection of α' back on α. If v' is the vanishing line in α' for the central projection, and w is the vanishing line in α, let v be the original of v' in the parallel projection. Then S is a mapping of α onto itself in which the points of v have no images, and the points of w have no originals. Having no more than two exceptional lines and clearly possessing all the other properties stated in V, 8.1, S is a projective transformation as we are using this term. The reader should verify that the resultant of a central projection of α on α' and a central projection of α' back on α could have as many as four exceptional lines and hence does not meet our specifications for a projective transformation.

Let us generalize these examples. Project a plane α on a plane α_1, then project α_1 on a plane α_2, and so forth, continuing in this way until n (> 1) projections have been achieved. Let *at most one* of these projections be a central projection of a plane on an intersecting plane. Then the resultant, T, of these projections is a transformation of α into the last plane α_n which has at most two exceptional lines, as well as the other properties stated in V, 8.1, and hence is a projective transformation in our sense of this term. If α_n coincides with α, then T is a projective transformation of α into itself.

In view of their great importance to us, the projective transformations of a plane into itself merit a separate definition.

2.1 Definition. *A projective transformation of a plane into itself is a transformation which* (*a*) *sends the plane into itself in a* 1-1 *way, except perhaps that the points of one line may have no images and the points of one line may have no originals,* (*b*) *preserves collinearity, and* (*c*) *preserves the cross-ratio of points.*

This definition uses only three of the projective properties stated in V, 8.1. From these three, however, the others can be deduced, as we shall see later.

To the list of transformations of a plane into itself considered in Chapters II–IV we now add the projective transformations. In those chapters we succeeded in gradually enlarging the class of transformations being considered, so that a new class often included all of the transformations of a prior class, together with others that were different. Thus the similarities include all the motions and also certain transformations that are not motions, while the affinities include all the similarities and, in addition, many transformations that are not similarities. The same relation holds between affine and projective transformations. First, according to 2.1 and what was learned in Chapter V, every affine transformation of a plane is a projective transformation of that plane into itself, the special kind in which the 1-1 correspondence of points holds without exception and ratio of division of points is preserved. Second, there are projective transformations of a plane into itself which are not affine, namely, those whose 1-1 correspondences of points involve exceptions. An example of such a projective transformation is the resultant of a central projection of a plane α on an intersecting plane α' and a parallel projection of α' back on α.

2.2 *The projective transformations of a plane into itself consist of all the* *affine transformations of that plane and also of certain transformations that are nonaffine.*

We have already noted that Definition 2.1 does not require that there be any exceptional elements, and that the case in which there are no exceptional elements occurs when the projective transformation is affine. On the other hand, when the transformation is nonaffine, 2.1 does not require that the exceptional lines be distinct. They may, indeed, be the same: the line which has no image may also have no original. Worth noting also is the fact that 2.1 says nothing about projections despite the fact that we were led to this definition by considering sequences of projections. Hence, although we know that the resultants of certain sequences of projections are projective transformations of a plane into itself, we do not yet know whether, conversely, every such transformation can be generated in this way. This question is answered in §3.

In the exercises that follow, and throughout the rest of this chapter, the term *projective transformation* will mean a projective transformation of a plane into itself. Although this plane could be any plane, it will be convenient to regard it as the plane dealt with in Chapters II–IV. The exceptional lines mentioned in Definition 2.1 and the points they contain will often be called *vanishing*, just as in the case of projections; nonexceptional points and lines will often be called *ordinary*.

EXERCISES

1. Verify that a projective transformation has an inverse and that it is also a projective transformation.

2. Verify that affine transformations meet the three conditions specified in Definition 2.1.

3. Using projections, describe how to generate a projective transformation (a) which has no vanishing elements, (b) which has vanishing elements.

4. Let α, α' be intersecting planes, T_1 a central projection of α on α' with vanishing lines $v, \overline{w}$ in α, α', respectively, and T_2 a parallel projection of α' on α such that $T_2(\overline{w}) = w'$. Show that v, w' are the only exceptional lines for T_1T_2, and hence that T_1T_2 is a projective transformation. Also show that v, w' may coincide.

5. In Ex. 4, if T_2 were a central projection, show that T_1T_2 would generally have four exceptional lines and hence not satisfy Definition 2.1.

6. If exactly one of the projections T_1, T_2, T_3 has vanishing lines, and $T_1T_2T_3$ transforms a plane into itself, verify that $T_1T_2T_3$ has two exceptional lines no matter which projection has vanishing lines, and hence is a projective transformation.

7. If a plane α is projected on an intersecting plane α' from a center O on neither plane, and α' is projected back on α from O, show that the resultant of these projections is a projective transformation of α into itself with only one exceptional line.

3. Some Implications of the Definition. Having given the definition of a projective transformation in terms of only three of its properties, let us now see how its remaining properties can be deduced from those three.

We see immediately that an ordinary line can contain at most one point with no image and at most one point with no original. Easily provable because of our work with affine transformations (see the proof of IV, 10.2) is the fact that if points A', B' are the images of A, B, then the set of all points of line AB (excluding at most one point) goes into the set of all points of line $A'B'$ (excluding at most one point), or, as we shall say more briefly, line AB goes into line $A'B'$. Familiarity with the proof of IV, 10.2 also helps one to show that a projective transformation establishes a 1-1 correspondence among the lines of a plane. We leave the details of these proofs to be covered in exercises.

Using the facts just cited, let us go on to show that a projective transformation T always sends a family of concurrent lines into a family of concurrent lines or parallel lines. Suppose, first, that T has no vanishing lines. It is then affine and we know from our study of affine transformations (IV, 6.3) that the image family is a concurrent one. Suppose that T is nonaffine, with vanishing lines v and w', such that v has no image and w' has no original. Let A be any point not on v. Then each line through A goes into a line through A', the image of A. Conversely, each line through A' must be the image of some line through A. Thus, the family through A goes into the family through A'. Now let A be on v. Each line through A except v goes into some line. No two of these image lines g', h' can meet in a point B' not on w'. For their originals would then have to meet in B, the original of B', which contradicts that they meet in A, a point with no image. Hence g', h' either meet in a point C' on w', or else are parallel. Similarly, the image of any other line through A either is parallel to g' or else meets it in a point of w'. If the latter case occurs, this point is necessarily C'. Thus, the family through A goes into a set of lines belonging either to the family through C' or to the family parallel to g'. It remains to show that this set is identical with one of the latter families. The student can do this by showing that any given line k' of these families other than g', h' (and of course w') is an image line.

Similarly, one can show that T sends a parallel family into either a parallel family or a concurrent family, and go on to show that it always sends a conic into a conic, and preserves cross-ratio of lines. Indeed, the properties of T that are provable on the basis of Definition 2.1 are precisely the properties common to all projections of one plane on another, and hence are the properties we called projective (V, 8.1). As this suggests, the answer to the question raised near the end of §2 is in the affirmative, and can be stated as follows:

3.1 *Any given projective transformation can be obtained as the resultant of a sequence of projections, not more than one of which has vanishing elements.*

We postpone the proof of this theorem until §6. As an immediate consequence of this theorem we have the proposition stated just prior to it, namely:

3.2 *The properties of projective transformations are precisely the same as the properties common to all projections of one plane on another, and hence are the properties we called projective.*

With Theorem 3.1 at our disposal, the proofs which we said could be given on the basis of Definition 2.1 become superfluous. However, they are instructive and are therefore included in the exercises. In Ex. 1(d), for example, the student is asked to show by use of 2.1 that a projective transformation sends a family of parallel lines into a family of parallel lines or concurrent lines. It is appropriate at this place to mention that families of parallel lines and families of concurrent lines are called **pencils of lines**, and to note that, in view of Ex. 1(d) and what was proved earlier, *the property of being a pencil of lines is projective.*

Before stating the next theorem on projective transformations let us recall from Chapter IV that any three noncollinear points can be sent into any three noncollinear points by some affine transformation (IV, 2.4). It may be surmised that more than this can be achieved by projective transformations since they include nonaffine, as well as affine, transformations. This is correct, and it will be shown later in the chapter that any four points, no three of which are collinear, can be sent into any four points, no three collinear, by some projective transformation. Without using this we now prove:

3.3 *There cannot be two projective transformations that send four given points A, B, C, D into four given points A', B', C', D', respectively, if no three points in either set are collinear.*

Proof. Let T_1, T_2 be distinct projective transformations that send A, B, C, D into A', B', C', D', respectively (Fig. VI, 1). Since these points are non-

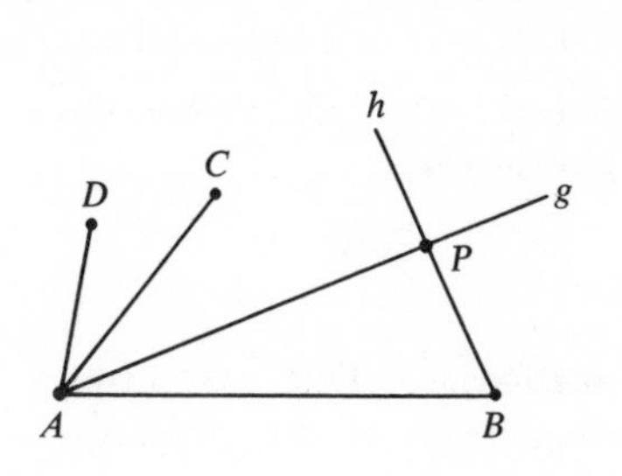

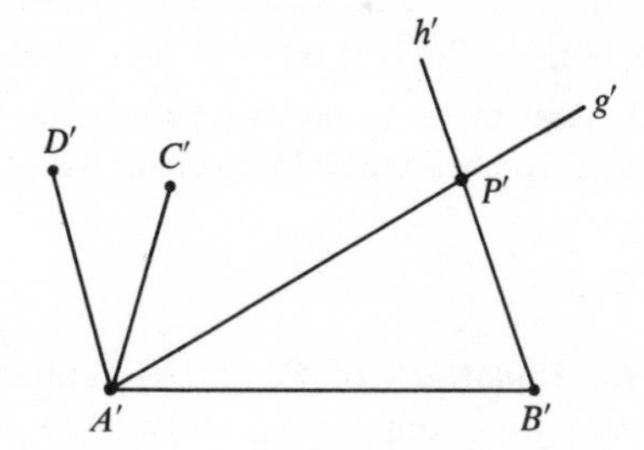

Fig. VI, 1

vanishing for T_1 and T_2, the same must be true of all lines through them. Hence T_1 sends the distinct lines AB, AC, AD into the distinct lines $A'B'$, $A'C'$, $A'D'$, and T_2 does likewise. Let g be any other line through A, and let T_1, T_2 send g into g_1', g_2', respectively. Then g_1', g_2' go through A' and we have

(1) $$(A'B'\ A'C',\ A'D'\ g_1') = (A'B'\ A'C',\ A'D'\ g_2')$$

since each of these cross-ratios equals $(AB\ AC, AD\ g)$ by 3.2. Hence $g_1' = g_2'$ (V, 11.1). Thus, T_1 and T_2 are identical as far as the lines through A are concerned, and clearly the same is true for the lines through B, C, and D.

Now let P be any point other than A, B, C, D which is ordinary for at least one of the transformations T_1, T_2. Joining A, B, C, D to P, we obtain at least three distinct lines. Suppose lines AP, BP are among them. If we denote AP, BP by g, h, respectively, then by the first part of the proof T_1, T_2 send g into the same line g' through A', and h into the same line h' through B'. Since g, h meet in P, and P is ordinary for at least one of the transformations, g', h' must meet also. If P' is their point of intersection, we then know that $T_1(P) = T_2(P) = P'$. In other words, a point which is ordinary for one of the transformations is also ordinary for the other, and its image is the same in both. Since the transformations must then have the same vanishing points, we conclude that $T_1 = T_2$. This contradicts the assumption that T_1, T_2 are distinct, and so the theorem is proved.

EXERCISES

1. Using 2.1, but not 3.1 or 3.2, show that a projective transformation: (a) preserves the noncollinearity of points; (b) sends a line into a line (except perhaps for one point on each line); (c) establishes a 1-1 correspondence among the lines of the plane; (d) sends a family of parallel lines into a family of parallel lines or concurrent lines; (e) preserves the cross-ratio of lines.

2. Find a sequence of projections whose resultant is a projective transformation in which the x-axis has no image.

3. Give the equations of the line v which has no image and the line w' which has no original in the projective transformations that meet the following conditions: (a) $(0, 0)$, $(1, 1)$ have no images, and $(1, 0)$, $(0, 1)$ no originals; (b) $(0, 0)$, $(1, 1)$ have no images, and $(1, 0)$, $(2, 1)$ no originals; (c) $(0, 0)$, $(1, 1)$ have no images and no originals.

4. Are there projective properties which are also affine? Are there affine properties which are also projective? Justify your answers.

4. Equations of Projective Transformations. Our next task, a comparatively long one, is to show that the equations of the projective transformations are of the form

$$x' = \frac{a_1x + a_2y + a_3}{c_1x + c_2y + c_3}, \qquad y' = \frac{b_1x + b_2y + b_3}{c_1x + c_2y + c_3}, \tag{1}$$

where

$$\begin{vmatrix} a_1 & a_2 & a_3 \\ b_1 & b_2 & b_3 \\ c_1 & c_2 & c_3 \end{vmatrix} \neq 0,$$

and that, conversely, equations of this form represent only such transformations. Until this is done we shall refer to the transformation represented by (1), under the specified condition, as a *fractional transformation*. The above determinant will be called the *determinant of the transformation* (1), and be denoted by Δ. Our procedure will be to show that (1) is affine when c_1, c_2 are both zero, that all affine transformations are so obtained, and that when c_1, c_2 are not both zero (1) gives all the nonaffine projective transformations and nothing but them. Attention is called to the fact that neither c_3 nor $a_1b_2 - a_2b_1$ can be zero when c_1, c_2 are both zero. To see this, expand Δ according to the elements of its last row and recall that $\Delta \neq 0$ (see Appendix, 3.1).

Letting $c_1 = c_2 = 0$ in (1) we get

$$x' = \frac{a_1}{c_3}x + \frac{a_2}{c_3}y + \frac{a_3}{c_3}, \qquad y' = \frac{b_1}{c_3}x + \frac{b_2}{c_3}y + \frac{b_3}{c_3},$$

where $c_3 \neq 0$. These are first-degree equations, whose determinant

$$\begin{vmatrix} a_1/c_3 & a_2/c_3 \\ b_1/c_3 & b_2/c_3 \end{vmatrix}, \quad \text{or} \quad \frac{1}{c_3^2}\begin{vmatrix} a_1 & a_2 \\ b_1 & b_2 \end{vmatrix},$$

is not zero since $a_1b_2 - a_2b_1 \neq 0$. Hence they represent an affine transformation. Conversely, any given affine transformation

$$\begin{cases} x' = a_1x + a_2y + a_3 \\ y' = b_1x + b_2y + b_3, \end{cases} \qquad \begin{vmatrix} a_1 & a_2 \\ b_1 & b_2 \end{vmatrix} \neq 0,$$

can be written in the form (1) by taking $c_1 = c_2 = 0$, $c_3 = 1$ and noting that in this case

$$\Delta = \begin{vmatrix} a_1 & a_2 & a_3 \\ b_1 & b_2 & b_3 \\ 0 & 0 & 1 \end{vmatrix} = \begin{vmatrix} a_1 & a_2 \\ b_1 & b_2 \end{vmatrix} \neq 0.$$

Thus, we get *all* the affine projective transformations from (1) by taking $c_1 = c_2 = 0$.

Let us now show that when c_1, c_2 are not both zero, (1) has the three defining properties of a projective transformation (2.1). If values are given to x, y in (1), unique corresponding values of x', y' can be computed if and only if $c_1x + c_2y + c_3 \neq 0$. Thus, each point (x, y) not on the line

$$c_1x + c_2y + c_3 = 0 \tag{2}$$

has a definite image (x', y'), whereas each point on the line has no image.

To show, conversely, that all points except those on a certain line have originals, we solve (1) for x and y, obtaining

$$x = \frac{A_1x' + B_1y' + C_1}{A_3x' + B_3y' + C_3}, \qquad y = \frac{A_2x' + B_2y' + C_2}{A_3x' + B_3y' + C_3}, \tag{3}$$

where $A_1, B_1, C_1, \ldots$ are the cofactors of $a_1, b_1, c_1, \ldots$ in Δ (see Appendix, §3). This algebraic work is valid only if the denominator in (3) is not identically zero, that is, zero for all values of x' and y'. If it were identically zero, then A_3, B_3, C_3 would all have to be zero. This cannot be, for the determinant

$$\delta = \begin{vmatrix} A_1 & B_1 & C_1 \\ A_2 & B_2 & C_2 \\ A_3 & B_3 & C_3 \end{vmatrix}$$

is not zero since $\delta = \Delta^2$ (see Appendix, 3.3). Thus, there are values of x', y' for which the denominator is not zero, and for each pair of them equations (3) enable us to compute unique corresponding values of x, y. In the transformation (1), then, each point (x', y') not on the locus of

$$A_3x' + B_3y' + C_3 = 0 \tag{4}$$

has a unique original, whereas each point on this locus has no original.

This locus is a line, for A_3, B_3 are not both zero. To see this let us denote the transformation (1) by T and recall that besides regarding (3) as the implicit form of T, which we just did, we may also regard it as the explicit form of T^{-1}. Now, T^{-1} is a fractional transformation, for, in addition to having the right form of such a transformation, its determinant δ is not 0, as we saw above. If A_3, B_3 were both zero, T^{-1} would be affine by the first part of the proof. Its inverse T would then also be affine. This contradicts our hypothesis that c_1, c_2 are not both zero.

Thus, assuming that c_1, c_2 are not both zero we have shown that (1) establishes a 1-1 correspondence among the points of the plane, excluding those on line (2), which have no images, and those on line (4), which have no originals. Now, with this same assumption, we show that (1) preserves collinearity. Let

$$ax + by + c = 0 \tag{5}$$

be any line other than (2). Then a, b are not both zero. On substituting from (3) we find the image of this line to be

$$a\frac{A_1x' + B_1y' + C_1}{A_3x' + B_3y' + C_3} + b\frac{A_2x' + B_2y' + C_2}{A_3x' + B_3y' + C_3} + c = 0,$$

or, after simplification,

$$(aA_1 + bA_2 + cA_3)x' + (aB_1 + bB_2 + cB_3)y' + (aC_1 + bC_2 + cC_3) = 0. \tag{6}$$

If P_1, P_2, P_3 are points on line (5) which have images, the coordinates of the images will satisfy (6). Hence if the coefficients of x' and y' in (6) were both zero, the constant term would also be zero. Another way of saying that these three numbers are zero is to say that the equations

$$A_1u + A_2v + A_3w = 0$$

$$B_1u + B_2v + B_3w = 0$$

$$C_1u + C_2v + C_3w = 0$$

in the unknowns u, v, w are satisfied when $u = a$, $v = b$, $w = c$. This is impossible, for 0, 0, 0 is the only solution of these equations since $\delta \neq 0$, whereas we saw above that a, b are not both zero. Hence the coefficients of x', y' in (6) are not both zero, the locus of (6) is a line, and the images of P_1, P_2, P_3 are collinear.

Finally, let us show that (1) preserves cross-ratio of points when c_1, c_2 are not both zero. Let A, B, C, D be distinct collinear ordinary points, with abscissas x_1, x_2, x_3, x_4. Their images A', B', C', D' are then distinct and collinear. Taking A', B', C' first, we have for their abscissas x_1', x_2', x_3' the equations

$$(7) \quad x_1' = \frac{a_1x_1 + a_2y_1 + a_3}{c_1x_1 + c_2y_1 + c_3}, \quad x_2' = \frac{a_1x_2 + a_2y_2 + a_3}{c_1x_2 + c_2y_2 + c_3},$$

$$x_3' = \frac{a_1x_3 + a_2y_3 + a_3}{c_1x_3 + c_2y_3 + c_3}.$$

If r is the ratio in which C divides $\overline{AB}$, then

$$x_3 = \frac{x_1 + rx_2}{1 + r}, \qquad y_3 = \frac{y_1 + ry_2}{1 + r}.$$

Substituting this in the equation for x_3' and simplifying, we get

$$x_3' = \frac{(a_1x_1 + a_2y_1 + a_3) + r(a_1x_2 + a_2y_2 + a_3)}{(c_1x_1 + c_2y_1 + c_3) + r(c_1x_2 + c_2y_2 + c_3)}.$$

Denoting $c_1x_1 + c_2y_1 + c_3$ by h, and $c_1x_2 + c_2y_2 + c_3$ by k, here and in the rest of (7), the above equation becomes

$$x_3' = \frac{hx_1' + krx_2'}{h + kr} = \frac{x_1' + (kr/h)x_2'}{1 + (kr/h)},$$

where neither h nor k is zero. Thus C' divides $\overline{A'B'}$ in the ratio kr/h.

If s is the ratio in which D divides $\overline{AB}$, we can show precisely as above that the ratio in which D' divides $\overline{A'B'}$ is ks/h. Since

$$(AB, CD) = \frac{r}{s} \quad \text{and} \quad (A'B', C'D') = \frac{(kr/h)}{(ks/h)} = \frac{r}{s},$$

it is seen that (1) preserves cross-ratio of points.

We have thus completely proved that:

4.1 *Every fractional transformation of the form* (1) *is a projective transformation. The latter is affine when* $c_1 = c_2 = 0$, *and only then.*

It remains to establish the converse. In order to do this we first prove:

4.2 *There is a unique fractional transformation of the form* (1) *that sends four given points into four given points, respectively, if no three points of either set are collinear.*

Let $(x_1, y_1), \ldots, (x_4, y_4)$ and $(x_1', y_1'), \ldots, (x_4', y_4')$ be the two given sets of points. To find a transformation (1) which maps the first four points onto the second four, respectively, we must find values of $a_1, a_2, a_3, b_1, b_2, b_3, c_1, c_2, c_3$ so that the eight equations

$$x_i' = \frac{a_1x_i + a_2y_i + a_3}{c_1x_i + c_2y_i + c_3}, \quad y_i' = \frac{b_1x_i + b_2y_i + b_3}{c_1x_i + c_2y_i + c_3}, \quad i = 1, 2, 3, 4,$$

are satisfied. It is easily seen after fractions are cleared in these equations that there result eight homogeneous linear equations in nine unknowns, the a's, b's and c's. According to algebraic theory, since there are more unknowns than equations these eight equations have at least one solution in which the values of the unknowns are not all zero. The determinant of the transformation (1) which employs these values is not zero, for such a transformation with zero determinant cannot send noncollinear points into noncollinear points (see Exs. 4, 5). Thus, there is at least one fractional transformation meeting the condition specified in 4.2. But there cannot be more than one such transformation in view of 3.3, for every fractional transformation is a projective transformation, as we saw in 4.1. This completes the proof of 4.2.

We are now ready to prove:

4.3 *Every projective transformation is a fractional transformation.*

Let T be a projective transformation, and A, B, C, D distinct ordinary points, no three of which are collinear. The images A', B', C', D' of these points are then distinct, and no three are collinear. There is a unique fractional transformation F that sends A, B, C, D into A', B', C', D', respectively, according to 4.2. Also, F is a projective transformation by 4.1. Therefore T must be identical with F by 3.3.

The following two theorems have now been completely established:

4.4 *Every projective transformation has equations of the form*

$$x' = \frac{a_1x + a_2y + a_3}{c_1x + c_2y + c_3}, \qquad y' = \frac{b_1x + b_2y + b_3}{c_1x + c_2y + c_3}, \qquad \Delta \neq 0,$$

and, conversely, such equations always represent a projective transformation. The latter is affine if and only if c_1, c_2 are both zero.

4.5 *There is a unique projective transformation that sends four given points into four given points, respectively, if no three points of either set are collinear.*

One matter remains in connection with 4.5. Consider, for example, the projective transformation

$$x' = \frac{x + y + 1}{x + 2y + 3}, \qquad y' = \frac{x - y + 1}{x + 2y + 3},$$

where $\Delta = -4$. Suppose we multiply the numerators and denominators by some nonzero number, say 3, thus obtaining

$$x' = \frac{3x + 3y + 3}{3x + 6y + 9}, \qquad y' = \frac{3x - 3y + 3}{3x + 6y + 9}.$$

These equations have different coefficients from those of the given equations, and a different determinant, with the value -108, but they represent the same projective transformation since each point has the same image it had before. This illustrates the fact that, although there is a unique projective transformation that sends four points, no three collinear, into four points, no three collinear, there is no unique set of coefficients associated with this transformation, and no unique determinant.

EXERCISES

1. Verify that $\Delta = 0$ for the transformation

$$x' = \frac{2x + 4y - 6}{x + 2y - 3}, \qquad y' = \frac{-3x - 6y + 9}{x + 2y - 3},$$

and that all points of the plane, excluding those on the line $x + 2y - 3 = 0$, go into the point $(2, -3)$.

2. Verify that $\Delta = 0$ for the transformation

$$x' = \frac{3x + 2y - 2}{8x + 7y - 5}, \qquad y' = \frac{2x + 3y - 1}{8x + 7y - 5},$$

and that all points of the plane, excluding those on the line $8x + 7y - 5 = 0$, go into points of the line $2x' + y' = 1$.

3. Show that the rank of the matrix of Δ is 1 in Ex. 1, and 2 in Ex. 2 (see Appendix, §1).

4. Show that in a transformation of the form (1), where the rank of the matrix of Δ is 1, all image points coincide.

5. Show that in a transformation of the form (1), where the rank of the matrix of Δ is 2, all image points are collinear. (*Hint*. Using (1), with c_1, c_2, c_3 not all zero, show that constants A, B, C exist, A and B not both zero, such that $Ax' + By' + C = 0$ holds for all image points (x', y').)

6. Show that the transformation $x' = 1/x$, $y' = y/x$ is projective, and find its exceptional lines. Show that it leaves $(-1, 0)$ and each point of line $x = 1$ fixed, and hence also each line through $(-1, 0)$.

7. Prove that a projective transformation is necessarily the identity if it leaves each of four points, no three collinear, fixed.

8. Show that the projective transformation that leaves each of the points (0, 0), (1, 0), (0, 1) fixed and sends (1, 1) into (2, 2) is not affine. Check by finding its equations.

9. Given $A(1, 0)$, $B(5, 0)$, $C(2, 0)$, $D(3, 0)$, find their images A', B', C', D' under the transformation of Ex. 6 and show that $(AB, CD) = (A'B', C'D')$.

10. Solve Ex. 9, using the transformation

$$x' = \frac{3x + y - 1}{x - y + 1}, \qquad y' = \frac{x - 4y + 2}{x - y + 1}.$$

11. (a) What line has no image in the transformation of Ex. 10? (b) What line has no original? (c) What point on line $3x + y - 1 = 0$ has no image? (d) Give an example of a line all of whose points have images.

12. See what happens when you try to find the image of the line $x - y + 1 = 0$ in the transformation of Ex. 10 by using the implicit form of the transformation to substitute for x and y.

5. The Projective Group. Among the things shown in the preceding section, although we did not call attention to it, is that the fractional transformation (1) has an inverse, and that the latter is the fractional transformation (3). In view of 4.1 and 4.3 we can therefore say that every projective transformation has an inverse and that it, too, is a projective transformation. This fact can also be established nonalgebraically (see §2, Ex. 1).

The identity

$$x' = x, \qquad y' = y$$

is a projective transformation since it is affine.

By the **resultant of two projective transformations**

$$(1) \qquad x' = \frac{a_1x + a_2y + a_3}{c_1x + c_2y + c_3}, \qquad y' = \frac{b_1x + b_2y + b_3}{c_1x + c_2y + c_3}, \qquad \Delta \neq 0,$$

and

$$(2) \qquad x'' = \frac{d_1x' + d_2y' + d_3}{f_1x' + f_2y' + f_3}, \qquad y'' = \frac{e_1x' + e_2y' + e_3}{f_1x' + f_2y' + f_3}, \qquad \Delta_1 \neq 0,$$

in that order, with determinants Δ and Δ_1, we shall mean the transformation $(x, y) \rightarrow (x'', y'')$ whose equations are obtained by substituting (1) in (2). Denote this resultant by R. Then if (1) sends a point P into P', and (2) sends P' into P'', $R(P) = P''$. This definition of resultant is therefore consistent with our previous usage. On the other hand, as we shall see, if (1) does not send a point P into anything, it can still happen that P has an image under R.

Substituting (1) in (2), and simplifying, we obtain the following equations of R:

$$(3)\qquad \begin{aligned} x'' &= \frac{(d_1a_1 + d_2b_1 + d_3c_1)x + (d_1a_2 + d_2b_2 + d_3c_2)y + (d_1a_3 + d_2b_3 + d_3c_3)}{(f_1a_1 + f_2b_1 + f_3c_1)x + (f_1a_2 + f_2b_2 + f_3c_2)y + (f_1a_3 + f_2b_3 + f_3c_3)} \\ y'' &= \frac{(e_1a_1 + e_2b_1 + e_3c_1)x + (e_1a_2 + e_2b_2 + e_3c_2)y + (e_1a_3 + e_2b_3 + e_3c_3)}{(f_1a_1 + f_2b_1 + f_3c_1)x + (f_1a_2 + f_2b_2 + f_3c_2)y + (f_1a_3 + f_2b_3 + f_3c_3)}. \end{aligned}$$

These equations have the form of a projective transformation. Also, their determinant Δ_2 is not zero, for

$$\Delta_2 = \begin{vmatrix} d_1a_1 + d_2b_1 + d_3c_1 & d_1a_2 + d_2b_2 + d_3c_2 & d_1a_3 + d_2b_3 + d_3c_3 \\ e_1a_1 + e_2b_1 + e_3c_1 & e_1a_2 + e_2b_2 + e_3c_2 & e_1a_3 + e_2b_3 + e_3c_3 \\ f_1a_1 + f_2b_1 + f_3c_1 & f_1a_2 + f_2b_2 + f_3c_2 & f_1a_3 + f_2b_3 + f_3c_3 \end{vmatrix}$$

$$= \begin{vmatrix} d_1 & d_2 & d_3 \\ e_1 & e_2 & e_3 \\ f_1 & f_2 & f_3 \end{vmatrix} \begin{vmatrix} a_1 & a_2 & a_3 \\ b_1 & b_2 & b_3 \\ c_1 & c_2 & c_3 \end{vmatrix} = \Delta_1 \cdot \Delta.$$

Thus R is a projective transformation and we have proved:

5.1 *The set of all projective transformations is a group.*

We call this group the **projective group.** Since the set of all affine transformations is also a group, we have:

5.2 *The affine group is a subgroup of the projective group.*

All the transformations of a plane into itself thus far considered in this book are projective and can now be classified as follows:

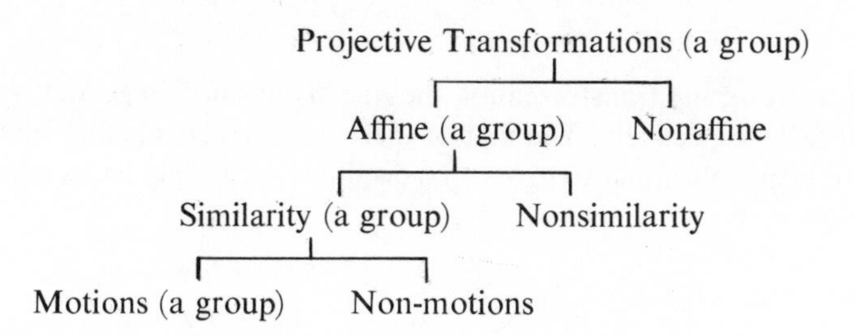

EXERCISES

1. Why do the nonaffine projective transformations not form a group?

2. Find the inverse of the transformation

$$x' = \frac{3x + y - 1}{x - y + 1}, \qquad y' = \frac{x - 4y + 2}{x - y + 1}$$

and show that it is projective.

3. Find the resultant of the given transformation in Ex. 2 and the transformation $x'' = 1/x'$, $y'' = y'/x'$ in that order. Check by using the relation among the three determinants.

4. Verify that equations (3) reduce to the identity when (2) is the inverse of (1).

5. If T is the transformation $x' = 1/x$, $y' = y/x$, find T^2. What does your result show about the nature of T, and about the nature of the set consisting of T and T^2?

6. Write the identity as a fractional transformation and find Δ.

6. Projective Transformations and Projections. Earlier we stated that every projective transformation can be obtained as the resultant of a sequence of projections (3.1) and postponed the proof until now. Before giving it we shall prove the following auxiliary theorem:

6.1 *There is a unique projective transformation in which a given line v has no image and which sends given noncollinear points A, B, C into given noncollinear points A', B', C', respectively.*

Proof. Let the equation of v be

$$ax + by + c = 0, \tag{1}$$

the coordinates of A, B, C be (x_1, y_1), (x_2, y_2), (x_3, y_3), and those of A', B', C' be (x_1', y_1'), (x_2', y_2''), (x_3', y_3'). Our task is to determine the coefficients in the equations

$$x' = \frac{a_1x + a_2y + a_3}{c_1x + c_2y + c_3}, \qquad y' = \frac{b_1x + b_2y + b_3}{c_1x + c_2y + c_3} \tag{2}$$

so that in the resulting transformation the line (1) has no image and A, B, C go into A', B', C', respectively. The first of these requirements is met by choosing c_1, c_2, c_3 to be proportional to a, b, c. To do this most simply let us take $c_1 = a$, $c_2 = b$, $c_3 = c$. We thus obtain

$$x' = \frac{a_1x + a_2y + a_3}{ax + by + c}, \qquad y' = \frac{b_1x + b_2y + b_3}{ax + by + c}, \tag{3}$$

and it remains to determine a_1, a_2, a_3, b_1, b_2, b_3.

Since (3) is to send A into A', we must have

$$x_1' = \frac{a_1x_1 + a_2y_1 + a_3}{ax_1 + by_1 + c}, \qquad y_1' = \frac{b_1x_1 + b_2y_1 + b_3}{ax_1 + by_1 + c}.$$

Likewise there are two equations corresponding to B, B', and two for C, C'. Thus, a_1, a_2, a_3, b_1, b_2, b_3 must have values satisfying these six equations. The latter, cleared of fractions, can be arranged as follows:

$$\begin{aligned}
x_1a_1 + y_1a_2 + a_3 \qquad\qquad\qquad\qquad &= x_1'(ax_1 + by_1 + c)\\
x_2a_1 + y_2a_2 + a_3 \qquad\qquad\qquad\qquad &= x_2'(ax_2 + by_2 + c)\\
x_3a_1 + y_3a_2 + a_3 \qquad\qquad\qquad\qquad &= x_3'(ax_3 + by_3 + c)\\
x_1b_1 + y_1b_2 + b_3 &= y_1'(ax_1 + by_1 + c)\\
x_2b_1 + y_2b_2 + b_3 &= y_2'(ax_2 + by_2 + c)\\
x_3b_1 + y_3b_2 + b_3 &= y_3'(ax_3 + by_3 + c).
\end{aligned}$$

These are six linear equations in the six unknowns a_1, a_2, a_3, b_1, b_2, b_3. The determinant of this system is

$$\begin{vmatrix} x_1 & y_1 & 1 & 0 & 0 & 0\\ x_2 & y_2 & 1 & 0 & 0 & 0\\ x_3 & y_3 & 1 & 0 & 0 & 0\\ 0 & 0 & 0 & x_1 & y_1 & 1\\ 0 & 0 & 0 & x_2 & y_2 & 1\\ 0 & 0 & 0 & x_3 & y_3 & 1 \end{vmatrix}, \quad \text{or} \quad \begin{vmatrix} x_1 & y_1 & 1\\ x_2 & y_2 & 1\\ x_3 & y_3 & 1 \end{vmatrix}^2$$

by Laplace's development, and hence is not zero since A, B, C are noncollinear. The system therefore has a unique solution; that is, there is only one set of values that the six unknowns can have so that the transformation (3) will meet all the requirements of the theorem.

Thus, when $c_1 = a$, $c_2 = b$, $c_3 = c$ there is a unique projective transformation meeting the requirements of the theorem. What happens if c_1, c_2, c_3 are proportional to a, b, c but not equal to them? If we took c_1, c_2, c_3 in (2) to be k times a, b, c, respectively, where $k \neq 0$, there would again be a unique solution for the six unknowns, but these values would merely be k times those first obtained. Cramer's rule for solving by determinants makes this clear. The resulting transformation would therefore be no different from the one in which $c_1 = a$, $c_2 = b$, $c_3 = c$. This completes the proof of 6.1.

We can now prove our main theorem:

6.2 *Any given projective transformation can be obtained as the resultant of a sequence of projections, not more than one of which has vanishing elements.*

Let T be the given transformation of a plane α. If T is affine, 6.2 is true by V, 4.2. Assume T is nonaffine. There is, then, a line v with no image. Let A, B, C be noncollinear points, none of which is on v, and let their images be A', B', C'. Using a central projection, project α on an intersecting plane α' so that v has no image. A, B, C will then have certain noncollinear images A_1, B_1, C_1.

If we now project α' back on α by a parallel projection, A_1, B_1, C_1 will go into three noncollinear points $\bar{A}$, $\bar{B}$, $\bar{C}$. Let R be the resultant of these projections in the stated order. Then R is a projective transformation of α into itself such that v has no image and A, B, C go into $\bar{A}$, $\bar{B}$, $\bar{C}$, respectively. Now, there is a sequence of parallel projections whose resultant R' maps α on itself so that $\bar{A}$, $\bar{B}$, $\bar{C}$ go into A', B', C', respectively (IV, 2.4 and V, 4.2). Since R' is affine, RR' is a projective transformation of α into itself such that v has no image and A, B, C go into A', B', C', respectively. By 6.1, then, $T = RR'$ and the present theorem is proved.

EXERCISES

1. Find the equations of the projective transformation which leaves each of the points (0, 0), (1, 0), (0, 1) fixed and in which the line $x = 2$ has no image.

7. Conic Sections. In view of 6.2 we know that the property of being a conic is preserved by all projective transformations. This had already been noted in 3.2. Of course, not every point of a conic c which is subjected to a projective transformation T need have an image. A conic, we recall, can meet a line in one or two points. Hence, if T is nonaffine, with vanishing lines v, w', c may meet v in one or two points and its image c' may meet w' in one or two points. Thus, one or two points of c may have no images, and one or two of c' may have no originals. Even when c and c' contain exceptional points, however, we shall continue to say, as a matter of convenience, that c goes into c'.

We now prove an important fact concerning the tangents to c, assuming it is nondegenerate. Let A, B, C, D, P be distinct ordinary points of c, and let A', B', C', D', P' be their images under a projective transformation. Then. since cross-ratio of lines is preserved, we have

$$(AP\ BP,\ CP\ DP) = (A'P'\ B'P',\ C'P'\ D'P'). \tag{1}$$

Let t be the tangent to c at D, and $\bar{t}$ the tangent to c' at D'. Then, by V, 14.2, $(AD\ BD,\ CD\ t)$ equals the left side of (1), and $(A'D'\ B'D',\ C'D'\ \bar{t})$ equals the right side. Hence

$$(AD\ BD,\ CD\ t) = (A'D'\ B'D',\ C'D'\ \bar{t}). \tag{2}$$

If t' is the image of t, then

$$(AD\ BD,\ CD\ t) = (A'D'\ B'D',\ C'D'\ t'). \tag{3}$$

The right sides of (2) and (3) are therefore equal, and we conclude that $\bar{t} = t'$ by V, 11.1. Thus t' is the tangent to c' at D', and we have proved the theorem:

7.1 *In mapping one conic section on another, a projective transformation sends the tangent at a point of the first conic into the tangent at the image point on the second conic.*

Thus, *the property of being a tangent to a conic is projective.* In checking this fact algebraically, and also for later purposes, it is useful to know that the equation of the tangent to the conic

$$Ax^2 + Bxy + Cy^2 + Dx + Ey + F = 0$$

at the point (x_0, y_0) is

$$2Ax_0x + B(x_0y + y_0x) + 2Cy_0y + D(x + x_0) + E(y + y_0) + 2F = 0.*$$

EXERCISES

All of these exercises involve the transformation $x' = 1/x, \quad y' = y/x$.

1. For each of the following conics c, find the image conic c' and the vanishing points on c and c'. Then describe the nature of the correspondence between the points of c and c'.

(a) $y = x^2$; (b) $xy = 1$; (c) $x^2 - y^2 = 1$;
(d) $y^2 - x^2 = 1$; (e) $b^2x^2 + a^2y^2 = a^2b^2$; (f) $x = y^2 + 1$.

2. Check 7.1 by using (a) the tangent to $y = x^2$ at (2, 4); (b) the tangent to $x = y^2 + 1$ at (1, 0).

3. (a) Find the image of $y = x^2$ and of the tangent to $y = x^2$ at (0, 0). Does your result contradict 7.1? Explain. (b) Do the same as in (a), using $y^2 - x^2 = 1$ and the tangent to it at (0, 1).

8. Projective Equivalence. Two geometric figures are said to be **projectively equivalent** if either figure can be mapped on the other by a projective transformation. Since affinely equivalent figures (IV, §9) are necessarily projectively equivalent, we can say immediately that any two lines or any two triangles are projectively equivalent.

Beyond this we can say, in view of 4.5, that any four points A, B, C, D, no three collinear, are projectively equivalent to any four points A', B', C', D', no three collinear. This is not the same as saying that any two quadrilaterals are projectively equivalent, for a projective transformation which sends A, B into A', B' need not send $\overline{AB}$ into $\overline{A'B'}$.

Also, it is easily seen that any two nondegenerate conics c, c' are projectively equivalent. For if they are of the same type (parabolic, elliptic, or hyper-

* For the derivation of this equation see, for example, L. P. Eisenhart, *Coordinate Geometry*, p. 228, or R. W. Brink, *Analytic Geometry*, Second Edition, p. 166.

bolic) they are affinely equivalent (IV, 9.1), and hence projectively equivalent. If they are of different types, suppose, for example, that c is a parabola and c' a hyperbola. An affine transformation T_1 will send c into $y = x^2$, and another, T_3, will send $xy = 1$ into c'. It is easily verified that the projective transformation $x' = 1/x, \quad y' = y/x$ sends $y = x^2$ into $x'y' = 1$. Denoting this transformation by T_2, we see that $T_1T_2T_3$, which is a projective transformation, sends c into c'. Also, the inverse of this transformation, which is $T_3^{-1}T_2^{-1}T_1^{-1}$, will send c' into c. A similar procedure will take care of the remaining cases, a parabola c and an ellipse c', and so on.

8.1 *Any two lines, any two triangles, or any two nondegenerate conics, respectively, are projectively equivalent. Any four points, no three collinear, are projectively equivalent to any four points, no three collinear.*

EXERCISES

1. Complete the proof that any two nondegenerate conics are projectively equivalent.

2. Prove that any two pairs of lines are projectively equivalent.

3. Assuming the proposition in Ex. 2, prove that any two pencils of lines are projectively equivalent.

9. Projective Geometry of the Euclidean Plane. Originally, we defined the projective properties of figures to be those properties that are preserved by all projections of one plane or another. Later, we saw that these properties are identical with those preserved by all projective transformations of a plane into itself, that is, by the projective group. Following the procedure of earlier chapters, then, we now designate as **projective geometry of the Euclidean plane** that part of Euclidean plane geometry which deals with projective properties. Thus, projective geometry is associated with the projective group in the same way as affine and metric geometries are associated with the affine and metric groups, respectively. Moreover, in view of the mutual relations of these groups we can say that all projective transformations preserve projective properties, only certain of these transformations, the affinities, also preserve affine properties, and only certain affine transformations, the similarities, also preserve metric properties. For this reason affine geometry may be regarded as a subgeometry of projective geometry, and metric geometry as a subgeometry of affine geometry.

Our work in this chapter has exhibited some of the most important projective properties: the collinearity of points, the cross-ratio of points and of lines, the property of being a pencil of lines, the property of being a conic, and the property of being a tangent to a conic. Not included among the projective properties are parallelism, concurrence, betweenness, ratio of division, and the other properties previously called affine. While parallelism and concurrence, regarded separately, are thus not projective, the property of being either parallel or concurrent is projective since a pencil of lines always projects into a pencil of lines.

In seeking, beyond what we have already done, for statements belonging to projective geometry of the Euclidean plane, the student would have to search among the statements of Euclidean geometry for those dealing with projective properties but not, at the same time, with affine or metric properties. He could reasonably expect to find comparatively few such statements, for geometric figures are usually far richer in metric and affine properties than in projective ones. Thus, on looking among the axioms of Euclidean geometry he would find only one statement of the desired sort, namely, that any two points are on one and only one line. His search among familiar theorems of Euclidean geometry would probably yield nothing, since projective theorems are not usually included in high school courses. We shall therefore assist him by mentioning a few such theorems, so-called classical ones. The proofs are omitted at this time.

Perhaps the earliest projective proposition of note is the following one, credited to Pappus of Alexandria, who was born about 340 A. D.

9.1 Theorem of Pappus. *If A, B, C are distinct points on a line and A', B', C' are distinct points on another line, the points of intersection of the lines AB' and $A'B$, BC' and $B'C$, AC' and $A'C$ are collinear, provided these intersections exist.*

This is illustrated in Fig. VI, 2, where L, M, N are the three intersections. The theorem is quite remarkable inasmuch as no condition is placed on the positions of the given points on their respective lines.

Unlike the Theorem of Pappus, which is somewhat isolated historically, the other theorems we shall state came at a time when considerable attention was being given to projections and their properties.

9.2 Theorem of Desargues (1593–1662). *If A, B, C and A', B', C'*

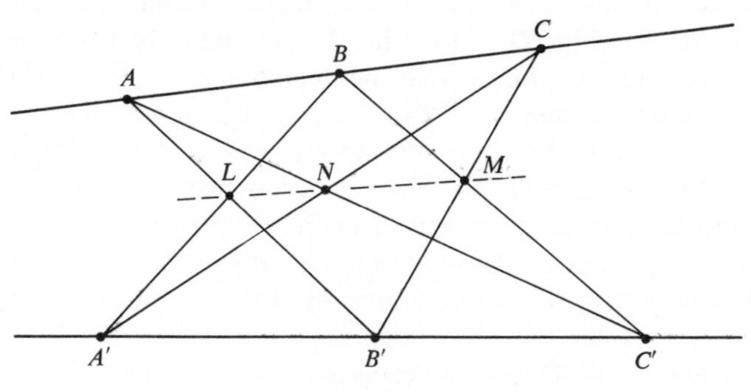

Fig. VI, 2

*are two sets of noncollinear points such that the lines AA', BB', CC' belong to a pencil, the intersections of the pairs of lines AB and $A'B'$, BC and $B'C'$, AC and $A'C'$ are collinear, provided these intersections exist.**

Fig. VI, 3 illustrates Desargues's theorem for the case in which AA', BB', CC' are concurrent. O is the point of concurrency and L, M, N are the three intersections which are collinear. It may be helpful in remembering the theorem to note that lines AA', BB', CC' join pairs of vertices of triangles ABC, $A'B'C'$,

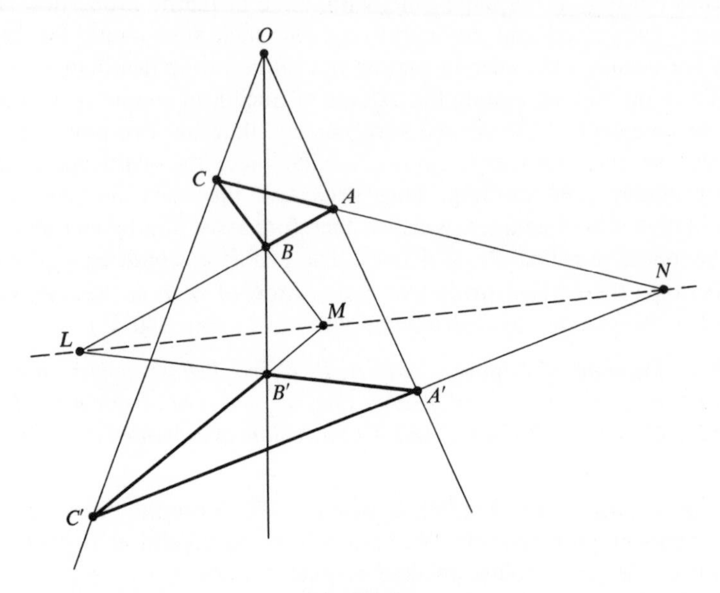

Fig. VI, 3

and that lines AB and $A'B'$, etc., contain pairs of sides of these triangles. In stating the theorem, however, we avoided using the word "triangle" since a triangle is an affine figure.

When two triangles ABC, $A'B'C'$ are situated so that lines AA', BB', CC' meet in a point, as in Fig. VI, 3, the triangles are said to be *perspective from the point*. When they are so situated that the intersections L, M, N of the pairs of lines AB and $A'B'$, AC and $A'C'$, BC and $B'C'$ are on a line, as in Fig. VI, 3, the triangles are said to be *perspective from this line*. As a consequence of the Theorem of Desargues, then, we can say that if two triangles are perspective from a point, they are also perspective from a line (provided, of course, that the specified intersections exist). This use of the word *perspective* in connection with a theorem of projective geometry reminds us of the close relation which existed

* The converse of the Theorem of Desargues is also true, and is sometimes included as a part of the theorem. We state it in Ex. 11.

historically between the study of projections and the discovery of the laws of perspective.

9.3 Theorem of Pascal (1623–1662). *If $A_1, A_2, \ldots, A_6$ are six points of a conic, the intersections of the pairs of lines A_1A_2 and A_4A_5, A_2A_3 and A_5A_6, A_3A_4 and A_6A_1 are collinear, provided these intersections exist.*

This is illustrated in Fig. VI, 4, where we have taken the conic to be a circle.

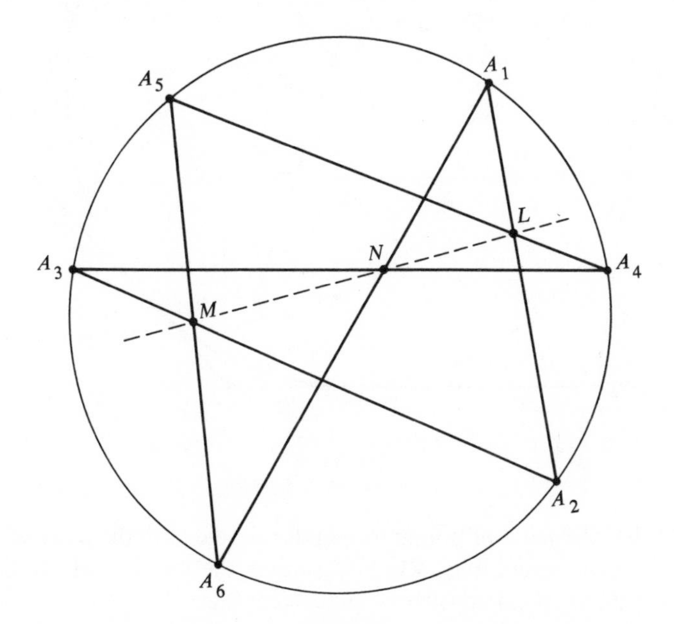

Fig. VI, 4

The three intersections are L, M, N. As an aid in remembering the theorem let us note that the six given points are the vertices of a hexagon $A_1A_2 \ldots A_6$ inscribed in the conic, and that each of the specified pairs of lines A_1A_2 and A_4A_5, etc., contains a pair of opposite sides of this hexagon. The hexagon of Fig. VI, 4, with sides $\overline{A_1A_2}$, $\overline{A_2A_3}$, etc., is not the type of hexagon most familiar to the reader, for some of its sides cut one another. However, the theorem holds regardless of the type of hexagon. We did not state the theorem in terms of a hexagon since this is an affine concept.

9.4 Theorem of Brianchon (1783–1864). *If $a_1, a_2, \ldots, a_6$ are six tangents to a conic, the three lines determined by the three pairs of points a_1a_2 and a_4a_5, a_2a_3 and a_5a_6, a_3a_4 and a_6a_1 belong to a pencil, provided these points exist.*

Whenever these points do exist, it is clear from Fig. VI, 5 that the six tangents determine a hexagon circumscribing the conic (which in Fig. VI, 5 is a

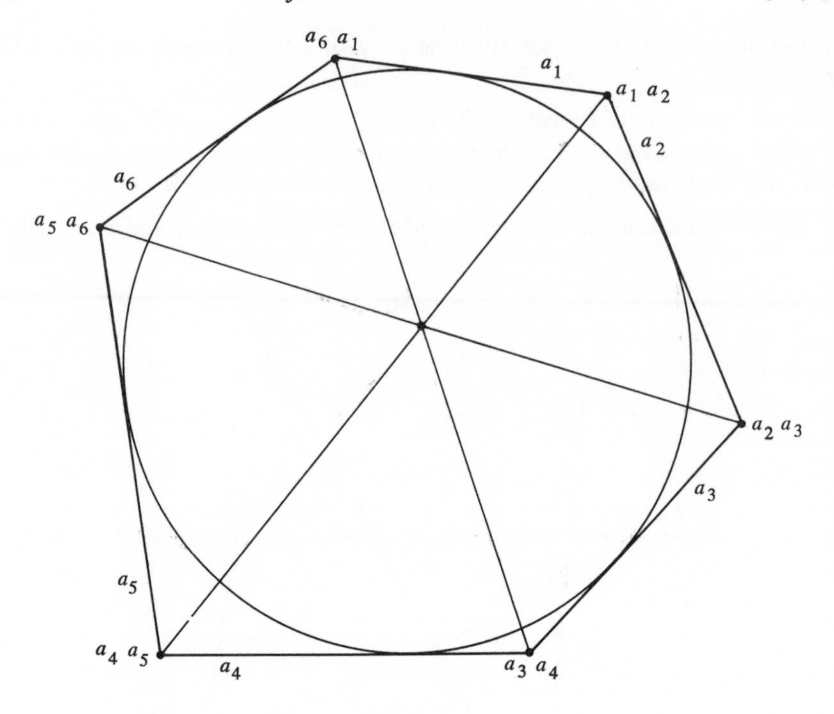

Fig. VI, 5

circle), and that the pairs of points a_1a_2 and a_4a_5, etc., are the pairs of opposite vertices of this hexagon. Fig. VI, 5 illustrates the case in which the pencil mentioned in the theorem consists of concurrent lines.

It cannot be too strongly emphasized that the preceding theorems hold only in case certain points exist, that is, only in case certain lines are not parallel. This provisional character of its theorems, this lack of complete generality, is typical of projective geometry of the Euclidean plane and constitutes a blemish. Even the properties we have called projective leave something to be desired in the way of completeness. Collinear points, we say, go into collinear points, but this is true only if the points have images. Of four lines belonging to a pencil it may be that only three have images, with the result that the statement "cross-ratio of lines is a projective invariant" is only provisionally true. Even in saying that a conic always goes into a conic we know that there may be one or two points on each which play no part in the mapping. By contrast, the properties we have called metric or affine involve no such exceptions. Never, for any reason, does a similarity transformation fail to send an angle into another of the same size, nor does an affine transformation fail to preserve ratio of division.

Thus we see that projective geometry of the Euclidean plane involves exceptional cases of a sort not encountered in metric or affine geometry. These exceptions, of course, were built in from the start since our definition of a projective transformation permitted them. But we had no choice in the matter. For we were led to this definition by considering the resultant of a sequence of projections of one plane on another, and if this resultant was to be a genuinely new transformation, that is, nonaffine, we could not exclude the projections which have vanishing points. The shortcomings in projective geometry of the Euclidean plane therefore stem from the nature of such projections.

In modern mathematics the presence of exceptional cases has often been the stimulus for the creation of some new idea to remove the exceptions. The introduction of negative numbers, we know, made subtraction possible in all cases, and as a result of the invention of imaginary numbers all algebraic equations could have roots. In much the same way, the exceptions arising in the study of projective properties have been removed. It was Kepler (1571–1630), a contemporary of Desargues, who first suggested how this could be done. In brief, by assuming the existence of new points we can enlarge the Euclidean plane so as to form a plane of a different sort, called a *projective plane*. When projections are studied in terms of this new plane, there are no longer any vanishing points or lines. The projective geometry of this plane, likewise, is free of exceptional cases. Thus, when the projective plane is used, the statement of Pascal's theorem simply ends with the word "collinear" since the specified intersections always exist in this plane. We shall explain these new ideas in Chapter VIII. In the meantime it will be well to keep in mind that in ordinary usage the term *projective geometry*, when referring to two-dimensional geometry, always means projective geometry of the projective plane.

EXERCISES

1. Verify the Theorem of Pappus for various orders of the points by means of diagrams.

2. Using diagrams, show that one or more of the intersections mentioned in Pappus's theorem can fail to exist.

3. Draw a diagram to illustrate Desargues's theorem when lines AA', BB', CC' are parallel.

4. In Desargues's theorem, if lines AA', BB', CC' are parallel, show by diagrams that one or two of the specified intersections can fail to exist. Prove that if two do not exist, the third cannot exist either.

5. Do Ex. 4, assuming that lines AA', BB', CC' are concurrent.

6. Illustrate Pascal's theorem, using a circle and a hexagon $A_1A_2 \ldots A_6$ whose sides do not cut each other.

7. Verify by a diagram that at least one of the intersections mentioned in Pascal's theorem can fail to exist.

8. Verify Pascal's theorem and Brianchon's theorem, using an ellipse other than a circle. (Draw the ellipse using a loop of string.)

9. Show by a diagram that one or more of the points mentioned in Brianchon's theorem may not exist.

10. When will a projective transformation always send (a) a triangle into a triangle; (b) a hexagon into a hexagon?

11. The **Converse of the Theorem of Desargues** is as follows: *If A, B, C and A', B', C' are two sets of noncollinear points such that the intersections of the lines AB and $A'B'$, BC and $B'C'$, AC and $A'C'$ are collinear, then the lines AA', BB', CC' belong to a pencil.* Draw a diagram to illustrate this.

CHAPTER VII

Topological Transformations

1. Introduction. By this time it is apparent that there is a close connection between a mapping, as we have been using this term, and a geographical map. In both cases we start out with certain objects, the originals, and establish a relation between them and other objects, which we call their images. Some features of the originals are preserved, in the sense that they also appear in the images, whereas other features are destroyed or distorted. A geographical map, serving an immediate practical end, should distort the original as little as possible. Our mappings or transformations, of course, have no such objective, otherwise we would not have gone beyond motions and similarities. Nonmetric properties of figures interest us as much as the metric ones, and we therefore proceeded to study transformations which can distort figures very considerably. Affinities, we have seen, can turn circles into elongated ellipses, while projectivities can do even more, transforming circles into hyperbolas.

Topological transformations, which we are about to study, are capable of still greater distortions in figures. The simple transformation used in I, §6 to map a circle on a square, for example, is topological, and so is the inverse of this transformation, which sends the square into the circle. We see from this example that topological transformations can fail to preserve collinearity or the property of being a conic section.

Although a circle and a square may seem quite different from one another, they do have a common property and it is preserved by the two transformations of the above example: each is a closed figure. In fact, the property of being a closed figure is preserved by all topological transformations, and hence is a *topological property*. In the course of the present chapter we shall become acquainted with other topological properties. A knowledge of topological properties is a key to deeper insights in geometry, and, in particular, can be helpful in understanding the projective plane, to which we made brief reference at the end of the preceding chapter.

The following ideas will be helpful as background for the chapter.

Let P be a point, and P_n, where $n = 1, 2, 3, \dots$, an infinite sequence of points, distinct from each other and from P. If the distance $PP_n \to 0$ when $n \to \infty$, that is, if PP_n approaches 0 when n becomes infinite, we say that the sequence P_n has the *limit point* P, or that P_n *approaches* P, and write $P_n \to P$.

If P and P_n are in the same plane, let their coordinates be (a, b) and (x_n, y_n), respectively, in which case $PP_n = \sqrt{(x_n - a)^2 + (y_n - b)^2}$. If $P_n \to P$ when $n \to \infty$, this radical approaches 0, and hence $x_n \to a$ and $y_n \to b$. Conversely, if $x_n \to a$ and $y_n \to b$ as $n \to \infty$, then $P_n \to P$. We shall understand the phrase Q *is a variable point which approaches* P, or, more briefly, Q *approaches* P, or $Q \to P$, to mean that Q denotes an infinite sequence of points which has the limit point P.

Let P' be the image of a point P in a mapping of a set S on a set S'. If, in the mapping, when a variable point Q approaches P, its image Q' approaches P', the mapping is called **continuous at P.** If the mapping is continuous at every point of S we call it a **continuous mapping of S on S'.** We shall accept without proof that the resultant of two continuous mappings is a continuous mapping.

A function $f(x, y)$ of the two variables x, y is said to be *continuous at a point* (a, b) of the plane if $f(a, b)$ is defined, that is, has a value, and if $f(x, y) \to f(a, b)$ when the variable point (x, y) approaches the point (a, b). The function is said to be *continuous on a set of points* of the plane if it is continuous at each point of the set. We accept without proof that if $f(x, y)$ is a rational function or the root of a rational function, then it is continuous at each point (x, y) for which it is defined.

2. Topological Transformations of the Plane.* It will be helpful in defining a topological transformation if we first look once more at affine transformations. As we know, they are 1-1 mappings of the plane onto itself. Another characteristic, to which we never called attention previously, is that they are continuous mappings. We shall prove this later. A third characteristic of an affine transformation is that its inverse is also a continuous mapping.

Affine transformations are not the only mappings of the plane onto itself with these three characteristics. The transformation $x' = x$, $y' = y^3$, for example, has them, and so do many other transformations with simple equations. These characteristics, in fact, distinguish the topological transformations.

2.1 Definition. *By a topological transformation of the plane, or briefly a topological transformation, we shall mean a mapping of the plane onto itself which is* 1-1, *continuous, and has a continuous inverse. A topological transformation is also called a homeomorphism.*

A projective transformation which has vanishing points does not satisfy this definition since it does not map the plane onto itself, but only a part of the plane onto a part of the plane. On the other hand, there exist 1-1 mappings of the plane onto itself which are not topological because they lack the necessary continuity. Consider, for example, the 1-1 mapping T of the plane onto itself

* Unless the contrary is stated, "plane" will mean "Euclidean plane."

defined as follows: T sends the origin O into itself, and any other point P into the point P' on the ray OP such that $OP' = 1/OP$ (Fig. VII, 1). If T were continuous, then, roughly speaking, since $T(O) = O$ points near O should go into points near O; but in fact, the nearer P is to O, the farther P' is from O. Stated more precisely, P' does not approach O when P approaches O.

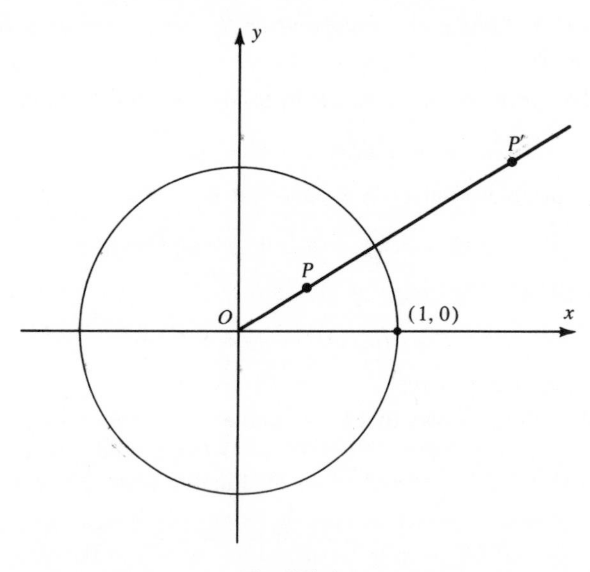

Fig. VII, 1

The inverse S^{-1} of a topological transformation S is clearly a topological transformation. The resultant ST of two topological transformations is likewise topological, for it is a 1-1 and continuous mapping of the plane onto itself since S and T are, and its inverse $T^{-1}S^{-1}$ is continuous since T^{-1} and S^{-1} are. Thus, the set of all topological transformations of the plane is a group. This group is called the **topological group** in the plane. The affine group is one of its subgroups, but the projective group is not.

Properties of geometric figures which are preserved by all transformations of the topological group are called **topological properties** or **topological invariants**, and the study of these properties is known as **topological geometry of the plane**, or simply **plane topology**. Two figures which can be mapped on each other by a topological transformation are said to be **topologically equivalent**, or **homeomorphic**, and each is called a **topological image**, or **homeomorph**, of the other. Figures which are affinely equivalent are therefore necessarily topologically equivalent.

The following two theorems enable us to prove that affine transformations are homeomorphisms and also are useful in other ways. Before stating them let us note that all the equations in rectangular coordinates that were used to represent transformations in the preceding chapters are of the form $x' = f(x, y)$, $y' = g(x, y)$, where each expression on the right denotes a function of x and y.

2.2 *If S, S' are any sets of points in the plane (including the plane itself), and S is mapped on S' by the equations*

$$x' = f(x, y), \qquad y' = g(x, y),$$

where the functions f and g are continuous on S, then the mapping is a continuous mapping of S on S'.

Proof. Let (a, b) be any point P in S, and (a', b') be its image P':

$$a' = f(a, b), \qquad b' = g(a, b); \tag{1}$$

let (ξ, η) be a variable point Q in S such that $Q \to P$:

$$PQ = \sqrt{(\xi - a)^2 + (\eta - b)^2} \to 0; \tag{2}$$

and let (ξ', η') be Q', the image of Q:

$$\xi' = f(\xi, \eta), \qquad \eta' = g(\xi, \eta). \tag{3}$$

We must show that $Q' \to P'$.

Since $Q \to P$, (2) shows that $\xi \to a$ and $\eta \to b$. From $\xi \to a$, $\eta \to b$ and the continuity of f and g, it follows that $f(\xi, \eta) \to f(a, b)$ and $g(\xi, \eta) \to g(a, b)$, or, as (1) and (3) show, $\xi' \to a'$ and $\eta' \to b'$. Finally, from $\xi' \to a'$ and $\eta' \to b'$ we see that

$$P'Q' = \sqrt{(\xi' - a')^2 + (\eta' - b')^2} \to 0,$$

that is, $Q' \to P'$.

Taking each of the sets S and S' in 2.2 to be the entire plane, we immediately obtain the following corollary:

2.3 *If T is a 1-1 mapping of the plane on itself with the equations*

$$x' = f(x, y), \qquad y' = g(x, y),$$

and its inverse has the equations

$$x = \varphi(x', y'), \qquad y = \psi(x', y'),$$

where the functions f, g, φ, ψ are continuous over the entire plane, then T is a homeomorphism of the plane.

An affine transformation meets all the conditions imposed on T in this theorem. For as we know from Chapter IV, (1) it is a 1-1 mapping of the plane on itself, (2) the functions f, g in its equations are polynomials of the first degree in x and y, and hence are continuous over the entire plane, and (3) the functions φ, ψ in the equations of its inverse are also polynomials of the first degree in x and y since the inverse is an affine transformation. Thus, ***affine transformations are homeomorphisms of the plane.***

EXERCISES

1. Find the equations of the transformation with fixed point O described above, and illustrated in Fig. VII, 1: (a) in polar coordinates; (b) in rectangular coordinates.

2. What is the relation between the set of all 1-1 mappings of the plane onto itself and the set of all homeomorphisms of the plane?

3. Give an example, not mentioned in the text, of a 1-1 transformation of the plane into itself which is not topological.

4. Prove that affine transformations are the only homeomorphisms of the plane which preserve collinearity.

5. Prove 2.3 in detail.

3. Example of a Nonaffine Topological Transformation. We shall begin our study of topological properties by considering in detail the simple nonaffine topological transformation

$$x' = x, \qquad y' = y^3$$

mentioned in §2. We denote it by T. Its inverse is

$$x = x', \qquad y = \sqrt[3]{y'}.$$

T is clearly a 1-1 transformation of the plane into itself. Also, by 2.2, T is continuous since the functions $f(x, y) = x$ and $g(x, y) = y^3$ that define it are continuous at every point of the plane. For example, consider (1, 2) and its image (1, 8). When a variable point approaches (1, 2), the equations of T show that the image of the point approaches (1, 8). The continuity of T^{-1} likewise follows from 2.2. Thus, T is topological. This could also have been proved immediately by use of 2.3.

Being nonaffine, T cannot always send lines into lines (see §2, Ex. 4). Let us check this. Although each vertical line remains fixed, and horizontal lines go into horizontal lines, the oblique lines $y = mx + b$, where $m \neq 0$, go into the cubic curves $y' = (mx' + b)^3$. These curves cross the horizontal axis, extending infinitely far above and below it. Fig. VII, 2 shows a line for which $m > 0$, $b < 0$ and the curve into which it goes. (The reader can quickly draw the curves in certain cases, for example, $m = 1$, $b = 0$.) Thus, although T does not always send lines into lines, it always sends them into curves which, like lines, are endless, unbroken, and enclose no area.

Let us now see the effect of T on a curve which encloses an area, say a circle. The circle $x^2 + y^2 = 1$, for example, is sent into the locus of

$$x'^2 + y'^{2/3} = 1, \qquad \text{or} \qquad y' = \pm(1 - x'^2)^{3/2}.$$

This curve, which is symmetrical to O and each axis, and tangent to the x-axis,

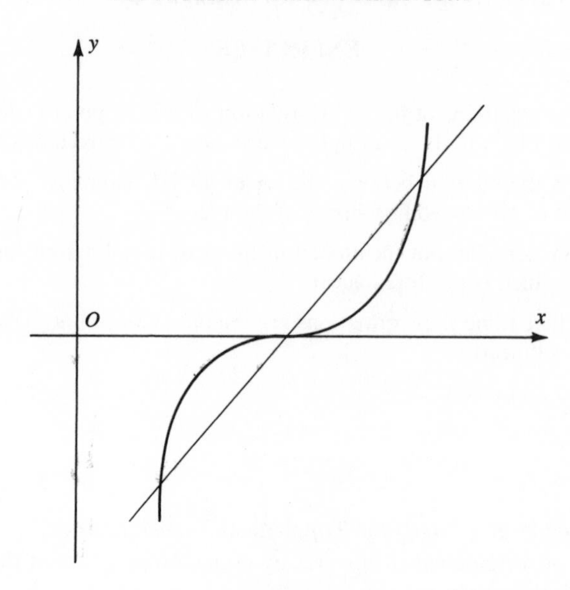

Fig. VII, 2

is shown in Fig. VII, 3, together with the circle, and is clearly not a conic. Although the curve is a considerable distortion of the circle, in common with the circle it encloses an area. If instead of a circle, we had started out with any other area-enclosing curve, its image would still have enclosed an area. Fig. VII, 3 brings out one further fact. Since T leaves the horizontal axis fixed and sends the circle, which is perpendicular to this axis, into a curve tangent to it, T^{-1} must leave the axis fixed and send the curve into the circle. Thus, T^{-1} carries the curve and a line tangent to it into the circle and an intersecting line. Since T^{-1} is topological, we see that topological transformations may fail to preserve tangency.

Next, consider a parabola, say $y = x^2$. T sends this into the curve $y' = x'^6$ (Fig. VII, 4), which is not a parabola, although it resembles the given one in certain ways. Like the parabola, for example, it is symmetrical to the vertical axis, has a minimum point at the origin, is endless, unbroken, and encloses no area. Only the last three of these properties, however, would be preserved by all topological transformations. If, instead of a parabola, we had used any other curve possessing these three properties, its image under T also would have possessed them. This has already been verified for the case in which the curve is a straight line.

Finally, consider a hyperbola, say $y = 1/x$. This goes into the curve $y' = 1/x'^3$, which, in common with the hyperbola, as the reader can easily show, consists of two separate parts or subcurves, one in the first quadrant and the other in the third quadrant, each endless, unbroken, and enclosing no area. Although there are other obvious common properties, for example, the fact

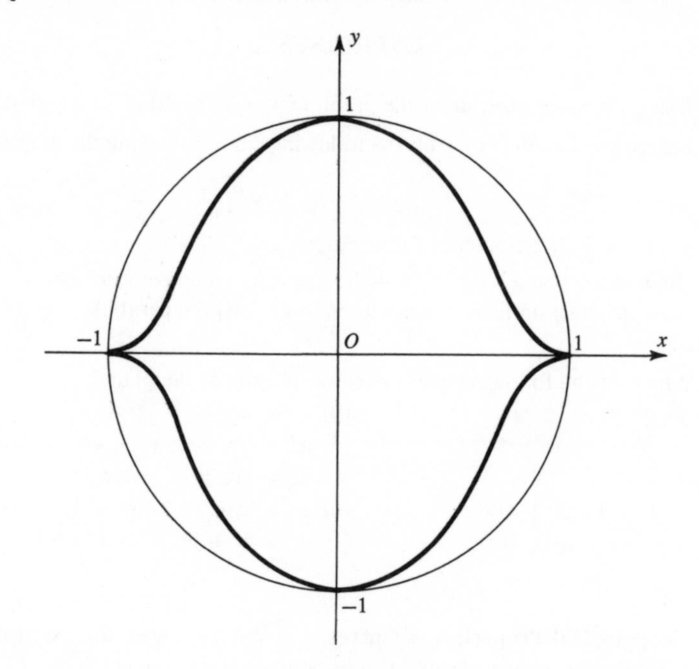

Fig. VII, 3

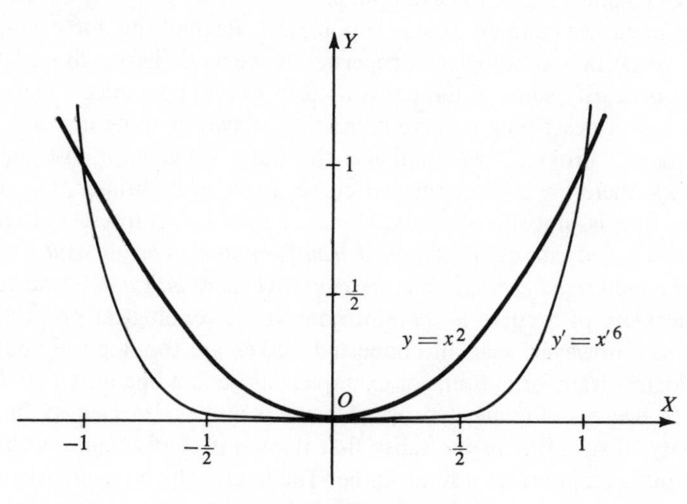

Fig. VII, 4

that the coordinate axes are asymptotes, only the foregoing would be preserved by all topological transformations. In fact, T would always send a curve consisting of two separate parts, each endless, unbroken, and enclosing no area, into another curve of this type.

EXERCISES

1. Using the same axes, draw the graph of $y = 1/x$ and of its image under T.

2. Determine the effect of T on the following curves, drawing the graphs in each case:
(a) $y = 2^x$; (b) $y = 1/x^2$; (c) $y = \sin x$;
(d) $x^{2/3} + y^{2/3} = 1$ (hypocycloid of four cusps).

3. Show that $x' = x, \quad y' = y^3 + 1$ represents a homeomorphism of the plane and determine its effect on lines, the circle $x^2 + y^2 = 1$, the parabola $y = x^2$, and the hyperbola $xy = 1$.

4. Which of the following are homeomorphisms of the plane?
(a) $x' = y^3,\ y' = x^3$; (b) $x' = x^2,\ y' = y^3$;
(c) $x' = e^x,\ y' = -e^y$; (d) $x' = 2x + 1,\ y' = y^3$;
(e) $x' = x + x^2,\ y' = 2y$; (f) $x' = \cos y,\ y' = \sin x$;
(g) $x' = x,\ y' = 1/(x^2 + 1)$; (h) $x' = x,\ y' = 1/(x^3 + 1)$;
(i) $x' = x + y^3,\ y' = x - y^3$; (j) $x' = x + y^2,\ y' = x - y^2$.

4. Topological Properties of Curves. We have used the example of the preceding section to prove certain things and suggest others. Since T, a topological transformation, did not always preserve collinearity, and sent conics into curves of other types, the example proved that no property called affine or projective in earlier chapters is also topological. Beyond this the example drew attention to certain topological properties of curves. Before discussing them further let us clarify some of the terms used in §3 and introduce a few others.*

Instead of describing a curve consisting of two or more completely separated parts as "broken," we shall use the more usual term **disconnected**. A hyperbola is therefore a disconnected curve. Each of its branches, however, is **connected**, that is, not disconnected. Lines, parabolas, ellipses, and circles are likewise connected curves. *Topological transformations always send a connected curve into a connected curve, and a disconnected curve into a disconnected curve.* The **connectivity** of a curve is therefore one of its topological properties. This is not to say, however, that all connected curves are topologically equivalent. No topological transformation, for example, can send a line into a circle. There are, in fact, degrees of connectedness. A circle is more connected, or has greater connectivity, than a line in the sense that it cannot be disconnected by the removal of just one point, as a line can be. The figure which results when a point is removed from a line, a circle, or a parabola is called a *punctured line*, a *punctured circle*, or a *punctured parabola*. Thus, a punctured circle is connected, and a punctured line disconnected.

The term **endless** was used in §3 to mean without endpoints or boundaries. Most familiar curves are of this sort. In traversing such a curve we can always

---

* This clarification, part of §7, and part of §8 are intuitive.

go beyond any point which has been reached. Topological transformations always send an endless curve into an endless curve. *The endlessness of a curve is therefore another of its topological properties.* Again, this does not mean that all endless curves are topologically equivalent, for curves can be endless in various ways. Lines and circles, for example, are both endless, yet are not topologically equivalent, as was mentioned above. An example of a curve which is not endless is $y = \sqrt{x}$. This curve is half of a parabola and has one end, the origin. The curve $y = \sqrt{1 - x^2}$, which is a semicircle, is also not endless; it has two ends, $(1, 0)$ and $(-1, 0)$.

We call a curve **closed** if the curve, in its entirety, is the boundary of some area, and **open** if no part of it encloses an area. Circles, two-leaved roses, or *lemniscates* (Fig. VII, 5a), and three-leaved roses are closed curves, for example, and lines, parabolas, hyperbolas are open curves. A *folium of Descartes* (Fig. VII, 5b) is neither open nor closed according to these definitions, for only a part of it encloses an area. Topological transformations always send a closed curve into a closed curve, and an open curve into an open curve. *The property of being a closed curve or an open curve, respectively, is therefore topological.* Since a line is connected and a hyperbola is disconnected we see that not all open curves are topologically equivalent. Nor are all closed curves topologically equivalent. No topological transformation, for example, can send a circle into a lemniscate. This appears reasonable when we note that a circle has the greater connectivity of the two, for the removal of a point cannot disconnect it, whereas the removal of the point in which the lemniscate intersects itself disconnects that curve.

A line, then, is an open, endless, connected curve. However greatly it might be distorted by a topological transformation, these properties would be preserved. Fig. VII, 2 is our first illustration of this. A parabola and a branch of a hyperbola also possess these three topological properties. In fact, *a line,*

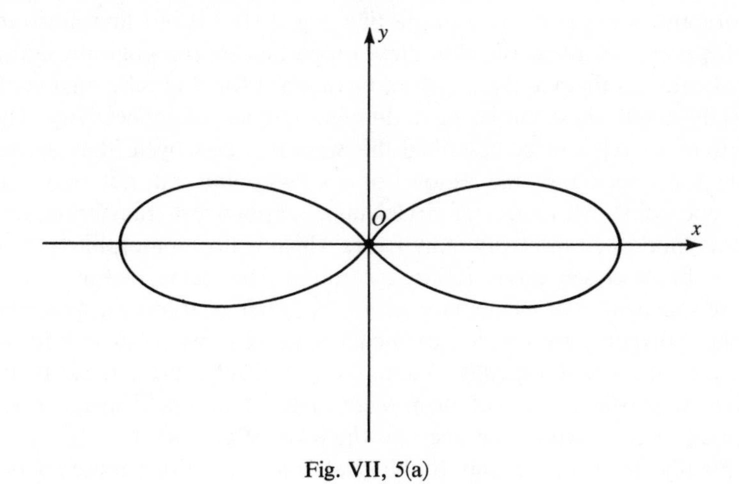

Fig. VII, 5(a)

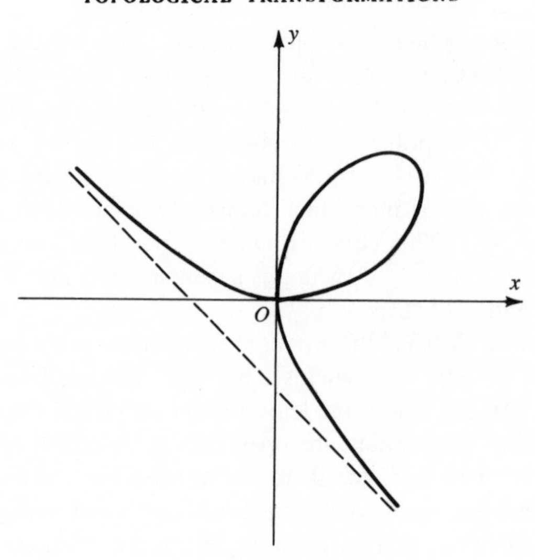

Fig. VII, 5(b)

a parabola, and a branch of a hyperbola are topologically equivalent, as we show in §6 by mapping each on the others by topological transformations. This means that from a topological viewpoint there is no difference between them, no way of distinguishing one from another, each topological property of one being possessed also by the others. If we had physical models of these curves made of a suitable material we could imagine bending each so as to obtain any of the others. Lines, parabolas, and hyperbolic branches belong to a class of topologically equivalent figures called *simple arcs*. To be precise, any figure which is homeomorphic, that is, topologically equivalent, to a line is called a **simple arc.**

Circles, we have seen, are closed, endless, connected curves. All topological transformations preserve these properties. Fig. VII, 3 is our first illustration of this. However, not all curves with these properties are topologically equivalent. A lemniscate has them and yet cannot be mapped topologically on a circle, for, as already noted, these curves have different degrees of connectivity. The connectivity of a circle can be described this way: it is destroyed (that is, the circle is made disconnected) by the removal of any two points, but not by the removal of any one point. All images of circles under topological transformations have this same kind of connectivity. Any figure which is homeomorphic to a circle is called a **simple closed curve.** Ellipses, cardioids, the curve of Fig. VII, 3, and many of the figures of elementary geometry, such as triangles, quadrilaterals, and other polygons, are simple closed curves, as we prove in §6 by transforming circles into them topologically. Viewed topologically, then, these figures are all alike. If plastic models of them were at hand we could imagine bending, stretching, or contracting any one into the form of any other.

Finally, let us note that topological transformations preserve the *inci-*

dence relation, that is, the meeting or nonmeeting of a point and a curve, or of two curves. If a point is on a curve, the image of the point always is on the image of the curve under a topological transformation, and if a point is not on a curve then the image of the point is not on the image of the curve. Suppose that we have two curves c_1, c_2 and a topological transformation T. Let $T(c_1, c_2) = c_1', c_2'$. Then each point which is on both c_1 and c_2 must go into a point which is on both c_1' and c_2'. Conversely, because of the 1-1 character of T, each point which is on both c_1' and c_2' must be the image of a point on both c_1 and c_2. Hence c_1', c_2' have the same number of intersections as c_1, c_2, and if c_1, c_2 do not meet, neither do c_1', c_2'. In particular, if c_1, c_2 are tangent at a common point P, then c_1', c_2' meet at the image of P, but they are not necessarily tangent there. This was illustrated in §3 in our mapping of the circle $x^2 + y^2 = 1$. If c_1, c_2 do not meet and are asymptotic, then c_1', c_2' do not meet, but they are not necessarily asymptotic.

EXERCISES

1. Can the figure consisting of a rectangle and one diagonal be disconnected by the removal of a point? If not, can two points be removed without disconnecting it?

2. Can the figure consisting of a circle and two diameters be disconnected by the removal of a point? If not, can as many as four points be removed without disconnecting it?

3. Describe the homeomorphs of each of the following figures: (a) a pair of tangent circles; (b) two parallel lines and a transversal; (c) a circle and a tangent; (d) a triangle and the inscribed circle; (e) the curve $y = e^x$ and the x-axis; (f) the curve $y = \sin x$, between $x = 0$ and $x = \pi$, and the coordinate axes.

4. Verify that a punctured circle is open, endless, and connected, like a line.

5. Verify that a punctured parabola has the same topological structure as a hyperbola.

5. More General Topological Transformations. Our understanding of topological equivalence will be improved if we take a broader view of a topological transformation. In addition to the topological transformations of the plane into itself which we have been considering, there are topological transformations which send the plane into a part of itself, or which send a part of the plane into another part of the plane, or which simply transform some geometric figure into another. We can achieve this broader view by adopting the following definition of a topological transformation.

5.1 Definition. *A transformation of a set of points S into a set of points S' is called topological, or is said to be a homeomorphism, if it is* **1-1,** ***continuous, and has a continuous inverse.***

No restriction is placed on the sets S and S'. They may, for example, form curves, or regions of the plane, or surfaces other than the plane, and so forth. In particular, each set may be the entire plane, which shows that the previous Definition 2.1 is a special case of the above.

If S can be sent into S' by a transformation meeting the conditions of 5.1, we say that S and S' are **topologically equivalent**, or **homeomorphic**, and that each set is the **topological image**, or **homeomorph**, of the other.

Previously we saw that a nonaffine projective transformation is not a topological transformation of the plane into itself. Now we can verify that it is a topological transformation of S into S', where S is the plane excluding one line, and S' is also the plane excluding one line. Consider, for example, the transformation T,

$$x' = \frac{x}{y}, \qquad y' = \frac{1 - x}{y},$$

with inverse

$$x = \frac{x'}{x' + y'}, \qquad y = \frac{1}{x' + y'}.$$

Here S consists of all points not on the line $y = 0$, and S' of all points not on the line $x' + y' = 0$. S and S' overlap, but this is irrelevant. T maps S on S' in a 1-1 way. Also, T is continuous by 2.2 since x/y and $(1 - x)/y$ are continuous functions in S. For example, $(1, 2)$ goes into $(\frac{1}{2}, 0)$, and clearly if a variable point of S approaches $(1, 2)$ the image of the point approaches $(\frac{1}{2}, 0)$. Similarly, T^{-1} maps S' on S continuously by 2.2 since $x'/(x' + y')$ and $1/(x' + y')$ are continuous functions in S'. Thus, T is topological.

Nonaffine projective transformations, being topological, preserve all the properties of curves that were called topological in §4. We now illustrate this, using the above transformation T. Consider the line $y = a$, where $a \neq 0$. Being parallel to the line $y = 0$, this line is entirely in S. Its image is the line $x' + y' = 1/a$, which, being parallel to the line $x' + y' = 0$, is entirely in S'. Thus, a complete line goes into a complete line, illustrating the fact that T preserves the property of being a simple arc. Now consider any line g that meets the line $y = 0$. A single point of g then fails to lie in S. The part of g situated in S thus consists of the two rays of g which are separated by that point, and hence is a disconnected figure, namely, a punctured line. To find its image, suppose that g has the equation $x = 2$. This is transformed into $x' + 2y' = 0$ by T. The latter line is not parallel to $x' + y' = 0$, so that the part of it situated in S' is also a punctured line. This is the image we were seeking. Had we used any other equation for g, we would have obtained a similar result: a punctured line always goes into a punctured line. This illustrates the fact that T sends a disconnected figure into a disconnected figure, and, moreover, into one of the same topological type. In this example each disconnected figure consists of two simple arcs with no common point.

As our final example, let us consider a conic, say $y = x^2$. This is a parabola tangent to the x-axis at the origin, so that the part of it in S is a punctured parabola, a disconnected figure consisting of two separate parts, each a simple arc. The given equation is changed by T to $y' = x'^2 - x'$. This is a parabola tangent to the line $x' + y' = 0$, so that the part of it in S' is likewise a punctured parabola. Thus, the conic in S and its image in S' have the same topological structure.

The above results can be generalized for all nonaffine projectivities. Complete lines of S always go into complete lines of S', and punctured lines into punctured lines. The image of a conic of S is always a conic of S' with the same topological structure, the image being a complete conic only if the original is. These are more precise statements than those made in VI, §§3, 7, where we said that a line goes into a line, excluding perhaps one point on each, and a conic goes into a conic, excluding perhaps one or two points on each.

EXERCISES

1. Prove that T always sends a punctured line into a punctured line.

2. Using the fact that T sends parallels to $y = 0$ into parallels to $x' + y' = 0$, prove nonalgebraically that it must send lines that meet $y = 0$ into lines that meet $x' + y' = 0$.

3. Given the nonaffine projectivity $x' = 1/x$, $y' = y/x$, (a) find the regions S and S'; (b) show that complete lines of S go into complete lines of S'; (c) show that the part of the line $y = mx + b$ which is in S, and its image in S', are punctured lines.

4. Using the transformation of Ex. 3, find the images of the parts of the following curves which are in S and verify that their topological properties are preserved:

(a) $x^2 + y^2 = 1$; (b) $xy = 1$; (c) $x = y^2 - 1$; (d) $y = x^3$.

5. Prove that a nonaffine projectivity,

$$x' = \frac{a_1x + a_2y + a_3}{c_1x + c_2y + c_3}, \qquad y' = \frac{b_1x + b_2y + b_3}{c_1x + c_2y + c_3},$$

sends lines parallel to $c_1x + c_2y + c_3 = 0$ into lines parallel to $A_3x' + B_3y' + C_3 = 0$, where A_3, B_3, C_3 are the cofactors of a_3, b_3, c_3 in the determinant of the transformation, and hence sends complete lines into complete lines, and punctured lines into punctured lines. (*Suggestion.* First consider the case in which c_1 or c_2 is zero.)

6. If M transforms S topologically into S', and N transforms S' into S'' topologically, does MN transform S into S'' topologically? Justify your answer.

6. Homeomorphs of Lines and Circles. Earlier we stated that parabolas and branches of hyperbolas are homeomorphic to lines, and that triangles, squares, and many of the other closed figures of elementary geometry are homeomorphic to circles. Using the broadened definition of a topological transformation given in §5 we can now show how these statements and others like them can be proved.

Let us first show that the parabola $y = x^2$ is homeomorphic to the x-axis. We do this by exhibiting a topological transformation that sends S into S', where S is the set of points forming the parabola, and S' the set of points forming the x-axis. Let P be any point of the parabola, and P' the foot of the perpendicular from P to the x-axis. Then the mapping $P \to P'$, which is simply the orthogonal projection of the parabola onto the x-axis, is topological. For it is clearly 1-1, continuous (when a point of the parabola approaches P, its image approaches P'), and has a continuous inverse (when a point of the x-axis approaches P', the point directly above it on the parabola approaches P). Since the coordinates of P are (x, x^2), and those of P' are $(x, 0)$, the mapping can be described algebraically as "the mapping $(x, x^2) \to (x, 0)$."

It now follows that any line g is homeomorphic to any parabola p. For, (1) g is affinely equivalent to the x-axis and hence topologically equivalent to it; (2) this axis is topologically equivalent to the parabola $y = x^2$, as shown above; and (3) this parabola is affinely equivalent to, and hence topologically equivalent to, p. The resultant of the three transformations expressing these equivalences is therefore a topological transformation of g into p.

What we have done for parabolas can be done in the same way for branches of hyperbolas and for other familiar curves of infinite extent which are endless, open, and connected. However, it is not necessary for a curve to be of infinite extent in order to be homeomorphic to a line. We now give some examples to prove this. A semicircle without its endpoints will be called a *deleted semicircle*, and a segment without its endpoints a *deleted segment*.

Let us map the x-axis onto the deleted semicircle BC shown in Fig. VII, 6. This semicircle is the lower half of the circle with center at $A(0, 1)$ and radius 1. If P is a point on the x-axis, its image will be the point P' in which line AP meets

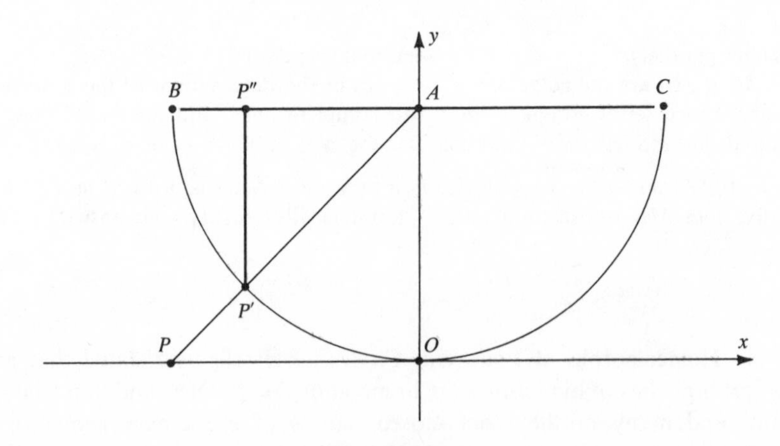

Fig. VII, 6

the semicircle. Clearly, each point of the x-axis has a unique image, and each point of the deleted semicircle is the image of a unique point of the x-axis. This mapping, which we call T, is therefore 1-1. It is, of course, simply the central projection of the x-axis onto the semicircle. If a point on the x-axis approaches P, the image of the point approaches P', and so T is continuous. T^{-1}, which maps the deleted semicircle onto the x-axis, is also continuous, for if a point of the semicircle approaches P', the image of that point under T^{-1} approaches P. Thus, T is topological. The deleted semicircle is therefore topologically equivalent to the x-axis, and hence to any line. Like a line, the deleted semicircle is endless, open, and connected. Instead of the semicircle, many other finite curves between B and C could have been used, for example, an arc of $y = x^2$, an arc of $y^3 = x^2$, or even the broken line BOC shown in Fig. VII, 7. Since these curves (without their endpoints) are homeomorphic to a line, they are necessarily homeomorphic to each other. It may be helpful to add that a semicircle or any of these curves cannot be mapped topologically on a line if its endpoints are included.

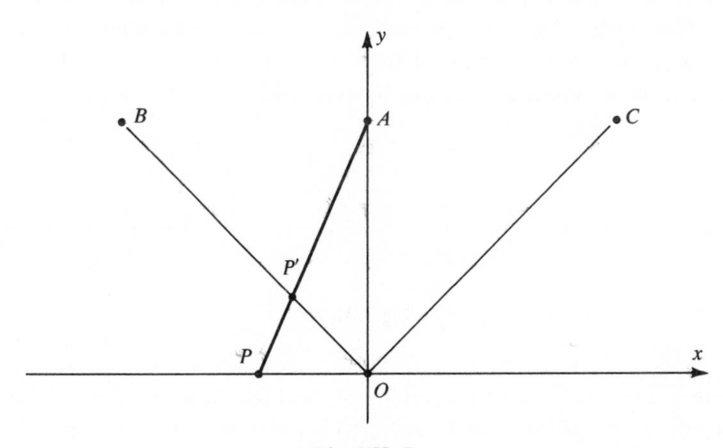

Fig. VII, 7

Let us now map the above deleted semicircle onto the deleted segment $\overline{BC}$ (Fig. VII, 6). If P' is a point of the semicircle, its image on the segment will be the point P'' with the same abscissa, that is, we are mapping by orthogonal projection. This mapping, which we denote by U, is clearly topological. Hence TU, where T is the mapping of the preceding paragraph, is topological and sends the x-axis into the deleted segment $\overline{BC}$. This deleted segment is therefore topologically equivalent to the x-axis. It follows that *any* deleted segment and *any* line are topologically equivalent. Combining this with what we saw in §4, we can then say that parabolas and branches of hyperbolas, too, are topologically equivalent to deleted segments. From a topological viewpoint, then, there is no difference between lines, parabolas, branches of hyperbolas, and deleted segments.

In §4 we suggested how physical models of lines, parabolas, and hyperbolic branches could be deformed into one another. This idea can be applied, more generally, to any simple arcs provided that we permit the models to be stretched or contracted, as well as bent. Then, for example, we could imagine deforming a model of a deleted segment so as to give a model of a parabola or a hyperbolic branch.

We have said nothing about the homeomorphs of ordinary segments. Such segments are homeomorphic, for example, to arcs of conics, with their endpoints included.

Ordinarily one thinks of a line as characterized by its straightness and infinite extent. We see now that neither property is topological. The first is related to the fact that lines are shortest paths, and hence is metric; the metric character of the second property is obvious. Viewed in terms of these properties, a line looks radically different from a parabola, a deleted semicircle, and a deleted segment; viewed topologically, it looks just like them—they are all simple arcs. For this reason, diagrams exhibiting topological properties often show lines as deleted segments. An ordinary segment and a segment from which only one endpoint is excluded will then not represent lines. Fig. VII, 8 shows how these three different figures can be exhibited: (a) shows a deleted segment,

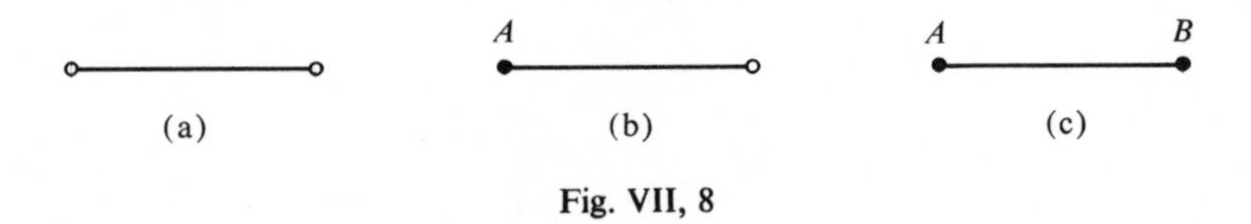

Fig. VII, 8

(b) shows a segment without one endpoint, and (c) shows an ordinary segment. The "empty dots" in (a) and (b) serve to suggest that the corresponding endpoints are omitted.

All the mappings considered in this section were achieved by projections. Projections can also be used in proving that many familiar closed figures are homeomorphs of circles. In I, §6, for example, we showed how a circle can be mapped in a 1-1 way on an inscribed square. It is now clear that a central projection was used and that the mapping is topological. This square is therefore homeomorphic to the circle. More generally, any square and any circle are homeomorphic. Similarly, by mapping a circle on a triangle inscribed within it, one can show that triangles are homeomorphic to circles. A different approach is generally called for in the case of polygons of more than three sides. We illustrate it by showing that there is a topological mapping T which sends a given circle into a given quadrilateral with consecutive vertices A', B', C', D'. Choose four points A, B, C, D of the circle which divide it into four equal arcs AB, BC, CD, DA. Let M_1 be the mapping which *merely* sends A, B, C, D into

A', B', C', D', respectively.* Let M_2 be a mapping which sends arc AB (endpoints included) topologically into $\overline{A'B'}$ so that A and B go into A' and B', respectively. Let M_3, M_4, M_5 be mappings which send arcs BC, CD, DA (endpoints included) topologically into $\overline{B'C'}$, $\overline{C'D'}$, $\overline{D'A'}$, respectively. Now let T be the mapping of the circle on the quadrilateral which is the composite of $M_1, M_2, \ldots, M_5$. That is, $T = M_1$ in so far as A, B, C, D are concerned, $T = M_2$ in so far as arc AB is concerned, etc. Then T is a mapping of the circle on the quadrilateral which is 1-1, continuous, and has a continuous inverse. Thus, the quadrilateral is homeomorphic to the circle. It has been assumed in the proof that $A'B'C'D'$ is a quadrilateral in which opposite sides do not meet.

In §4 we suggested how physical models of simple closed curves could be deformed so as to duplicate one another. We had only plane curves in mind, for we had been considering homeomorphisms of the plane. A simple closed curve, however, may also be a space curve, which is a curve not situated in a plane. If space curves are included among the models, then it is not always possible to deform one of these models, that is, bend, stretch, or contract it, so as to obtain another. The knotted curve in Fig. VII, 9, for example, cannot be obtained

Fig. VII, 9

by deforming a circle in this way. We could imagine it obtained, however, by cutting the circle in one place, deforming the resulting figure into the shape of the knotted one, and then bringing the two ends together.

It was seen in Fig. VII, 8 that the diagram of a segment can symbolize three topologically different types of figures depending on the number of endpoints that are included. By use of a certain device, which we shall now describe, the diagram of a segment can also be made to symbolize a simple closed curve. Let us take an ordinary segment $\overline{AB}$ and imagine bending it until A and B coincide, thus producing a simple closed curve. As this suggests, we may think of a simple closed curve as a segment whose endpoints are the same point, and hence represent the curve as in Fig. VII, 10(a). This device, in which physically distinct points in a diagram are regarded as the same or identical points mathematically, is called **identification of points.** The figure symbolized in Fig. VII,

---

* A mapping of a finite set on a finite set is constructed, or determined, simply by specifying what is to be the image of each element in the first set.

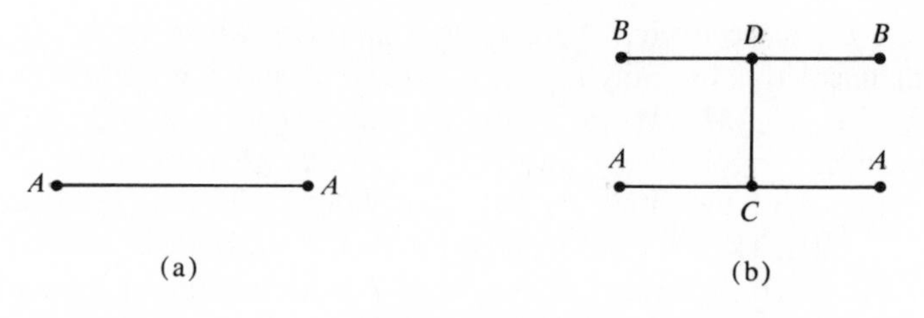

Fig. VII, 10

10(b), for example, is a pair of simple closed curves joined by a segment. This device of identifying points can be very useful in making diagrams of surfaces, as we shall see.

EXERCISES

1. Prove that the following curves are homeomorphs of a line:

(a) a branch of a hyperbola; (b) $y = \sin x$;

(c) $y = \log x$; (d) a punctured circle.

2. Prove that the following are simple arcs:

(a) the part of the curve $y = \sin x$ between $(0, 0)$ and $(2\pi, 0)$, excluding these points;

(b) the positive side of the x-axis;

(c) a quadrant of a circle, endpoints excluded;

(d) a circle from which one of the quadrants and the endpoints of this quadrant are excluded.

3. Show that the equations of our mapping of the x-axis on the deleted semicircle BC are

$$x' = \frac{x}{\sqrt{1 + x^2}}, \qquad y' = 1 - \frac{1}{\sqrt{1 + x^2}},$$

where x is the abscissa of P, and x', y' are the coordinates of P'.

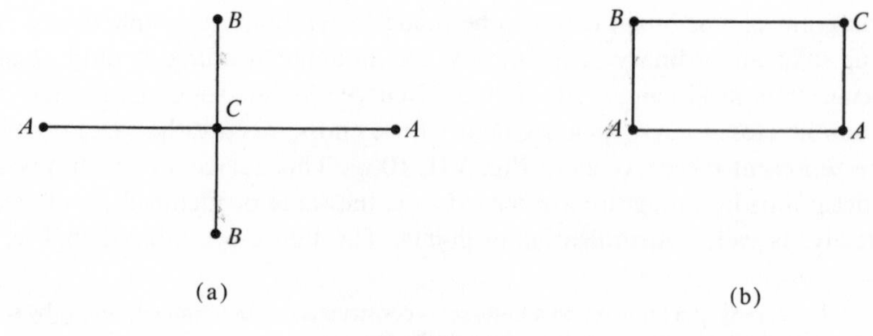

Fig. VII, 11

4. Show how to map the x-axis topologically on the broken line BOC in Fig. VII, 7 and find the equations of the mapping.

5. Where would our proof that a quadrilateral is the homeomorph of a circle break down if two opposite sides of the quadrilateral meet?

6. Prove that a triangle and a quadrilateral (in which opposite sides do not meet) are homeomorphic.

7. Prove that the following are simple closed curves: (a) an ellipse; (b) the boundary of a sector of a circle; (c) a hypocycloid of four cusps.

8. Give topological descriptions of the geometrical objects symbolized in Fig. VII, 11.

7. Topological Transformations and Order. Since topological transformations do not in general send lines into lines, betweenness is not a topological property. However, such transformations do preserve the type of order occurring on a line or on any simple arc, namely, *linear order*. If, for example, A, B, C are points of a line g, with B between A and C, these points are said to be in the **linear order** ABC and also in the **linear order** CBA. This can be interpreted intuitively to mean that we encounter A first, then B, then C when traversing g in one direction, and C first, then B, then A when traversing it in the other direction. In neither traversal are the points in any of the orders ACB, BCA, BAC, or CAB. If g is mapped topologically on g', with A, B, C going into A', B', C', then B' may not be between A' and C' since g' may not be a line; but it is always true that in traversing g' one way we shall encounter A' first, then B', then C', and in traversing it the other way we shall first encounter C', then B', then A'.

To generalize the preceding discussion, let g be any simple arc. If A, B, C are points of g, and we traverse g in one direction, we shall encounter one of these points first, say A, and one of the remaining ones second, say B. The points are thus in the order ABC when g is traversed this way, and hence they are in the order CBA when g is traversed the other way. In neither case are the points in any of the orders ACB, BCA, BAC, or CAB. Thus the type of order on g is the same as on a line. Now let g be mapped topologically on g', with A, B, C going into A', B', C'. The latter points will then be in the order $A'B'C'$ when g' is traversed one way, and in the order $C'B'A'$ when it is traversed the other way, but in neither case are they in any of the orders $A'C'B'$, $B'C'A'$, $B'A'C'$, or $C'A'B'$. In other words, *topological transformations preserve linear order.*

Like a line, a circle can be traversed in either of two directions. If we encounter A first when going in one of these directions, then B, then C, we can say that these three points are in *the order ABC on the circle* (Fig. VII, 12). Clearly, though, by changing our starting place, but not our direction, we can meet B first, then C, then A, with the result that the points are in the order BCA. Simi-

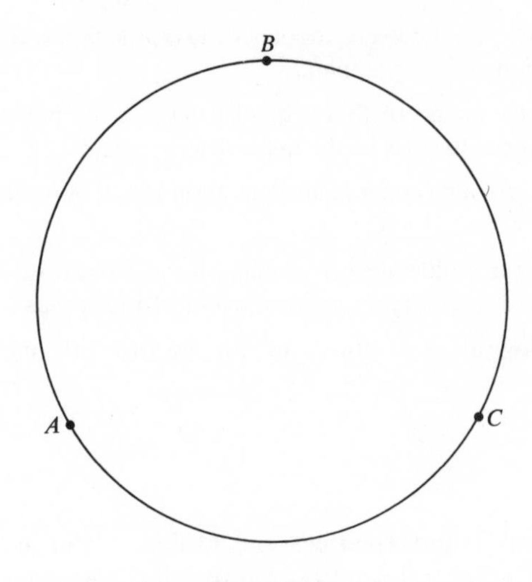

Fig. VII, 12

larly, one can also see that they are in the order *CAB*. On traversing the circle in the opposite direction, we find, in like manner, that the points are in all of the orders *CBA*, *ACB*, *BAC*. Thus, if three points on a circle are in the order *ABC*, they are also in five other orders. But on a line, as we have seen, if three points are in the order *ABC*, then they are also in just one other order, *CBA*. The type of order encountered on a circle is therefore different from linear order. It is called **cyclic order**. The order on any simple closed curve is also cyclic.

Neither type of order is restricted to three points. Four points on a line, for example, may be in the order *ABCD* (Fig. VII, 13a), in which case they are also in the order *DCBA*, but in no other order. Traversing the line one way, we meet *A* first, then *B*, then *C*, then *D*; traversing it the other way, we meet *D* first, then *C*, then *B*, then *A*. To say that four points on a circle are in the order *ABCD* (Fig. VII, 13b) can be interpreted in this same way. Then, however, the

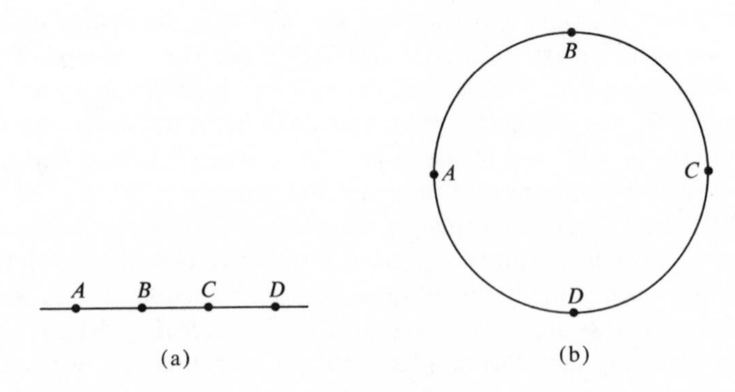

Fig. VII, 13

points are also in the seven orders *BCDA*, *CDAB*, *DABC*, *DCBA*, *CBAD*, *BADC*, *ADCB*, but not in any of the other sixteen orders in which these four letters can be written.

Topological transformations preserve cyclic order. If, for example, the points A, B, C, D are in the order *CADB* on a circle γ, and γ is mapped topologically on γ', then the image points A', B', C', D' are in the order $C'A'D'B'$ on γ'.

There is another useful way of viewing both types of order when there are four points. In each type, when four points are in the order *ABCD* (Fig. VII, 13) it will be noticed that the pairs A, C and B, D *separate* each other. This term was explained in V, §9 for the case in which the points are collinear. Its meaning for the case in which the points are on a circle is similar.

EXERCISES

1. We have just seen that if four points are in the order *ABCD*, linear or cyclic, the pairs A, C and B, D separate each other. Verify by a diagram that the converse statement holds for cyclic order but not for linear order.

2. Verify by a diagram that if A, B, C, D are points on a simple closed curve, exactly one of the following holds: either A, B separate C, D, or A, C separate B, D, or A, D separate B, C (that is, the points are in just one of the cyclic orders *ABCD*, *ABDC*, *ACBD*).

3. If three points are in the order *ABC*, verify that it would be reasonable to say that B separates A and C in the case of linear order but not in the case of cyclic order.

8. Homeomorphs of the Plane. Having made good use of the broadened definition of a topological transformation in proving facts about curves, let us now apply the definition to the plane as a whole and thereby learn something of its topological structure. Since we have already mapped the plane onto itself, let us now map it onto something else. When we say that a point is in a certain quadrant we shall mean that it is an interior point of that quadrant and not on an axis.

A simple transformation which maps the plane on something other than itself is

$$x' = e^x, \qquad y' = e^y.$$

Each point of the plane has an image which is in the first quadrant; conversely, each point in this quadrant is the image of a unique point. The transformation is continuous by 2.2 since e^x, e^y are continuous functions for all values of x and y. The inverse transformation, $x = \log x'$, $y = \log y'$, where x', y' are positive, is also continuous, for $\log x'$ and $\log y'$ are continuous functions in the first quadrant. Thus we have a topological mapping of the plane onto the first quadrant. This region is therefore homeomorphic to the plane and possesses the same topological properties. We do not immediately say what these properties are, but give examples to show that the flatness and infinitude of the plane are not among them.

In the above example we mapped the plane on a part of itself. We now map it on a hemispherical surface. The method is essentially the same as that used in §6 to map the x-axis on a deleted semicircle, and, just as we excluded the endpoints of the semicircle, we now exclude the boundary c of the hemisphere. Imagine the plane horizontal. The hemisphere to be used is the lower half of the sphere with radius 1 whose center M is 1 unit above the origin O (Fig. VII, 14).

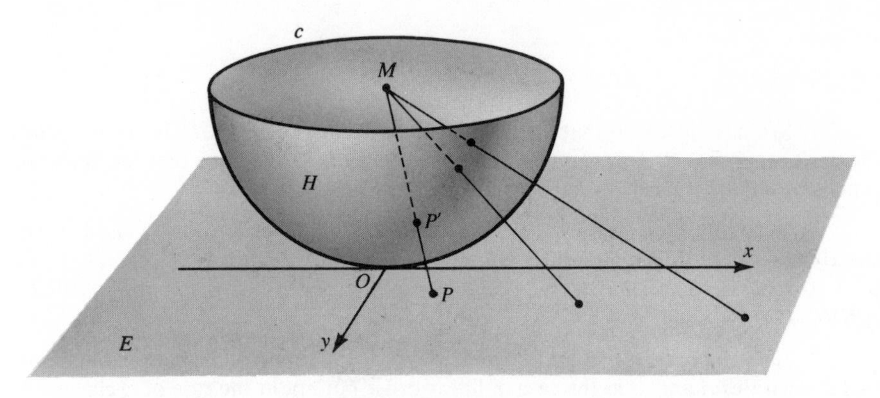

Fig. VII, 14

If P is any point of the plane, let its image be the point P' in which the line MP meets the hemisphere. This clearly gives a topological mapping of the plane onto the hemisphere (excluding c). It will be noted that points very far from O have images very close to c.

Instead of the hemisphere we could have used a portion of a paraboloid or cone with vertex at O and bounded by c, or many another familiar surface. Hence each of these surfaces (excluding its boundary c), the first quadrant region of the earlier example, and the plane all have the same topological properties. These properties are the two-dimensional analogues of those of lines and can be described by the same terms used previously. Each of these surfaces is **open** in the sense of enclosing no volume. Each consists of a single piece and hence is **connected** (the surface consisting of two concentric spheres, for example, is disconnected). Each is **endless** in the sense of containing no boundary. In the case of the plane there is no boundary at all, whereas for each of the other surfaces there is a boundary which we have agreed to exclude.

We are in a position to exhibit one more homeomorph of the plane, of particular importance to us. Having mapped the plane on the hemisphere by a central projection, which we denote by T, we now go one step further and map the hemisphere back on the plane by an orthogonal projection U. Thus, if $T(P) = P'$, then $U(P') = P''$, where P'' is the foot of the perpendicular from P' to the plane (Fig. VII, 15). U is topological and maps the hemisphere onto the interior of the circle γ, in the plane, with radius 1 and center at O. The resultant TU therefore maps the plane topologically onto the interior of γ, an arbitrary point P of the plane going into a point P'' on ray OP as shown in Fig. VII, 15. Only O remains fixed in this mapping; otherwise P'' is closer to O than is P. In effect, then, we have a compression of the plane into the interior of γ, points "moving" straight toward O, and circles with center O going into smaller such circles. The disk-shaped part of the plane interior to γ thus has the same topological structure as the whole plane. Like the latter, it is a surface which is open, endless, and connected. These properties are also possessed by the plane region interior to any simple closed curve, although we shall not show this.

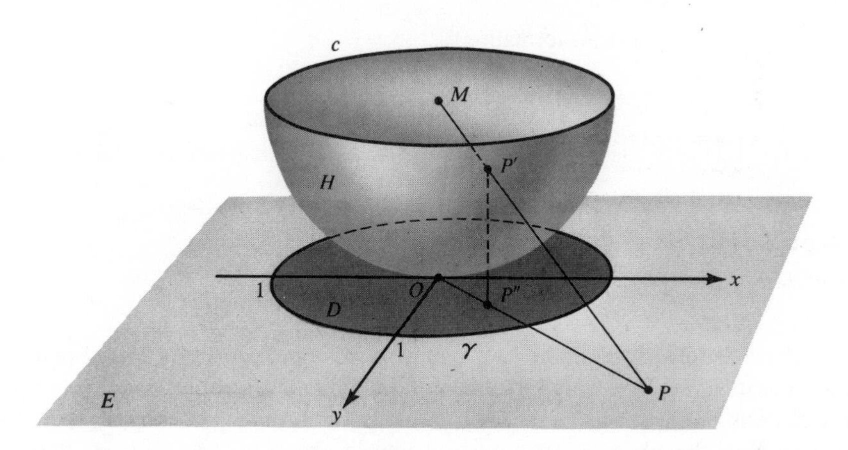

Fig. VII, 15

Most of the homeomorphs of the plane which we have mentioned are of finite extent. Some familiar surfaces of infinite extent, too, are homeomorphic to the plane, for example, a paraboloid of revolution (obtained by rotating a parabola about its axis), one sheet of a two-sheeted hyperboloid of revolution (obtained by rotating a hyperbola about its transverse axis), and one nappe of a cone (including the vertex). More generally, any surface of infinite extent which we can imagine to be obtained from a plane by bending the latter, as if it were made of a plastic material, is homeomorphic to the plane. Similarly, any surface of finite extent which we can imagine to be obtained by bending, stretching, or contracting a circular disk (without its boundary) is homeomorphic to the disk, and hence to the plane. The hemisphere, for example, can be imagined as obtainable this way.

EXERCISES

1. Find the region into which the plane is sent by each of the following transformations and determine whether the transformation is topological:

(a) $x' = x^2, y' = y^3$; (b) $x' = x^2, y' = y^2$;

(c) $x' = e^{-x}, y' = y^3$; (d) $x' = e^y, y' = 2/(e^x + 1)$.

2. Find the equations of a transformation that maps the plane topologically onto the interior of the square bounded by the axes and the lines $x = 1, y = 1$.

3. Prove that the interior of each of the following curves is homeomorphic to the interior of the circle γ, and hence to the plane: (a) any circle; (b) any ellipse.

4. Show that a *punctured sphere* (a sphere from which one point has been removed) is topologically equivalent to a plane.

5. Let g' be the image of a line g in our mapping of the plane onto the hemisphere, and g'' be the image of g' in our mapping of the hemisphere onto the interior of γ. What kind of curve is g'; g''?

6. Show that the transformation

$$x' = \frac{x}{\sqrt{x^2 + y^2 + 1}}, \qquad y' = \frac{y}{\sqrt{x^2 + y^2 + 1}}$$

maps the plane onto the interior of γ in a 1-1 way. Is the mapping topological? Justify your answer.

7. Prove that a paraboloid of revolution is homeomorphic to a plane.

9. Models of the Plane. We are accustomed to think of the plane as a perfectly flat surface which extends infinitely far in all directions. From what was done in the previous section it is clear that this view is not topological. The infinitude of the plane is, of course, a metric property, and so is its flatness, which is closely tied up with distance. As we well know, the plane cannot be seen in its entirety when visualized in such metric terms. Viewed topologically, however, the plane can be seen completely. For, since there is no topological difference between the plane and its homeomorphs, any of the latter can serve as a topological representative or model of the plane, and in particular we can choose one of limited extent, for example, the hemisphere or the circular disk. In looking at such a *bounded model* (or *finite model*) topologically, we are seeing the entire plane topologically. Since it is very useful on occasion to have such a complete visual grasp of the plane, it is worthwhile to study these models more closely. The hemisphere and circular disk, without their boundaries, will be denoted by H and D, respectively, and the plane by E. When H and D are regarded as the images of E in the mappings used in §8, we call them the **hemispheric model of the plane** and the **circular model of the plane**, respectively.

Let us return to the mapping of E on H (Fig. VII, 14). Lines, we know, go into curves which are open, connected, and endless. What is the nature of these curves? If g is any line in E, all the lines joining its points to M lie in a plane. Hence the image of g is the curve in which this plane meets H. Since this plane goes through M, the center of the original sphere, it meets the sphere in a circumference, and H in half of this circumference. Thus the lines of E go into the curves of H which are semicircumferences of the sphere without their endpoints. We shall refer to these curves as *semicircumferences*. The lines through O, for example, go into the semicircumferences through O. Fig. VII, 16 shows x', the image of the x-axis, and y', the image of the y-axis.

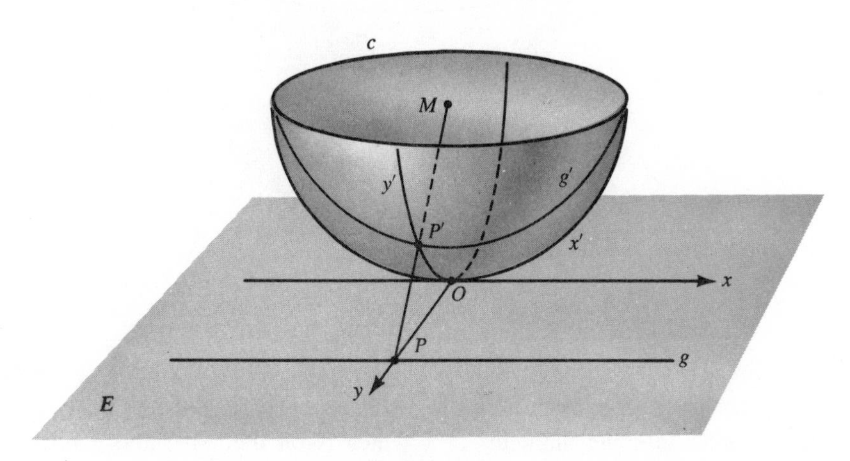

Fig. VII, 16

Individually and in their mutual relations the semicircumferences have the same topological properties as the lines of E. In particular they have the same incidence relations. The following comparison makes this clear. A unique line goes through any two points of E; a unique semicircumference goes through any two points of H. Two lines meet in one point or not at all; two semicircumferences meet in one point or not at all. Through a point of E not on a given line there passes a unique line that does not meet the given one; through a point of H not on a given semicircumference there passes a unique semicircumference that does not meet the given one. Fig. VII, 16 shows two parallel lines g, x and their images g', x'. Since g, x do not meet, neither do g', x'. The nonmetric character of the mapping is well illustrated in the fact that g', x' get very close to each other near c, whereas g, x are everywhere equidistant.

Let us return now to the mapping of E on D and determine the curves of D which correspond to the lines of E. These curves are simply the figures into which the orthogonal projection U of §8 sends the above-mentioned semicircumferences. A semicircumference passing through O projects into a diameter of γ, and one not passing through O projects into an arc of an ellipse, endpoints excluded in both cases. It is clear from the nature of U that the chord subtending this arc is a diameter of γ, and hence is 2 units long, and that no chord of this ellipse exceeds 2. The chord subtending this arc is therefore the major axis of the ellipse and the arc is a semiellipse, one of the two symmetrical parts into which this axis divides the ellipse. Fig. VII, 17 shows the diameter x'' and semiellipse

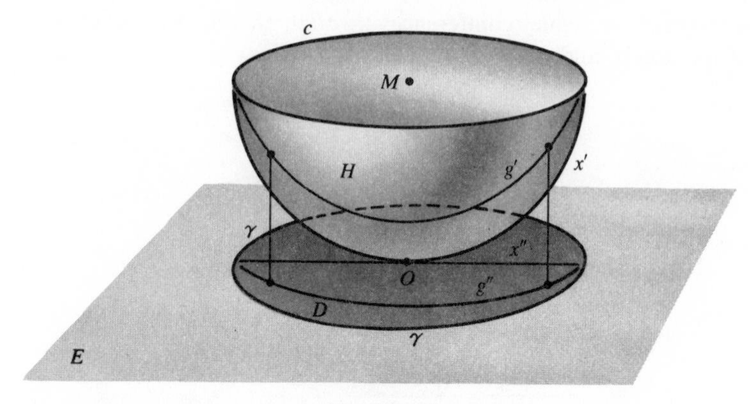

Fig. VII, 17

g'' into which the semicircumferences x', g' of Fig. VII, 16 project. x'' is the major axis of the ellipse containing g''. In the mapping of E on D, then, a line through O goes into the diameter of γ which is on that line, and a line not through O goes into a semiellipse whose related major axis is a diameter of γ. Since a diameter of γ is the limiting position of a semiellipse which becomes narrower as its eccentricity approaches 1, we shall also refer to diameters as semiellipses. We can then simply say that in the mapping of E on D lines go into semiellipses (without endpoints) whose related major axes are diameters of γ.

The set of all these semiellipses, then, is in 1-1 correspondence with the set of all the lines of the plane. Individually and collectively these semiellipses have the same topological properties as the lines, and in particular they have the same incidence relations. Thus, through any two points of D there passes a unique semiellipse, two semiellipses meet in one point of D or not at all, and through a point of D not on a given semiellipse there passes a unique semiellipse not meeting the given one.

Fig. VII, 18 shows three semiellipses a'', b'', c'' which do not meet. They come arbitrarily close to the same two diametrically opposite points of γ. Since they do not meet, the lines a, b, c corresponding to them are parallel. We can now be more precise than we were earlier about the image of a line not through O: it is a semiellipse whose related major axis is the diameter of γ which is parallel to the line, and it is between the line and the diameter.

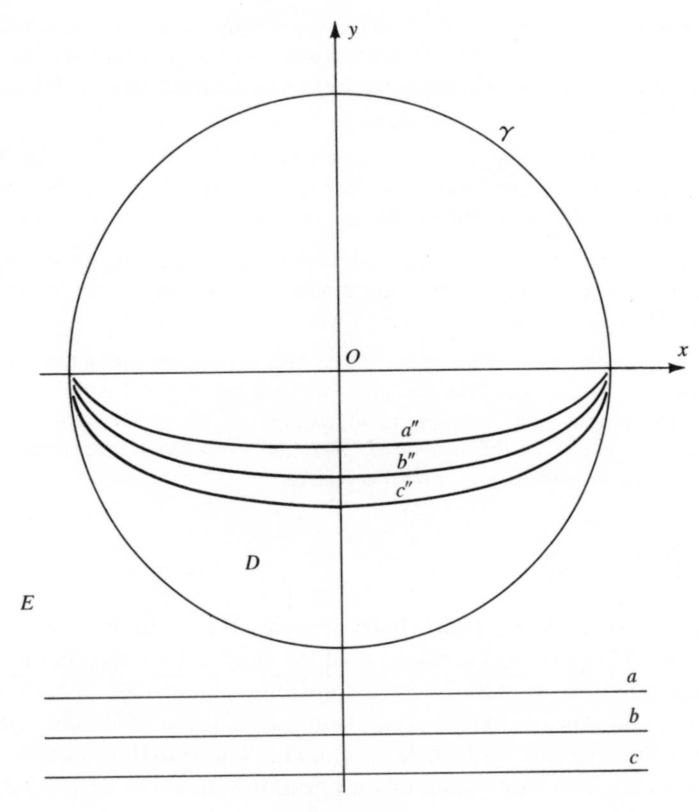

Fig. VII, 18

EXERCISES

Exercises 1–6 deal with the mapping of E on D discussed above.

1. Find the excluded endpoints of the semiellipses corresponding to the following lines:

(a) $x = 0$; (b) $y = 0$; (c) $y = x$; (d) $y = 2x$;
(e) $x = 1$; (f) $y = 2$; (g) $y = x + 3$; (h) $y = 2x + 5$.

2. Give the polar coordinates (θ, r) of the points which are approached as we come arbitrarily close to γ along the semiellipses corresponding to the following lines:

(a) $y = x$; (b) $y = x + 2$; (c) $y = 3x$;
(d) $y = 3x + 2$; (e) $ax + by = 0$; (f) $ax + by + c = 0$.

3. Prove that a circle in E with center O goes into a smaller circle in D with center O.

4. Prove that the semiellipses have the incidence relations described in the discussion.

5. Describe the images in D of the following figures in E: (a) the vertex and axis of a parabola; (b) the asymptotes of a hyperbola; (c) two parallel lines one of which goes through O; (d) two parallel lines, neither of which goes through O; (e) two parallel lines symmetrical to O.

6. Describe the topological character of the image in D of each of the following figures in E, and state whether the image comes arbitrarily close to γ: (a) a very large circle with center far from O; (b) an ellipse; (c) a parabola; (d) a hyperbola.

7. In §8 we obtained an unbounded model of E by mapping it topologically on the first quadrant. Find the curves in this model corresponding to the following lines: (a) $x = k$; (b) $x + y = 0$; (c) $x + y = 1$; (d) $x - y = 0$; (e) $x - y = 1$.

8. A *spherical model of the plane* can be obtained by projecting the plane onto a tangent sphere from the point B of the sphere which is opposite the point of tangency O. (a) If this mapping is to be topological which point of the sphere must be excluded? (b) Describe the curves on this *punctured sphere* that correspond, respectively, to lines, to lines through O, and to circles with center O.

10. More on the Circular Model of the Plane. In discussing the mapping of E on D in §9 we said nothing about the images of conics other than circles with center O. Now we wish to learn something about these images when the conics are parabolas and hyperbolas. These curves, being of infinite extent, have images which come arbitrarily close to γ, and it will be useful to know precisely how this occurs. In investigating this we need the equations of the mapping of E on D, which are

$$x' = \frac{x}{\sqrt{1 + x^2 + y^2}}, \qquad y' = \frac{y}{\sqrt{1 + x^2 + y^2}}, \tag{1}$$

and, in implicit form,

$$x = \frac{x'}{\sqrt{1 - x'^2 - y'^2}}, \qquad y = \frac{y'}{\sqrt{1 - x'^2 - y'^2}}, \tag{2}$$

where we now use single primes instead of the double primes used previously. The derivation of (1), and the obtaining of (2) from (1), are left as exercises. The student can also show that the polar equations of the mapping are

$$\theta' = \theta, \qquad r' = \frac{r}{\sqrt{1 + r^2}}.$$

These are useful in some of the exercises.

Let us find the image k of the parabola $y = x^2$. One sees, by considering the two projections which make up the mapping, that k is a curve tangent to the horizontal diameter of γ at O, symmetrical to the vertical diameter, and opening upward. Fig. VII, 19(a) gives a three-dimensional view of k and Fig. VII, 19(b) gives a two-dimensional view. The equation of k is

$$y' = \frac{x'^2}{\sqrt{1 - x'^2 - y'^2}}.$$

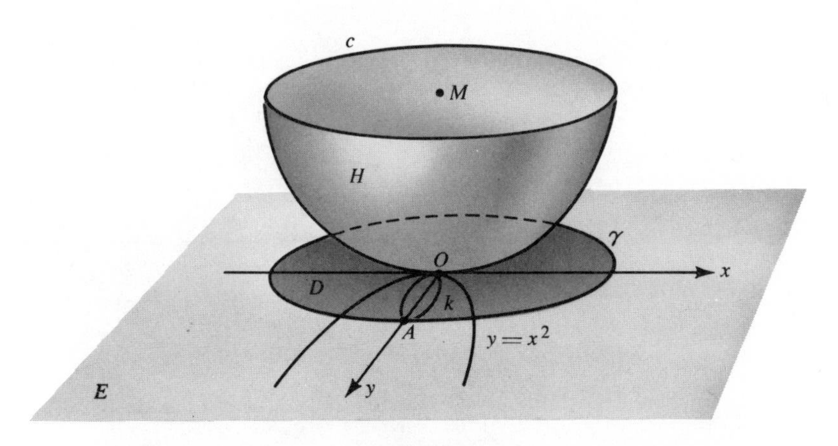

Fig. VII, 19(a)

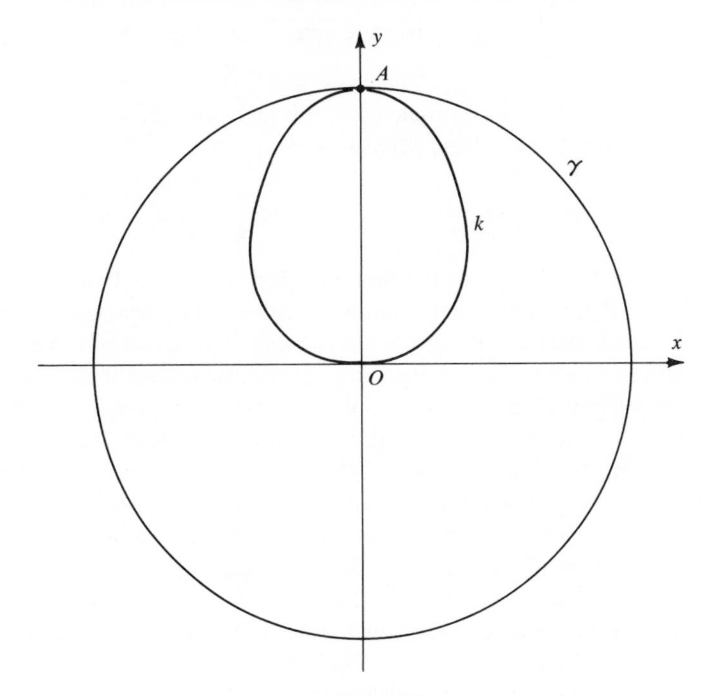

Fig. VII, 19(b)

Being the topological image of a parabola, k is endless, connected, and open. Also, it lacks but a single point of being closed, namely, the point A on γ with coordinates $x' = 0$, $y' = 1$. Hence, if we proceed along k toward γ, either in the first or second quadrant, we approach A. The latter is an endpoint of the vertical diameter of γ, which is the semiellipse corresponding to the axis of the parabola.

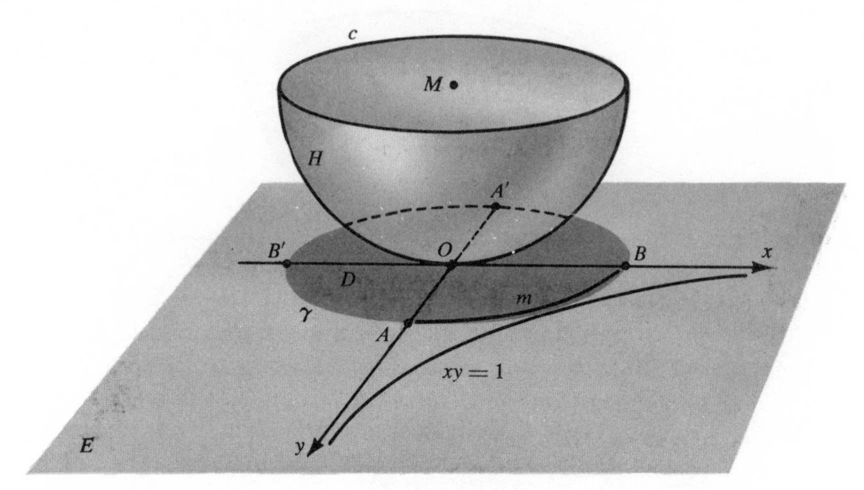

Fig. VII, 20(a)

To prove that (0, 1) is approached in the specified way we substitute x^2 for y in (1) and rewrite the resulting equations as

$$(4) \qquad x' = \pm \frac{1}{\sqrt{1 + x^2 + x^{-2}}}, \qquad y' = \frac{1}{\sqrt{1 + x^{-2} + x^{-4}}},$$

where we choose the + sign in the first equation when $x > 0$, and the − sign when $x < 0$. If P is a point on the parabola with nonzero abscissa x, equations (4) give the coordinates (x', y') of the image point P' in terms of x. Let P recede from O without bound. Then x becomes infinite, positively if P is in the first quadrant, negatively if in the second. When x becomes positively infinite, (4) shows that $y' \to 1$ and $x' \to 0$ through positive values; hence P approaches the point (0, 1) from the first quadrant. When x becomes negatively infinite, (4) shows that $y' \to 1$ and that $x' \to 0$ through negative values; hence P' approaches (0, 1) from the second quadrant. In view of this result and the fact that (0, 1) is not on the locus of (3), we may say that the locus is punctured at this point. It is not punctured anywhere else, for then it would be disconnected. Thus our earlier statement, that it lacks but a single point of being closed, is substantiated.

This result can be generalized. *The image of every parabola is a curve lacking only one point of being closed. This point is on γ and is an endpoint of the* *semiellipse corresponding to the axis of the parabola.* The proof of this in case the axis is vertical or horizontal can be handled by the method just used for $y = x^2$ and is covered in exercises. When the axis is vertical the missing point is always $(0, 1)$ or $(0, -1)$ according as the parabola is concave upward or concave downward; when the axis is horizontal the point is $(1, 0)$ for a parabola concave to the right, and $(-1, 0)$ for one concave to the left. We omit the proof for the case in which the axis is an oblique line.

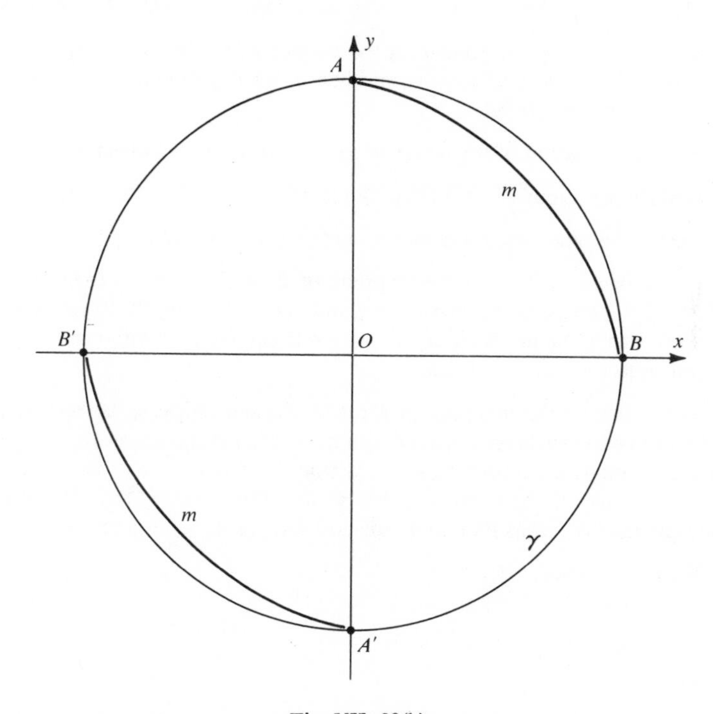

Fig. VII, 20(b)

We merely state the facts in the case of hyperbolas and leave their verification to the exercises. Consider the hyperbola $xy = 1$. Its asymptotes, the coordinate axes, are transformed into the horizontal and vertical diameters of γ. The hyperbola goes into a two-branched curve m coming arbitrarily close to the four points A, B, A', B' of γ which are the ends of these diameters. Fig. VII, 20(a) gives a three-dimensional view showing one branch of m; Fig. VII, 20(b) gives a two-dimensional view showing all of m. In the general case we have the following statement. *The image of a hyperbola is a two-branched curve that comes arbitrarily close to the four points of* γ *which are the ends of the semiellipses corresponding to its asymptotes.* Since the branches do not meet the semiellipses nor one another, they approach the four points as indicated in Fig. VII, 20(a), that is, each branch approaches two points of γ associated with different semiellipses.

EXERCISES

1. As a check on Fig. VII, 19(b), use the equation of the image of $y = x^2$ to find x' when y' is 0.2, 0.5, 0.9, 0.99, and plot the corresponding points.

2. The images of the following curves come arbitrarily close to points of γ. Find these points by the method used for $y = x^2$.

(a) $y = -x^2$; (b) $x = y^2$; (c) $x = -y^2$; (d) $y = ax^2 + bx + c$;
(e) $xy = 1$; (f) $xy = -1$; (g) $xy = k$; (h) $x^2 - y^2 = 1$.

3. If P' is the image of point P in the mapping of E on D, and we let $d = OP$, $d' = OP'$, show that $d' = d/\sqrt{1 + d^2}$, and hence that P' gets arbitrarily close to γ as P recedes from O without bound.

4. Find the equation of the image of $xy = 1$ and use it to check Fig. VII, 20.

5. Deduce equations (2) from equations (1).

6. Derive the polar equations of the mapping of E on D.

7. Prove algebraically that the mapping of E on D (a) leaves O fixed and sends every other point P into a point interior to γ and between O and P; (b) sends each circle with center O into a smaller such circle; (c) sends each ellipse with center O and foci on the x-axis into a smaller such ellipse.

8. Verify that in the mapping of E on D, the semiellipse g' corresponding to a given line g at positive distance d from O can be located in the following way: (1) draw diameter $\overline{AB}$ through O parallel to g; (2) denote by C the foot of the perpendicular from O to g; (3) take C' between O and C so that $OC' = d/\sqrt{1 + d^2}$; (4) g' is then the semiellipse having $\overline{AB}$ as its major axis and $\overline{OC}'$ as its semiminor axis.

9. Derive equations (1).

11. Surfaces Not Homeomorphic to the Plane. If, to the interior of the circle γ, we adjoin γ itself, there results a flat surface not homeomorphic to the plane, namely, a circular disk with its boundary included. This surface therefore contains the previously excluded endpoints of each semiellipse. These new semiellipses, with endpoints included, therefore do not possess all of the topological properties of lines, either individually or in their mutual relations. If, for example, two such semiellipses meet in a point P on γ, they also meet in the diametrically opposite point Q on γ.

Most of the surfaces in elementary geometry are likewise not homeomorphic to the plane. Spheres, ellipsoids, cubes, and other regular polyhedra, for example, differ from the plane in being closed surfaces, and are all alike topologically. We can imagine them deformed into one another, but they cannot be deformed into the plane or any of its homeomorphs. Cylinders and one-sheeted hyperboloids (obtained by rotating hyperbolas about their conjugate axes) represent a third topological type, and cones represent a fourth (Fig. VII,

21). A sphere, for example, cannot be deformed into a cylinder, nor the latter into a cone.

Like the plane, the surfaces mentioned in the preceding paragraph are all connected. Also, as with the plane, every simple closed curve on such a surface divides it into two separate parts. Stated differently, if any simple closed curve is excluded from the surface, what remains is disconnected. This can be illustrated physically by cutting a paper model of the surface along a simple closed curve, in which case the model will fall into two pieces.

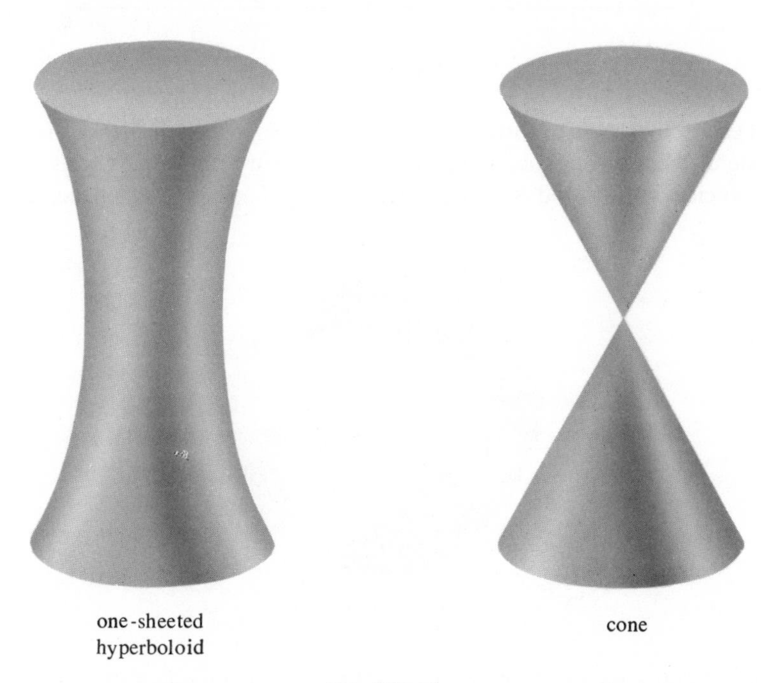

Fig. VII, 21

One might be tempted to say that any surface would be severed in two when cut completely around in this way, but this is not so. The *torus* or doughnut-shaped surface shown in Fig. VII, 22, for example, does not fall into two

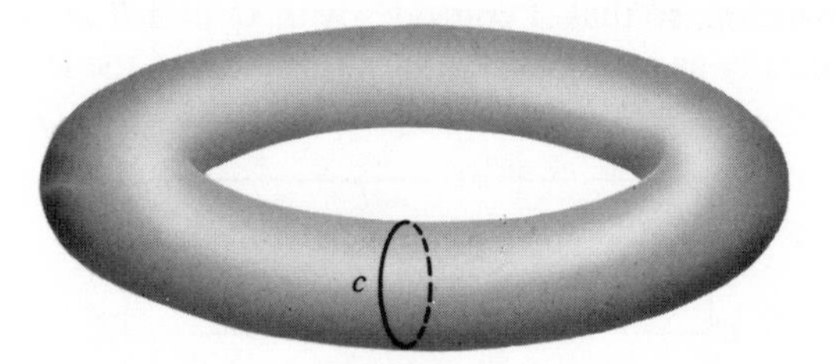

Fig. VII, 22

pieces when cut along the curve c. To obtain another example of such a surface, and one that is important for us, consider the following construction. Take a long rectangular piece of paper $ABCD$ (Fig. VII, 23) and bring the edges AB, CD

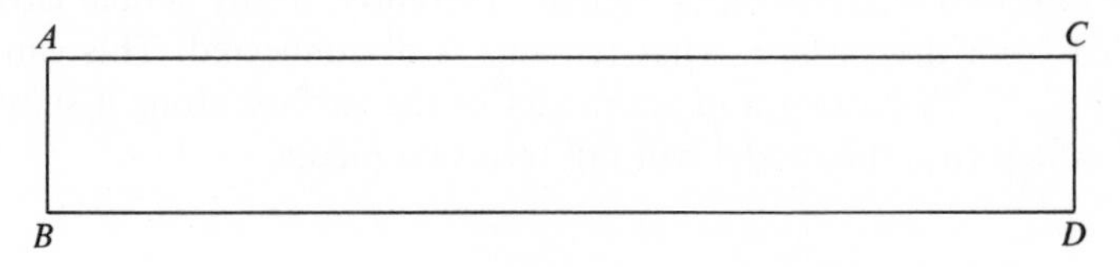

Fig. VII, 23

toward each other as if to form a cylinder. Just before the edges meet, however, twist one of them through an angle of 180°, and then bring them together, making A coincide with D, and B with C (Fig. VII, 24). The resulting surface is called

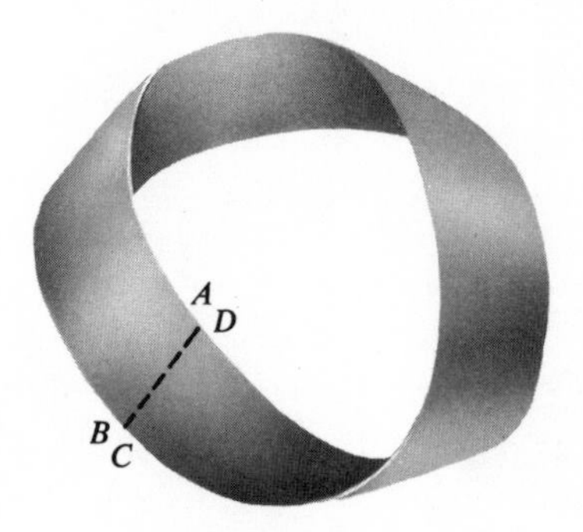

Fig. VII, 24

a *Möbius band*, after the 18th century German mathematician who first called attention to its interesting properties. If one punctures the paper at any point P and cuts completely around the band so as to return to P, he will find, very likely to his surprise, that the piece of paper is still intact.

It is helpful in explaining this if we first show how to represent a cylinder and a Möbius band symbolically. Since a cylinder can be formed from the original strip of paper $ABCD$ (Fig. VII, 23) by bringing the edges AB, CD together, without twisting, so that A coincides with C, and B with D, we represent a cylinder symbolically as in Fig. VII, 25. In this symbol the left and right edges

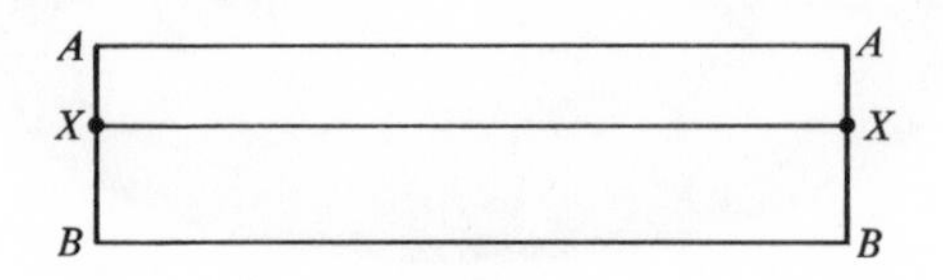

Fig. VII, 25

are *identified*, that is, regarded as identical. More precisely, an arbitrary point X at the left is identified with the point at the right which is at the same height above B. Hence the horizontal line joining X at the left to X at the right symbolizes a simple closed curve, namely, a certain circle encompassing the cylinder.

If we now recall that in constructing the Möbius band we twisted one of the edges AB, CD of the original strip before bringing them together, it is clear that the band can be symbolized as in Fig. VII, 26. The left and right edges are

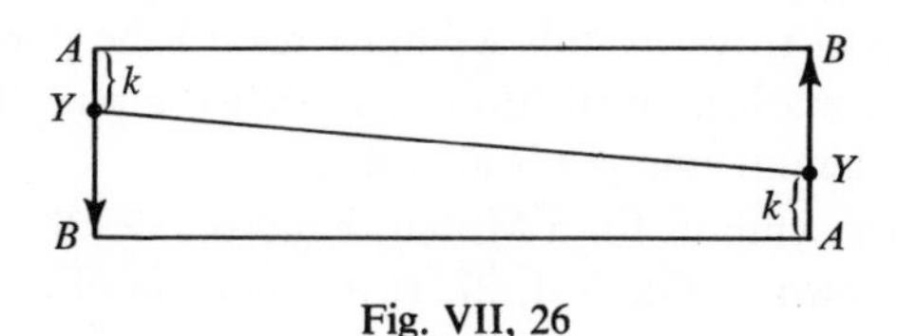

Fig. VII, 26

again identified, but this time in a reversed way, points near the top at the left being identified with points near the bottom at the right. More precisely, a point on the left k units below A is identified with the point on the right k units above A. (The top and bottom edges of the symbol are not identified.) As a result, the line in Fig. VII, 26 joining Y on the left to Y on the right symbolizes a simple closed curve encompassing the band. To say that cutting the paper model along such a curve does not sever the model in two means, geometrically, that this curve does not divide the band into two separate parts. Fig. VII, 26 helps us to verify this. Although the line YY seems to divide the symbol into two separate parts, this division actually does not occur; for, in view of the identification, the left edge above Y is regarded as joined to the right edge below Y, and the left edge below Y to the right edge above Y. On the other hand, in the symbol of the cylinder (Fig. VII, 25), where line XX also seems to divide the symbol into two separate parts, the division really occurs, for the part of the symbol above line XX and the part below it have no common point. Cutting a paper model of the cylinder along the circle symbolized by this line therefore severs the model in two.

A Möbius band differs from all the other surfaces we have mentioned in that it is a *one-sided surface*, whereas they are *two-sided surfaces*. These terms are easily understood intuitively if physical models are used. Thus, a hollow spherical globe has an outer side and an inner side, and a horizontal plane has an upper side and a lower side. The two sides in each case could be painted in different colors and then referred to, for example, as the red side and the blue side. This procedure would not work in the case of a Möbius band, for if we started painting in blue and always remained on what we believed to be the same side, we would soon find that the whole surface was painted blue. This is only as it should be, for the rectangle from which the band can be imagined to be formed had two sides and they became merged as a result of the 180° twist.

There are various mathematical ways of distinguishing between one-sided and two-sided surfaces, and we shall describe one, using the Möbius band and the plane as representative surfaces. With any point A as center, take a small circle c in the plane and choose a direction on it. Let A describe a simple closed curve in the plane, and, as it does so, let the directed circle c (with constant radius) vary so that A is always its center. When c returns to the starting position, its direction is the same as the original no matter what closed curve is used. On a Möbius band, however, when we follow the same procedure we find that there are simple closed curves (YY in Fig. VII, 26, for example) such that when c returns to the starting position its direction is the opposite of the original. In using the paper model to verify this we must regard the paper as if it had no thickness, just like a mathematical surface, so that if a pin pierces the paper at A, the place where the pin emerges is also A.

There are other symbols for a Möbius band besides the one we have been using. Another is shown in Fig. VII, 27. It is a portion of a circular disk, with

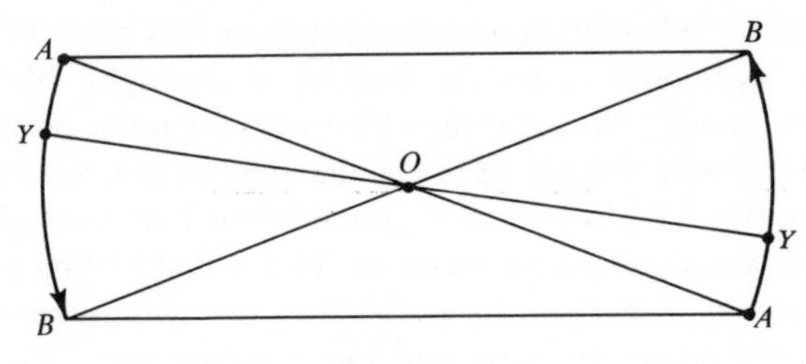

Fig. VII, 27

center O, in which any two diametrically opposite points on the sides (but not on the top and bottom except for A and B) are regarded as the same point. We can imagine this symbol as having been obtained by deforming the one shown in Fig. VII, 26. In both cases any two identified points are symmetrical to the center of the symbol.

One reason for our discussing a Möbius band and contrasting it with the familiar surfaces of elementary geometry is that *the one-sidedness or two-sidedness of a surface is a topological property of the surface*. Another reason is that familiarity with a Möbius band is helpful in understanding a projective plane, which is also a one-sided surface.

EXERCISES

1. (a) Find a simple closed curve which does not encompass the torus in Fig. VII, 22 the way c does, and which also does not divide the torus into two parts. (b) What restriction must be placed on a simple closed curve if it is to divide the torus in two?

2. Construct a paper model of a Möbius band and determine whether the boundary is a connected curve or a disconnected curve.

3. To further clarify what is meant by saying that a simple closed curve c divides or does not divide a surface in two, let us do the following. Take a point X on the surface, very close to c, draw a simple arc from X to a nearby point Y on c, and extend this arc slightly beyond c to a point Z so as to obtain a simple arc XYZ which crosses c. If c divides the surface in two, then every simple arc joining X to Z crosses c; if c does not divide the surface in two, then there exist simple arcs joining X to Z which do not cross c. Illustrate this for (a) a sphere; (b) a plane; (c) a torus; (d) a Möbius band.

12. The Projective Plane. At the beginning of §11 we obtained our first example of a surface not homeomorphic to the plane by adding to the points inside the circle γ those on γ itself. That disk-shaped surface with a boundary is shown in Fig. VII, 28, where A, B, C, D, E, F are six specially chosen points of the boundary, D, E, F being diametrically opposite A, B, C, respectively. Let us now convert this surface into another by identifying each two diametrically opposite points. The new surface, whose symbol appears in Fig. VII, 29, is our last and, for our purposes, most important example of a surface not homeomorphic to the plane. It is a **projective plane**.

As its symbol shows, a projective plane is the surface which would result if we could bring the edges of a disk-shaped surface together so that they are joined at each two diametrically opposite points. Any attempt to make a paper model of a projective plane by cutting out a disk and joining its edges as required is obviously futile. Nor would we fare any better if the disk were made of a

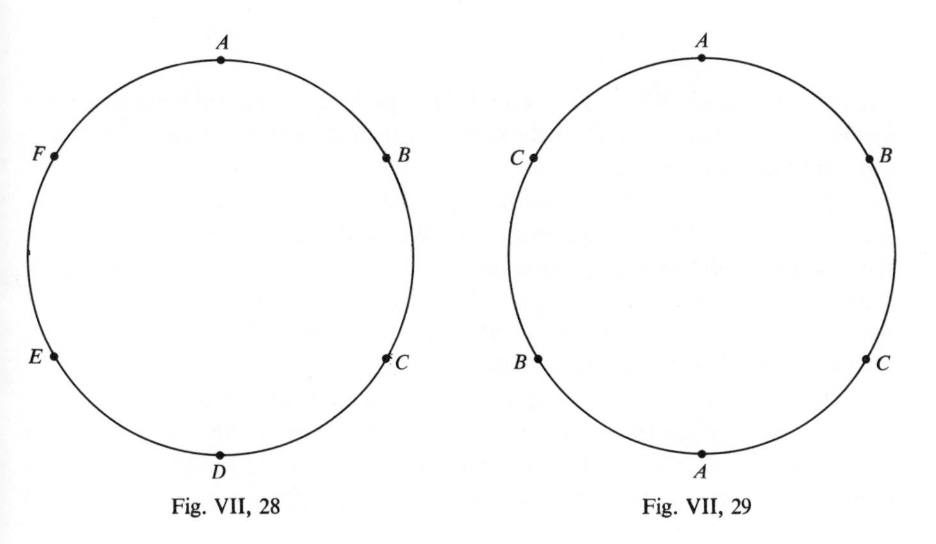

Fig. VII, 28

Fig. VII, 29

flexible material which could be deformed as much as we pleased, without tearing. In other words, a projective plane is an abstract or conceptual surface in the sense that a complete physical model of it cannot be constructed. Studying the surface is therefore best done in terms of its symbol. For example, since a centrally located zone in this symbol, such as each shaded area of Fig. VII, 30, represents a Möbius band, we see that a Möbius band is part of a projective plane. Like the band, then, a projective plane is one-sided, and is not divided in two by every one of its simple closed curves. A projective plane resembles a sphere in possessing no boundary; that is, there is no edge or frontier beyond which we cannot go and still remain on the surface. Hence, a point tracing a path on a projective plane will, so to speak, never encounter a sign saying "the

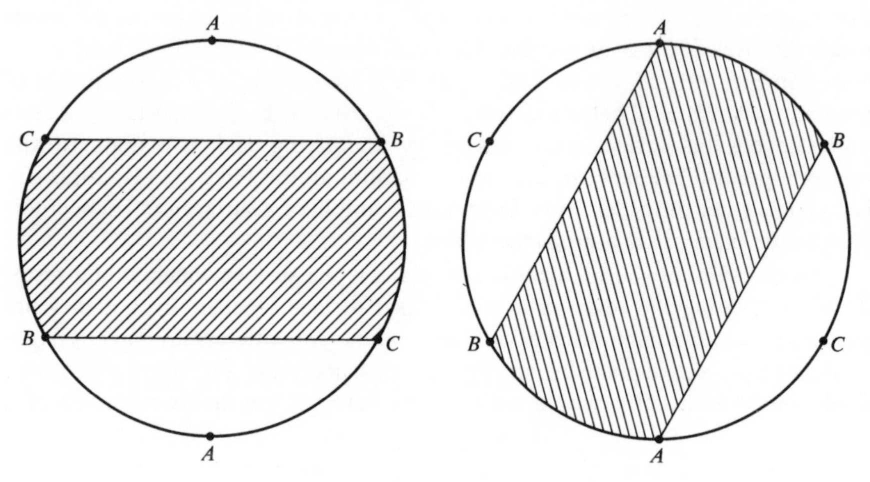

Fig. VII, 30

surface ends here." When, for example, the path leads from the center of the symbol in Fig. VII, 29 to A at the top, it is then also at A on the bottom and can continue on from there.

Just as a Möbius band can be symbolized in various ways, so there are numerous symbols for a projective plane. We shall call the one discussed above the **circular model of a projective plane** and denote it by $\overline{D}$. Any ellipse, for example, could serve as a symbol provided that every two points of its perimeter which are symmetrical to its center are identified. A square, with like identification of points, is also a symbol.

Before leaving $\overline{D}$ let us examine in more detail its apparently close relation to D, the circular model of a Euclidean plane. The latter model, we recall from §9, consists of all the points interior to the unit circle γ; by adding to these points those on γ, and identifying diametrically opposite points on γ, we obtained $\overline{D}$. What effect does this addition of points have on the semiellipses of D studied in §9? The effect is clearly to convert them into simple closed curves of $\overline{D}$. Stated more precisely, if we add to each semiellipse of D the two endpoints which had previously been excluded, and identify these endpoints, the result is a simple closed curve of $\overline{D}$. We shall call these simple closed curves *modified semiellipses*. Fig. VII, 31 shows two of them, AKA and BLB. Although γ, too, is a simple closed curve, there is no semiellipse that corresponds to it. However, the structure of $\overline{D}$ will appear more orderly if we broaden the meaning of the term "modified semiellipse" to include γ. *We shall therefore regard γ as belonging to the set of modified semiellipses.*

As a result of this agreement, for example, we find that the modified semiellipses have simpler incidence relations than do the semiellipses. Two semi-

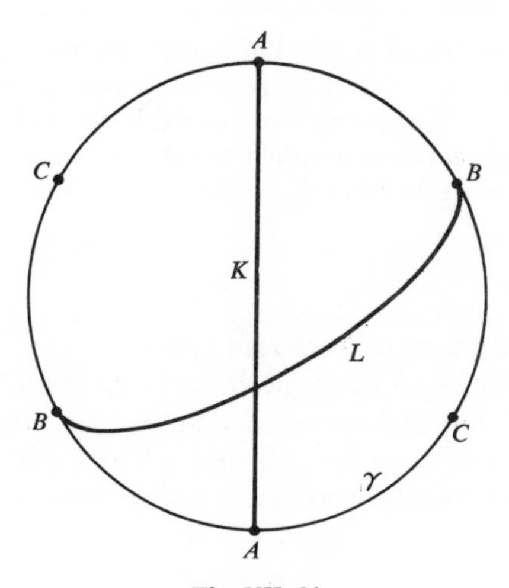

Fig. VII, 31

ellipses, we know, meet in one point or not at all, but two modified semiellipses meet in one point without exception. On the other hand, a unique modified semiellipse goes through each two points of $\overline{D}$, just as a unique semiellipse goes through each two points of D. We can therefore say that two modified semiellipses always determine one point, and, conversely, that two points always determine one modified semiellipse. No such statement can be made for the unmodified semiellipses.

When, in the following chapter, we turn to the problem of perfecting projective geometry by studying it within the framework of the projective plane, we shall find that a new kind of line, the *projective line*, has replaced the straight line in basic importance. The circular model $\overline{D}$ of the projective plane will then serve us well, for the topological structure and incidence relations of the modified semiellipses are precisely those of the projective lines.

EXERCISES

1. Draw a curve in $\bar{D}$ which divides it into two separate parts. Draw another which does not.

2. Prove that a unique modified semiellipse goes through each two points of $\bar{D}$. Consider these cases: (a) both points interior to γ; (b) one point interior to γ, one on γ; (c) both points on γ.

3. Prove that any two modified semiellipses of $\bar{D}$ meet in a unique point. Consider the cases: (a) neither curve is γ; (b) one of the curves is γ.

4. In §10 we determined the curves in D which correspond to parabolas and hyperbolas. What happens to those curves when D is converted into $\bar{D}$? More precisely, what results when we add to these curves the points of γ which are arbitrarily close to them, and identify diametrically opposite ones?

5. The hemispheric model H of a Euclidean plane discussed in §9 can be converted into a *hemispheric model* $\bar{H}$ *of the projective plane* by adding to H the points on its boundary, diametrically opposite points on the boundary being identified. Which curves in $\bar{H}$ have the same topological structure and incidence relations as the modified semiellipses of $\bar{D}$, and hence as the projective lines?

13. A Bounded Model of Euclidean Space. In much the same way as we mapped a Euclidean plane topologically onto the interior of a circle, we can map three-dimensional Euclidean space topologically onto the interior of a sphere. We take the latter to be the unit sphere S with center at the origin O of a three-dimensional rectangular coordinate system, and use

$$(1)\quad x' = \frac{x}{\sqrt{1 + x^2 + y^2 + z^2}}, \quad y' = \frac{y}{\sqrt{1 + x^2 + y^2 + z^2}}, \quad z' = \frac{z}{\sqrt{1 + x^2 + y^2 + z^2}}$$

as the equations of the mapping, where the point $P'(x', y', z')$ is the image of the point $P(x, y, z)$. The implicit form of the mapping is

$$(2)\quad x = \frac{x'}{\sqrt{1 - x'^2 - y'^2 - z'^2}}, \quad y = \frac{y'}{\sqrt{1 - x'^2 - y'^2 - z'^2}}, \quad z = \frac{z'}{\sqrt{1 - x'^2 - y'^2 - z'^2}}.$$

It is left to the student to show that a 1-1 correspondence is established between the points of space and the points interior to S. (In particular, he should show that P' approaches the boundary of S when P recedes indefinitely far from O.) That the mapping is topological then follows from the continuity of the functions on the right side of (1) throughout three-dimensional space, and of the functions on the right side of (2) interior to S.

The mapping being topological, the straight lines and Euclidean planes in space go, respectively, into curves and surfaces interior to S with the same topological properties, in particular, the same incidence relations. We state the following facts without proof. (1) Each space family of parallel lines goes into a space family of curves consisting of the diameter of S parallel to those lines and all semiellipses in space having this diameter as their major axis. No curves of this family meet, but all come arbitrarily close to the same two diametrically opposite points of S. (2) Each family of parallel Euclidean planes goes into a family of surfaces consisting of the circular disk with radius 1 and center O that is parallel to these planes, and all semiellipsoids "based on" this disk. (By this we mean the semiellipsoids obtained by rotating through 180° about their major axes the above-mentioned semiellipses that lie in this disk.) No surfaces in this family meet, but all come arbitrarily close to the boundary of the disk.

Whenever one wishes to view three-dimensional Euclidean space in its entirety, one may do so in terms of this model, which will be called a *spherical model of Euclidean space*. We shall find this useful in the following chapter.

EXERCISES

All the exercises refer to the mapping (1).

1. Show that the mapping establishes a 1-1 correspondence between the points of space and the points interior to S.

2. Show that: (a) each sphere with center O goes into a smaller such sphere; (b) each point P and its image P' are related by the fact that P' is between O and P, and $(OP')^2 = (OP)^2/[1 + (OP)^2]$.

3. Deduce from Ex. 2 (b) that (a) each line through O goes into the diameter of S which is on the line; (b) when a point P recedes indefinitely far from O, its image P' approaches the boundary of S.

CHAPTER VIII

The Projective Plane

1. Introduction. Fortified with a few topological ideas, we can now resume the study of projective geometry begun in Chapter VI. There we saw that, because of the presence of vanishing points and lines, projective geometry in the Euclidean plane was full of exceptional cases and hence lacked the generality of metric and affine geometry. Anticipating the work of the present chapter, we remarked that these exceptions could be removed by enlarging the Euclidean plane so as to form a new plane, the *projective plane*. Let us now see how this is done.

2. Ideal Points. In our study of projections we saw that parallelism and vanishing elements go hand in hand—if there were no parallel lines, there would be no vanishing elements, and hence no exceptional cases. We cannot, of course, eliminate parallelism from Euclidean geometry without changing that subject drastically, but we can do the next best thing. By adding new points to the Euclidean framework we can convert straight lines into lines of a new type in such a way that each two new lines meet. If projections are then studied in the enlarged framework, and in terms of the new lines, they will be free of exceptions. A similar procedure is common in developing the number system. Root-taking, we recall, is not always possible in the real number system; but when the latter is supplemented by new numbers, the imaginaries, to form the enlarged system, of complex numbers, there are no longer any exceptions.

The new points are called *ideal points*, and the "old" ones, with which we have dealt exclusively until now, will be called *Euclidean points*. No significance other than a historical one is to be attached to the term "ideal". The two types of points are on an equal footing logically and later we exhibit them both in the same diagram. The new lines will be called *extended lines*; the "old", or Euclidean lines are called, as usual, straight lines.

In what follows we continue to use "space" to mean "three-dimensional space." Although we are concerned mainly with plane geometry, it is convenient, first, to add the new points to Euclidean space, and later to see the effect of this on Euclidean planes. We therefore make the following agreements:

(1) *To each straight line g in Euclidean space we add a single ideal point. The geometrical object which consists of g and this point is called an extended line and denoted by $\bar{g}$. More precisely, we speak of $\bar{g}$ as the extended line associated with g.*

(2) *The ideal points which are added to two parallel straight lines are the same.*

(3) *The ideal points which are added to two nonparallel straight lines are distinct.*

(4) *The geometrical object which consists of Euclidean space and all ideal points is called extended space.*

Part of what was promised earlier has now been achieved. For, by the above agreements, two parallel straight lines g, h are converted into two extended lines $\bar{g}$, $\bar{h}$ with a common ideal point. Thus, g and h do not meet in Euclidean space, whereas $\bar{g}$ and $\bar{h}$ meet in extended space. We do not call $\bar{g}$, $\bar{h}$ parallel, nor shall we ever apply this term to extended lines.

We can also show without delay how one of the projections involving exceptions can be perfected by the use of ideal points. Consider the projection of a straight line g on an intersecting straight line g' from a center O in their plane (Fig. V, 7). Previously we saw that the vanishing point V on g had no image since straight lines OV and g' are parallel, and that the vanishing point W' on g' had no original since straight lines OW' and g are parallel. Now, since extended lines $\overline{OV}$, $\bar{g}'$ meet in an ideal point V', we can say that V projects into V'. Similarly, if W is the ideal point common to $\overline{OW'}$ and $\bar{g}$, we can say that W projects into W'. Thus the projection of $\bar{g}$ on $\bar{g}'$ is perfect: each point of $\bar{g}$ has an image on $\bar{g}'$, and each point of $\bar{g}'$ has an original on $\bar{g}$. At present we offer no diagram to illustrate this.

The above agreements also enable us to perfect the remaining troublesome projection, the central projection of a plane on an intersecting plane, but we postpone that discussion until §3.

Since we have offered no diagram to illustrate the relation between Euclidean space and extended space, the following description of the relation may be helpful. Let us think of Euclidean space as consisting of families of parallel straight lines, one family for each direction in space. According to our agreements, one and the same ideal point is added to all the straight lines of each such family, different families acquiring different ideal points. Extended space can therefore be thought of as consisting of the resulting families of extended lines. All of the extended lines in each such family have the same ideal point, but no other common point; extended lines in different families have different ideal points.

EXERCISES

1. If a, b are distinct straight lines in Euclidean space, state whether $\bar{a}$, $\bar{b}$ meet and, if they do, in what kind of point or points, given that

(a) a, b are parallel,
(b) a, b are coplanar but not parallel,
(c) a, b are skew.

2. Show that two points in extended space determine a unique extended line if

(a) both points are Euclidean,
(b) just one point is Euclidean.

3. Let v be the vanishing line in plane α for a central projection of α on an intersecting plane (Fig. V, 10). If V is any point of v, show that there is an ideal point which could reasonably serve as the image of V.

3. Extended Planes. Let α be any Euclidean plane. By agreement (1) of §2 one ideal point has been added to each straight line in α. The geometrical object consisting of α and all of these ideal points is called an **extended plane** and denoted by $\bar{\alpha}$. More precisely, we speak of $\bar{\alpha}$ as **the extended plane associated with** α.

We can also regard $\bar{\alpha}$ as consisting of all the extended lines obtained from the straight lines of α. According to agreements (2) and (3), these extended lines are grouped in families: the extended lines in each family have the same ideal point, and those in different families have different ideal points, each family being associated with a certain family of parallel lines of α.

Two extended lines of $\bar{\alpha}$ meet in a unique point. (This point is ideal when the lines belong to the same family, and Euclidean when they do not.) The set of ideal points of $\bar{\alpha}$ and each extended line of $\bar{\alpha}$ likewise meet in a unique point, namely, the ideal point on that line. For this reason, among others, the set of ideal points of an extended plane $\bar{\alpha}$ is regarded as a line, and hence is called the **ideal line** in $\bar{\alpha}$. Another reason is that this set is in 1-1 correspondence with the set of points on an extended line (Ex. 3). A third reason is that the set of ideal points of $\bar{\alpha}$ is the locus of intersection of $\bar{\alpha}$ and any extended plane $\bar{\beta}$ such that α, β are parallel. To show this, let I be any ideal point of $\bar{\alpha}$. Then I is on some extended line in $\bar{\alpha}$, say $\bar{g}$. Let h be the straight line in β which is parallel to g. Then I is on $\bar{h}$, and hence in $\bar{\beta}$. Similarly, we can show that each ideal point of $\bar{\beta}$ is also in $\bar{\alpha}$. Thus $\bar{\alpha}$ and $\bar{\beta}$ have all their ideal points in common, and, of course, have no other common point.

Since $\bar{\alpha}$ is any extended plane, the relation between Euclidean space and extended space can now be described in terms of planes rather than in terms of lines, as was done at the end of §2. Let us think of Euclidean space as consisting of families of parallel planes.* Corresponding to each such family f there is a family $\bar{f}$ consisting of the associated extended planes; all of the latter have the same ideal points, and no two have any other common point. Extended space consists of all such families $\bar{f}$.

We saw above that the ideal line and the extended lines of $\bar{\alpha}$, taken collectively, have the property that each two of them meet in a unique point. They also have the further property that exactly one of them passes through any two points of $\bar{\alpha}$. For if both points are Euclidean, there is a unique straight line through them, and hence a unique extended line. If one point, A, is Euclidean and the other, B, ideal, there is likewise a unique extended line through them; for B was added only to a certain family of parallel lines and just one member of this family contains A. If both points are ideal, no extended line contains them, but the ideal line does.

Later we shall also see that ideal lines and extended lines behave similarly in projections. In recognition of this common property and of the others already noted we apply a common name to ideal lines and extended lines, calling them **projective lines.** Thus we can state the following theorem:

3.1 *In every extended plane a unique projective line passes through any two points and any two projective lines meet in a unique point.*

It is now easy to see how to perfect the projection, from a point O, of one Euclidean plane α on another, α', which intersects it (Fig. V, 10). Previously we had to accept the fact that each point V on v, the vanishing line in α, projects into nothing since straight line OV does not meet α'. Now let us consider the extended line $\overline{OV}$. Its ideal point V' lies in the extended plane $\overline{Ov}$, and hence in the extended plane $\bar{\alpha}'$. Therefore extended line $\overline{OV}$ meets $\bar{\alpha}'$ in V', and we may take this point as the image of V. Thus the vanishing points in α are regarded as projecting into ideal points of $\bar{\alpha}'$. In like manner, the vanishing points in α' can be regarded as being the images of ideal points in $\bar{\alpha}$.

EXERCISES

1. Prove that two extended planes which are not associated with parallel planes meet in an extended line.

2. Prove that a unique projective line passes through any two points of extended space.

3. Prove that the set of ideal points in an extended plane is in 1-1 correspondence with the set of points on any extended line of the plane. (*Hint.* Since this set of ideal points is in 1-1 correspondence with the set of directions in a Euclidean plane, one need

---

* *Parallel planes* will always mean *parallel Euclidean planes.*

only show that the latter set is in 1-1 correspondence with the set of points on an extended line. Our mapping of the x-axis onto a deleted semicircle in VII, §6 is helpful in this connection.)

4. Show that in the central projection of a Euclidean plane α on an intersecting Euclidean plane α', images can be found for the vanishing points of α'.

4. Model of an Extended Plane. We have made no attempt to visualize the ideal points used in building up an extended plane. If the student has made such an attempt he has perhaps said to himself: "Two parallel lines are everywhere equidistant; the same point is added to each; the result is two new lines meeting in this point. How can this possibly be imagined?"

Probably it cannot be imagined if one thinks of Euclidean lines in such metric terms. But it is not necessary to do this. After all, our purpose in introducing ideal points is to perfect projective geometry, and metric concepts are of no importance in that subject. In Chapter VII we saw that Euclidean lines can also be thought of topologically and that there is a large variety of topological models of the Euclidean plane. Moreover, in some of these models the curves corresponding to parallel lines actually approach each other more and more closely. It would therefore seem reasonable, in seeking to visualize the introduction of ideal points, to use such a Euclidean model. In the *circular model of the Euclidean plane* (VII, §9), for example, a family of parallel lines appears as a certain family of nonintersecting semiellipses, all completely interior to the circle γ and coming arbitrarily close to the same two diametrically opposite points of γ. If we regarded these points as one point and added it to each semiellipse of the family, the resulting curves would then meet in just this point. The latter could therefore be interpreted as an ideal point, and the curves as the extended lines meeting in this ideal point. Thus, by adding the points of γ in the specified way to the circular model of a Euclidean plane we would obtain a model of an extended plane. This is precisely what we did in obtaining the *circular model of a projective plane* (VII, §12). In other words, if in the latter model we interpret the points of γ as ideal points and the modified semiellipses other than γ as extended lines, we have a model of an extended plane. In this model, then, γ represents the ideal line.

In §2 we showed how the central projection of a straight line g on a coplanar straight line g' is perfected by the use of ideal points, but offered no illustrative diagram. Now we can do so, using the above model of an extended plane. Thus, Fig. VIII, 1 shows all the extended lines involved in the projection as modified semiellipses, the center of projection O and the vanishing points V, W' as points inside γ, and the ideal points V', W as points of γ. Also shown is the case in which a Euclidean point A projects into a Euclidean point A', the ideal point on their extended line being Z.

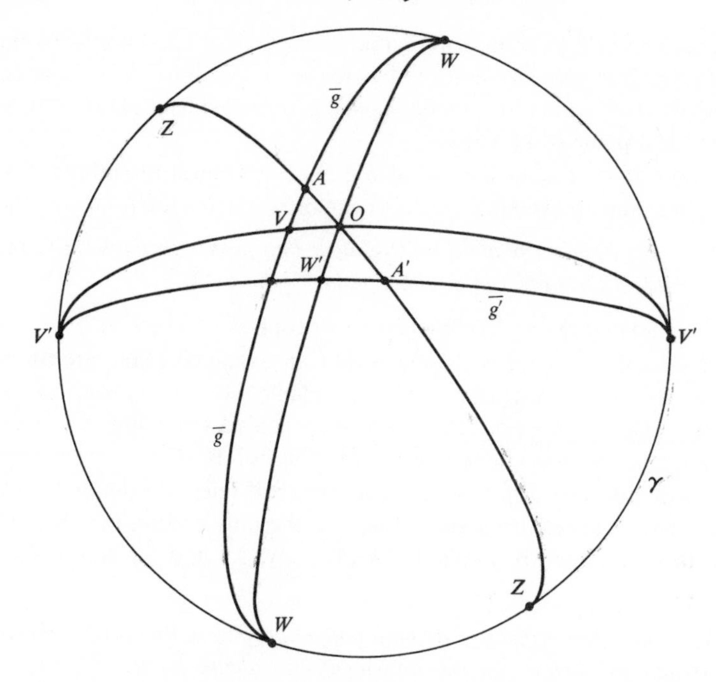

Fig. VIII, 1

EXERCISES

1. Show that in an extended plane it is possible to project one extended line on another, using an ideal point as center. Illustrate this in the model of an extended plane.

2. Show that in an extended plane it is possible to project the ideal line on an extended line, using a Euclidean point as center. Illustrate this in the model of an extended plane.

5. The Ideal Plane; Projective Planes. In §3 we saw that two extended planes which are associated with parallel planes meet in an ideal line. Also, two extended planes not associated with parallel planes meet in an extended line (§3, Ex. 1). Thus, each extended plane meets any other in a projective line. Now, the set of all ideal points in extended space likewise meets each extended plane in a projective line since this set has an ideal line in common with any extended plane. This is one reason for thinking of the set as being like an extended plane. Another reason is that this set can be shown to be in 1-1 correspondence with the set of points in any extended plane (Ex. 3). A third reason, which is verified in a later section, is that the set of all ideal points in extended space behaves just like any extended plane in projections. For these reasons, the set of all ideal points in extended space is regarded as a plane, and hence is called the **ideal plane** of that space.

Beyond these reasons there is also a close structural resemblance between the ideal plane and an extended plane, as can be seen from the following theorem:

5.1 *In the ideal plane a unique ideal line passes through any two points and any two ideal lines meet in a unique point.*

This is analogous to 3.1. We leave its proof as an exercise.

The structures of the ideal plane and an extended plane are, in fact, so much alike that our model of an extended plane becomes a model of the ideal plane if we regard all its points as ideal, its modified semiellipses as ideal lines, and continue to identify diametrically opposite points of γ.

In recognition of their resemblances we shall refer to extended planes and the ideal plane as **projective planes.** They are the only entities in extended space to which this name will be applied. Accordingly, 3.1 and 5.1 can be combined as follows:

5.2 *In every projective plane a unique projective line passes through any two points and any two projective lines meet in a unique point.*

The structure of a projective plane, as regards its points and projective lines, is seen to be more orderly than that of a Euclidean plane, as regards its points and straight lines, for two coplanar straight lines do not always meet.

The circular model and the other models of the projective plane mentioned in Chapter VII all involve the identification of points. Now that Theorem 5.2 is available we can describe a model of that plane which does not employ this procedure. It should first be emphasized that any two sets of mathematical objects can serve as a model of the projective plane if these sets are in 1-1 correspondence, respectively, with the points and projective lines of that plane, and if there is an incidence relation between the objects of the two sets just like that between these points and lines. For with these basic properties of the projective plane as a start, one could go on to develop all the others, working only with the two sets. Consider, then, the set of straight lines through any point O in Euclidean space and the set of Euclidean planes through O. In the following chapter we show by algebraic means that these lines and planes are in 1-1 correspondence, respectively, with the points and lines of the projective plane (IX, §3). In Exercise 5 of the present section we suggest how these correspondences can also be established geometrically.

Turning to the incidence relations of these Euclidean lines and planes, let us note (1) that any two of these planes meet in one of these lines (this is logically equivalent to two projective lines meeting in one point), and (2) that through any two of these Euclidean lines there passes one of these planes (this is logically equivalent to the fact that through any two points of a projective plane there passes one projective line). It follows then, in view of what was said above, that the Euclidean lines and planes through O constitute a model of the projective plane, the lines playing the role of the points of the projective plane, and the planes playing the role of the projective lines. The points on these Euclidean lines play no part in this model, nor do any of the other lines in these Euclidean planes.

EXERCISES

1. Prove 5.1. (*Hint*. Keep in mind that the ideal plane necessarily consists of all the ideal lines in extended space.)

2. Prove that a projective line not lying in a given projective plane meets the latter in a unique point.

3. Prove that the set of ideal points in extended space is in 1-1 correspondence with the sets of points in an extended plane. (*Hint*. See §3, Ex. 3.)

4. Verify that in the circular models of an extended plane and the ideal plane, projective lines are represented by simple closed curves, none of which divides the plane into two separate parts.

5. Using the *hemispheric model of the projective plane* (VII, §12, Ex. 5), show that the Euclidean lines and planes through O are in 1-1 correspondence, respectively, with the points and lines of the projective plane.

6. Projective Space.

Extended space is also called **projective space,** and the points of extended space, that is, the Euclidean and ideal points, are known as **projective points**. Thus, projective space can be regarded as consisting of projective points, or of projective lines, or of projective planes. The incidence relations of these basic elements are useful in the next section, where we complete the study of projections. We therefore state them now. Some have already been established; the student can prove the others.

(a) *A unique projective line passes through any two points of projective space* (§3, Ex. 2).

(b) *Any two projective planes meet in a unique projective line* (§5).

(c) *A projective line not lying in a given projective plane meets the latter in a unique point* (§5, Ex. 2).

(d) *Through any point and any projective line not containing the point there passes a unique projective plane.*

(e) *A unique projective plane passes through any three noncollinear points.*

(f) *A projective plane which contains two points contains all points of their projective line.*

(g) *Three projective planes which have no common projective line meet in a unique point.*

Previously we suggested that one can obtain an understanding of projective space by thinking of it as consisting of all the families of extended lines which are associated with the families of parallel lines. Now we offer a diagram, or model, for visualizing this. In VII, §13 we obtained a *spherical model of Euclidean space* by mapping the latter topologically onto the interior of a sphere S so that a family f of parallel lines goes into the family f' of curves consisting of the diameter of S parallel to f and all semiellipses having that diameter as their major axes. No curves of f' meet, but all come arbitrarily close to two diametrically opposite points of S. If we now regard these two points as one, add it to each curve of f', and do likewise for every other diametrical pair and the curves related to it, we obtain a model of projective space in which the added points correspond to the ideal points, and the families of modified curves correspond to the families of extended lines. In this model, which will be called a *spherical model of projective space*, the surface of S, with its points identified in the specified way, corresponds to the ideal plane.

EXERCISES

1. Prove incidence relation (*c*).
2. Prove incidence relation (*d*).
3. Prove incidence relation (*e*).
4. Prove incidence relation (*f*).
5. Prove incidence relation (*g*).
6. What corresponds to a family of extended planes in our model of projective space?
7. When the surface of a sphere is taken as a model of the ideal plane, which curves on it correspond to the ideal lines? Justify your answer.

7. Projections Viewed More Broadly. Having seen how projections involving vanishing points are perfected by the use of ideal points, let us now show that ideal points can make for even greater simplification in the study of projections.

The variety of projections considered in Chapter V was considerable. There were three kinds of projections of a line g on a coplanar line g': central projection for the case in which g, g' meet, central projection for the case in which they do not meet, and parallel projection. Similarly, there were three kinds of projections of one plane on another, two central and one parallel. An important distinction that had to be made among these six types was that some possessed vanishing elements and some did not. We broke down this distinction by introducing ideal points. But these points also enable us to banish the distinction between central and parallel projections, as we shall now show.

Consider, for example, a parallel projection of a straight line g on a coplanar straight line g' (Fig. V, 3). The straight lines AA', BB', CC', etc., joining pairs of corresponding points are parallel and hence are associated with the same ideal point I. Since the extended lines $\overline{AA'}$, $\overline{BB'}$, $\overline{CC'}$, etc., all go through I, the parallel projection can be regarded as a central projection with I as center. Fig. VIII, 2 illustrates this, using the circular model of an extended plane. Simi-

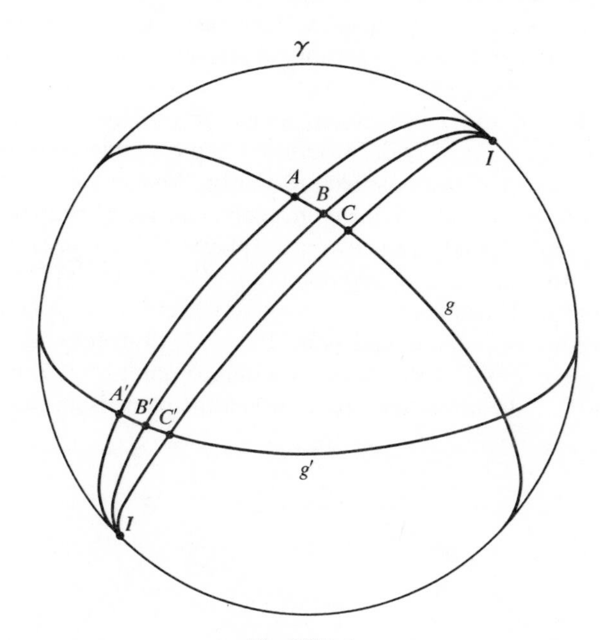

Fig. VIII, 2

larly, a parallel projection of one Euclidean plane on another can be regarded as a central projection in which the center is the ideal point associated with a certain family of parallel lines in space. Thus, when ideal points are used, all projections can be regarded as central projections. Under these conditions a standard notation for a projection becomes possible. To express in symbols that points A, B, C go into points A', B', C' in a projection with center J we write

$$ABC \overset{J}{\underset{\wedge}{=}} A'B'C'.$$

If A', B', C' then go into A'', B'', C'' in another projection with center K, we write

$$ABC \overset{J}{\underset{\wedge}{=}} A'B'C' \overset{K}{\underset{\wedge}{=}} A''B''C''.$$

Since so much simplicity is achieved by the use of ideal points, we shall henceforth suppose that projections employ them to the fullest extent possible. By this we mean, first, that every projection will be regarded as a central projection. Second, any projective point can be the center of a projection. Third, the lines joining pairs of corresponding points in a projection will be understood to be projective lines. Fourth, in projecting one line on another, or one plane on another, it will be understood that these terms refer to projective lines and projective planes. In brief, projections will be regarded as involving projective points, lines, and planes, in contrast to the projections of earlier chapters, which involved only Euclidean points, lines, and planes. These earlier *projections in Euclidean space* are therefore just parts of the *projections in projective space* with which we shall be concerned.

By working in the more ample arena of projective space with this broadened concept of a projection we can develop a simpler and more orderly projective geometry than was possible previously. We can see immediately, for example, that the projection of a line g on a coplanar line g' establishes, without exception, a 1-1 correspondence between the points of these lines. For let O be the center of projection, and P be any point of g. (We must keep in mind that the points, lines, and plane dealt with are projective.) There is then a unique line OP by 5.2. This line meets g' in a unique point P' by 5.2. Conversely, 5.2 also shows that each given point of g' is the image of a unique point of g. Thus, P projects into P'. Fig. VIII, 3 illustrates the case in which the plane is an extended one, g

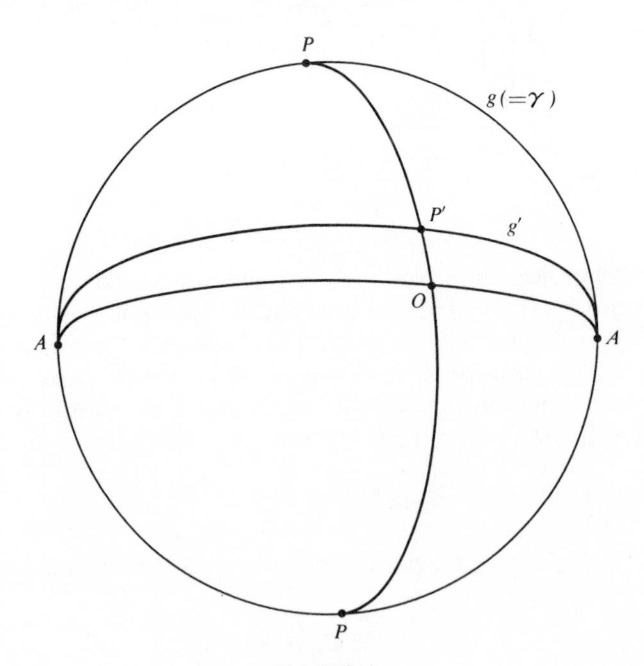

Fig. VIII, 3

is ideal, and O is Euclidean. A, the point of intersection of g and g', projects into itself.

Similarly, it can be shown that the projection of one plane on another establishes, without exception, a 1-1 correspondence between the points of the planes and a 1-1 correspondence between the lines of the planes. In showing this one would use the incidence relations for projective space stated in §6.

When working in Euclidean space, we defined a projective property to be a property that is preserved by all projections of one Euclidean plane on another, and later showed that these properties are the same as those preserved by the projective transformations of a Euclidean plane into itself. We proceed in the same way in working with projective space.

7.1 Definition. *The properties preserved by all projections of one projective plane on another are called projective properties.*

Later, we shall see that these properties are the same as those preserved by all projective transformations of a projective plane into itself.

EXERCISES

1. Show in detail how the parallel projection of one Euclidean plane on another becomes a central projection by the use of ideal points.

2. Prove that the projection of one projective plane on another establishes a 1-1 correspondence among (a) the projective points of the planes, (b) the projective lines of the planes.

8. Collinearity, Concurrence, Duality. One of the important projective properties considered in earlier chapters was the collinearity of points. The presence of vanishing points, however, had kept this property from being universal. Now that the image of every line in a projection is a line we know that collinearity is preserved without exception. Moreover, since the point-correspondences established by projections are 1-1, n collinear points always go into n collinear points.

Working in Euclidean space, we did not find that the concurrence of lines was a projective property, for the presence of vanishing points made it possible for a family of concurrent lines to project into a family of parallel lines, and vice versa. A family of concurrent lines in a projective plane, however, must project into a family of concurrent lines. For, (1) the image of each line is a line, (2) the point common to the original lines goes into a point common to their images, and (3) these images can have no other common point without implying that the original lines had two common points. Thus, the concurrence of lines in a projective plane is preserved by projections of that plane on another. In particular, since the line-correspondences established by such projections are 1-1, n concurrent lines always go into n concurrent lines. It is to be noted that in a projective plane the term *pencil of lines* means only a family of concurrent lines.

We have already noticed that a projective plane possesses a certain symmetry of structure as regards its points and lines which does not occur in a Euclidean plane. This was brought to our attention by the pair of theorems: *through two points there passes a unique line* and *two lines meet in a unique point.* The symmetry of these statements becomes even more apparent if we rephrase them, using the word "on": *two points are on a unique line* and *two lines are on a unique point.* Later in the chapter we shall encounter other examples of such symmetrical pairs of theorems. They illustrate what is known as the *principle of duality in the projective plane.*

Before describing this principle let us mention that the two statements above are called *dual statements*, or *duals*, because replacing "point" by "line" and "line" by "point" in either statement yields the other. In general, if a given statement involving collinearity, concurrence, or both, is expressed in the *"on" language*, and we replace "point" by "line" and "line" by "point" in the given statement, the resulting statement is called the **dual statement** or the **dual** of the given one, and vice versa. Although the "on" language makes for simplicity, it need not be used. If we do not use it, then, besides making the above-mentioned replacements, we must also replace "collinear" by "concurrent" and "concurrent" by "collinear" in order to obtain the dual statement.

8.1 Principle of Duality in the Projective Plane. *If one of two dual statements is a theorem, so is the other.*

This theorem, whose proof we omit, is very useful in developing the geometry of the projective plane. Because of the part which *point* and *line* play in the definition of a dual statement, and hence in the theorem, they are called **dual elements in the projective plane.**

It may be helpful if we illustrate the dualizing process a bit more. The dual of the statement, "this figure consists of three collinear points and four concurrent lines," is the statement, "this figure consists of three concurrent lines and four collinear points." The two figures are called **dual figures** (each being regarded as the dual of the other) since their descriptive statements are duals. Similarly, the dual of the figure consisting of four points, no three of which are collinear, is the figure consisting of four lines, no three of which are concurrent. In the "on" language we would say that the first of these figures is made up of four points, no three on a line, and that the second is made up of four lines, no three on a point.

EXERCISES

1. A figure consists of three collinear points, the line containing them, and three other lines, one on each point. Describe the dual figure.

2. A figure consists of four points, no three collinear, and the six lines joining them in pairs. (a) Describe the dual figure; (b) describe the given figure in the "on" language.

3. A triangle in a projective plane is defined to be the figure consisting of three noncollinear points and the lines joining them in pairs. Dualize this definition.

9. Cross-Ratio and Ideal Elements. Like the collinearity of points, the invariance of cross-ratio also had its limitations when discussed previously. As long as four collinear points projected into four collinear points, for example, it was true that cross-ratio was preserved, but one or more of the four points might not project into anything. This difficulty, of course, has been removed since n collinear points now always project into n collinear points. But a new difficulty appears in the fact that some of these points may be ideal, whereas cross-ratio was defined only for Euclidean points. This definition involves distances and hence cannot be applied to ideal points, with which we have not associated the idea of distance. Since we do want to retain the invariance of cross-ratio as a projective property, our first task is to say what the cross-ratio of four points means when they are not all Euclidean. Similar remarks apply to the cross-ratio of lines. Except when the contrary is indicated, all points and lines considered in this section are understood to lie in the same extended plane.

First, let us see how to assign a value to the symbol (AB, CD) when A, B, C, D are distinct collinear points, D alone being ideal. Let P be any other point on their line (Fig. VIII, 4). Since P is Euclidean,

$$(AB, CP) = \frac{AC}{CB} \div \frac{AP}{PB}.$$

A B C P

Fig. VIII, 4

If we keep A, B, C fixed and let P recede indefinitely far along their straight line in either direction, AP/PB approaches -1, and hence (AB, CP) approaches $-AC/CB$. When P recedes in this manner, the point (also labeled P) in the circular model which corresponds to it approaches the point (also labeled D) corresponding to D (Fig. VIII, 5). This suggests that we attach the value $-AC/CB$ to the symbol (AB, CD).

9.1 Definition. *If A, B, C, D are distinct collinear points, and D alone is ideal,*

$$(AB, CD) = -\frac{AC}{CB}.$$

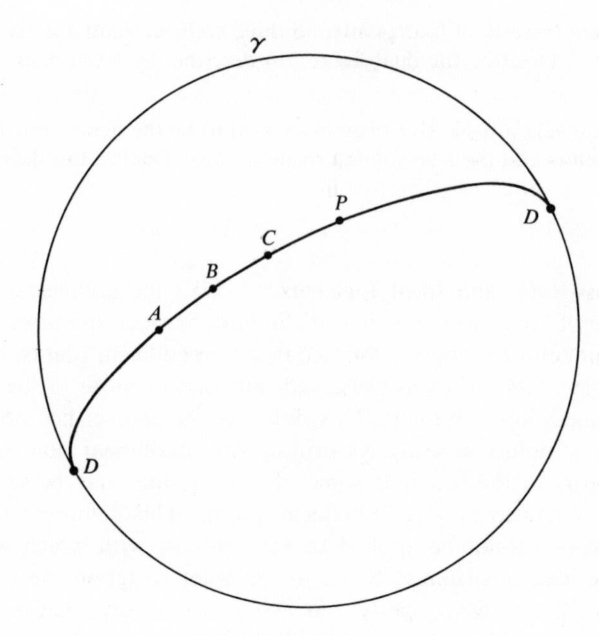

Fig. VIII, 5

In other words, (AB, CD) is the negative of the ordinary ratio in which C divides segment $\overline{AB}$. If C is the midpoint of $\overline{AB}$, then $AC/CB = 1$ and $(AB, CD) = -1$. Retaining the earlier terminology concerning harmonic sets, we can then state:

9.2 *Two Euclidean points are separated harmonically by the midpoint of their segment and the ideal point of their line.*

Before the introduction of ideal points we saw that if A, B, C are distinct collinear points and C is not the midpoint of $\overline{AB}$, there is a unique point D such that $(AB, CD) = -1$ (V, §9, Ex. 7). In view of 9.1 a more general statement can now be made:

9.3 *If A, B, C are distinct collinear Euclidean points, there is a unique point D on their line such that $(AB, CD) = -1$. D is ideal if and only if C is the midpoint of segment $\overline{AB}$.*

The other cases in which just one of the four points is ideal are covered by the following definition, suggested by V, 9.2 and V, 9.4.

9.4 Definition. *Let A, B, C, D be distinct points on an extended line. Then*

$$(AB, CD) = 1/(AB, DC) \qquad \textit{if } C \textit{ is ideal,}$$

and

$$(AB, CD) = (CD, AB) \qquad \textit{if } A \textit{ or } B \textit{ is ideal.}$$

Before defining the cross-ratio of four ideal points it will be helpful to consider the cross-ratio of four lines a, b, c, d through a common point P. If none of these lines is ideal, then they are associated with four Euclidean lines, which meet in P or are parallel according as P is Euclidean or ideal. Cross-ratio of these Euclidean lines having been defined in V, §§11, 13, it is natural to proceed as follows for the projective lines a, b, c, d:

9.5 Definition. *The cross-ratio of four concurrent lines, none ideal, is the same as that of the associated Euclidean lines.*

From this and V, 11.1, then, we have at once:

9.6 *If a, b, c are distinct lines meeting in a Euclidean point, and R is any number not* 0 *or* 1, *there is a unique line d through this point such that* $(ab, cd) = R$.

It remains to define (ab, cd) when one of the lines a, b, c, d, say d, is ideal. Since this corresponds to nothing considered previously, we need a clue to help us. Let a, b, c, d meet in P, which is therefore ideal, and suppose m, m' are any lines coplanar with a, b, c, d, but not containing P (Fig. VIII, 6). Then m meets a, b, c, d in distinct points A, B, C, D, m' meets them in distinct points A', B', C', D', and D, D' alone are ideal. Hence, by 9.1,

$$(1) \qquad (AB, CD) = -AC/CB \qquad \text{and} \qquad (A'B', C'D') = -A'C'/C'B'.$$

Since a, b, c have a common ideal point, the Euclidean lines AA', BB', CC' associated with them are parallel and cut off proportional segments on m and m'. Hence, the ratio in which C divides $\overline{AB}$ equals the ratio in which C' divides $\overline{A'B'}$. It then follows from (1) that $(AB, CD) = (A'B', C'D')$. Thus, the cross-

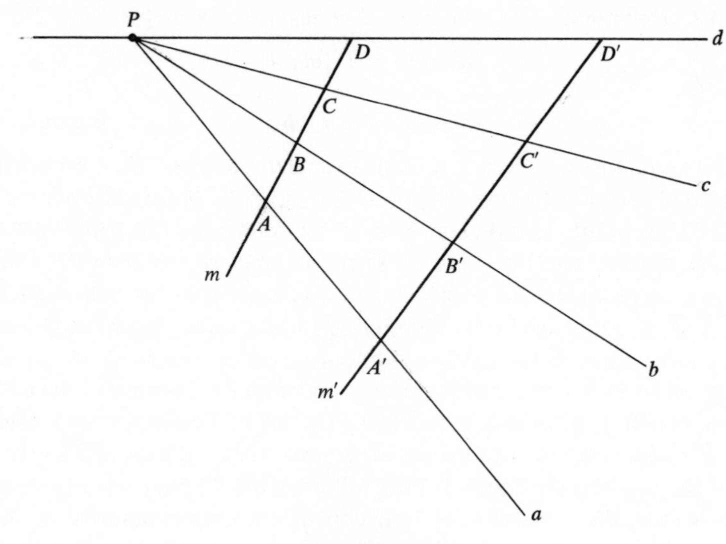

Fig. VIII, 6

ratio of the points in which a, b, c, d are cut by a transversal is independent of the transversal used. This is our clue, for we have found a number related to a, b, c, d which might serve as the value of (ab, cd). Using this number is, in fact, very reasonable in view of a previous theorem on cross-ratio of Euclidean lines (V, 11.2). Accordingly we make the following definition:

9.7 Definition. *If a, b, c, d are distinct concurrent lines and d is ideal, the cross-ratio (ab, cd) is defined to be the same number as (AB, CD), where A, B, C, D are the distinct points in which a, b, c, d are met by any transversal.*

Since a, b, c, d in this definition meet in an ideal point, the Euclidean lines α, β, γ associated with a, b, c are parallel. Suppose γ is midway between α and β. Then C is the midpoint of segment $\overline{AB}$. It follows from 9.3 that $(AB, CD) = -1$, and hence from 9.7 that $(ab, cd) = -1$. If V, §13, Ex. 2 is now consulted, it will be seen that the excluded case in which $R = -1$ and one line is midway between two others can now be taken care of by use of the ideal line, and that we can state the following more general proposition:

9.8 *If a, b, c are distinct extended lines meeting in an ideal point, and R is any number not 0 or 1, there is a unique line d concurrent with them such that $(ab, cd) = R$. Line d is ideal if $R = -1$ and the Euclidean line associated with c is midway between those associated with a and b.*

Combining this with 9.6 we then have:

9.9 *If a, b, c are distinct concurrent lines and R is any number not 0 or 1, there is a unique line d concurrent with them such that $(ab, cd) = R$.*

The other cases in which one of the four lines is ideal are covered by the following definition, suggested by our work in Chapter V:

9.10 Definition. *Let a, b, c, d be distinct concurrent lines. Then*

$$(ab, cd) = 1/(ab, dc) \qquad \text{if } c \text{ is ideal,}$$

and

$$(ab, cd) = (cd, ab) \qquad \text{if } a \text{ or } b \text{ is ideal.}$$

Still confining ourselves to a single extended plane, we shall now define the cross-ratio of four ideal points. Suppose that A, B, C, D are ideal points, that P is a Euclidean point, and that a, b, c, d are four lines through P containing A, B, C, D, respectively (Fig. VIII, 7). Then (ab, cd) has a value by 9.5. Similarly, if Q is any other Euclidean point, and e, f, g, h are four lines through Q containing A, B, C, D, respectively, then (ef, gh) has a value. Since a and e contain the same ideal point A, the Euclidean lines associated with them are parallel, as are those associated with b and f, c and g, and d and h. Because of the equalities of angles resulting from this parallelism, the set of Euclidean lines associated with a, b, c, d is congruent to the set associated with e, f, g, h, and so the cross-ratios of the two sets are equal. It then follows from 9.5 that $(ab, cd) = (ef, gh)$. In other words, the cross-ratio of four concurrent lines containing A, B, C, D, respectively, is the same, regardless which lines are used. In order that these

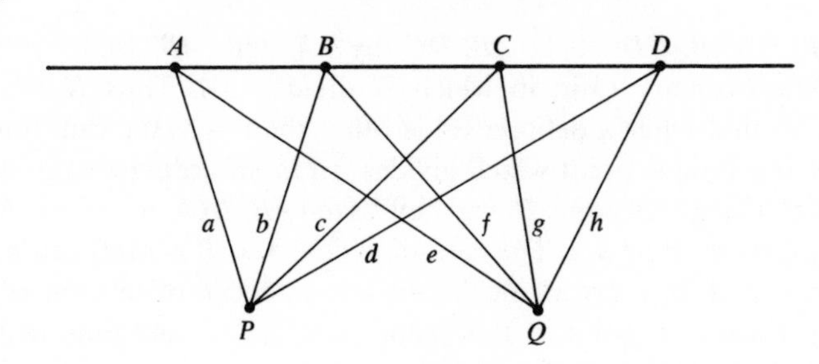

Fig. VIII, 7

points have the same property as four Euclidean points (see V, 11.2, 11.3), we therefore agree to the following:

9.11 Definition. *If A, B, C, D are distinct ideal points in an extended plane, and a, b, c, d are distinct concurrent lines in that plane which contain these points, respectively, then (AB, CD) is defined to be the same number as (ab, cd).*

Cross-ratio of points and lines in an extended plane has now been defined in all possible cases and in such a way that the following basic propositions now hold without exception. (We have not defined cross-ratio of four ideal lines, but it can be done so that these propositions hold in this case, too. We assume this.)

9.12 *If distinct concurrent (coplanar) lines a, b, c, d are intersected by a line in distinct points A, B, C, D, respectively, then $(ab, cd) = (AB, CD)$.*

9.13 *The cross-ratio of points and lines is preserved by every projection.*

9.14 *If α, β, γ, δ are distinct collinear points, or concurrent coplanar lines, then $(\alpha\beta, \gamma\delta) = (\gamma\delta, \alpha\beta) = 1/(\alpha\beta, \delta\gamma)$.*

9.15 *If A, B, C are distinct collinear points and R is any number not 0 or 1, there is a unique point D such that $(AB, CD) = R$. Hence the harmonic conjugate of C with respect to A and B is uniquely determined.**

9.16 *If a, b, c are distinct concurrent lines in the same plane, and R is any number not 0 or 1, there is a unique line d such that $(ab, cd) = R$. Hence the harmonic conjugate of c with respect to a and b is uniquely determined.*

9.17 *If A, B, C, D, E are distinct collinear points, then*

$$(AB, DE)\,(BC, DE) = (AC, DE).$$

These theorems can be proved in the order in which they are stated. We shall prove 9.15. The others are covered in exercises.

Proof of 9.15. If A, B, C are Euclidean, the theorem follows from 9.3 and V, §9, Ex. 7. To get a clue for the proof when C alone is ideal, let us see what follows if we assume that a point D exists such that $(AB, CD) = R$. Then D is

---

* In all cases where we have not said so explicitly it is understood that the term *harmonic* refers to cross-ratios which equal -1.

Euclidean, $R = (AB, CD) = 1/(AB, DC)$ by 9.4, and $(AB, DC) = -r$, by 9.1, where r is the ordinary ratio in which D divides $\overline{AB}$. Thus $R = -1/r$, or $r = -1/R$, so that r has a definite value other than -1. We can now give the proof. There is a unique point which divides $\overline{AB}$ in an ordinary ratio having the value $-1/R$. Calling this point D, we then have $(AB, DC) = -(-1/R) = 1/R$, so that $(AB, CD) = R$ by 9.4. The case in which A or B is ideal can be handled in like manner. If A, B, C are all ideal, take any extended plane containing them, let P be any Euclidean point in this plane, and a, b, c any lines in it that go through P and contain A, B, C, respectively. There is a unique Euclidean line d' such that $(a'b', c'd') = R$, where a', b', c' are the Euclidean lines associated with a, b, c (see V, 11.1), and hence a unique extended line d such that $(ab, cd) = R$ by 9.5. For the point D in which d meets the ideal line we have $(AB, CD) = R$ by 9.11. Also, D is unique since d is unique.

EXERCISES

1. Prove 9.12 in these cases regarding a, b, c, d: (a) they are associated with parallel lines; (b) they meet in a Euclidean point; (c) just one of them is ideal.

2. Prove 9.13 in these cases: (a) for points; (b) for lines.

3. Prove 9.14 in these cases: (a) for points; (b) for lines.

4. Prove 9.16 in these cases regarding a, b, c: (a) they meet in a Euclidean point; (b) they meet in an ideal point but none is ideal; (c) just one of them is ideal.

5. Prove 9.17 in these cases: (a) E alone is ideal; (b) D alone is ideal; (c) C alone is ideal; (d) all are ideal.

10. Order on a Projective Line. At the time we introduced ideal points our only concern was to bring in enough of them to banish parallelism, and to do this in the simplest way. Hence the agreements we made then tell us nothing about the structures of projective lines and planes other than that they possess certain incidence properties and certain relations to Euclidean lines and planes. On the other hand, the models that we introduced to aid us in visualizing these properties and relations actually go far beyond meeting these requirements. In the circular model of the projective plane, for example, the curves which represent projective lines are simple closed curves. Also, because of the way in which this model is built up from these curves, we may say that it represents the projective plane as a type of surface which is not divided in two by any of its projective lines.

As we said above, there is nothing in our previous agreements about ideal points that implies these things about the projective line and plane. Yet, it would be hard to imagine a simpler diagram than the one we have used which illustrates the incidence of points and lines in an extended plane and at the same time exhibits the relation of this plane to a Euclidean plane. *Let us therefore agree that the model correctly exhibits the topological structures of a projective plane and all of the loci on it, and that, in particular, it correctly exhibits the relation between an extended plane and the associated Euclidean plane.** This agreement, which has the effect of an axiom, implies, for example, that a locus in a projective plane is a simple closed curve if its image in the model is such a curve. Accordingly, *a projective line is a simple closed curve.*

Since a projective line is a simple closed curve, its points are necessarily in cyclic order (VII, §7). If A, B, C, D are four of its points, then, they are in exactly one of the orders $ABCD$, $ABDC$, $ACBD$, each implying that the points are also in seven other orders. The order $ABCD$, for example, implies the orders $BCDA$, $CDAB$, $DABC$, $DCBA$, $CBAD$, $BADC$, $ADCB$. Reference to the model shows that our agreement, insofar as it involves order, can be spelled out in detail for A, B, C, D as follows:

(1) *if they are Euclidean we say that they are in the cyclic order ABCD on their projective line when they are in the linear order ABCD on the associated Euclidean line*;

(2) *if only D is ideal we say that they are in the cyclic order ABCD on their projective line when A, B, C are in the linear order ABC on the associated Euclidean line*;

(3) *if all are ideal we say that they are in the cyclic order ABCD when they project into Euclidean points A', B', C', D' which are in the linear order $A'B'C'D'$* (Fig. VIII, 8).

Suppose that A, B, C, D are in the cyclic order $ABCD$. Then the pairs A, B and C, D do not separate each other (V, §9 and VII, §7). If the points are Euclidean, these pairs also do not separate each other on the associated Euclidean line, and hence $(AB, CD) > 0$ by V, 9.6. Suppose just one of the points is ideal: (a) if it is D, then $(AB, CD) > 0$ by VIII, 9.1 and the fact that C is outside of $\overline{AB}$; (b) if it is C, then $(AB, CD) = 1/(AB, DC) > 0$ since $(AB, DC) > 0$ by (a); (c) if it is B, then $(AB, CD) = (CD, AB) > 0$ by (a); (d) if it is A, then $(AB, CD) = 1/(BA, CD) = 1/(CD, BA) > 0$ since $(CD, BA) > 0$ by (a). Finally, if A, B, C, D are ideal, they project into Euclidean points A', B', C', D' such that A', B' and C', D' do not separate each other by agreement (3) above. Then $(A'B', C'D') > 0$. Since cross-ratio is a projective invariant, $(AB, CD) > 0$. Similarly, if we assumed A, B and C, D separate each other, we could show that $(AB, CD) < 0$. Thus V, 9.6 can be generalized:

10.1 *Two pairs of points A, B and C, D on a projective line separate each other if $(AB, CD) < 0$ and do not separate each other if $(AB, CD) > 0$.*

---

* A projective plane and its model are, of course, not the same thing, any more than a ship and a model of it are the same.

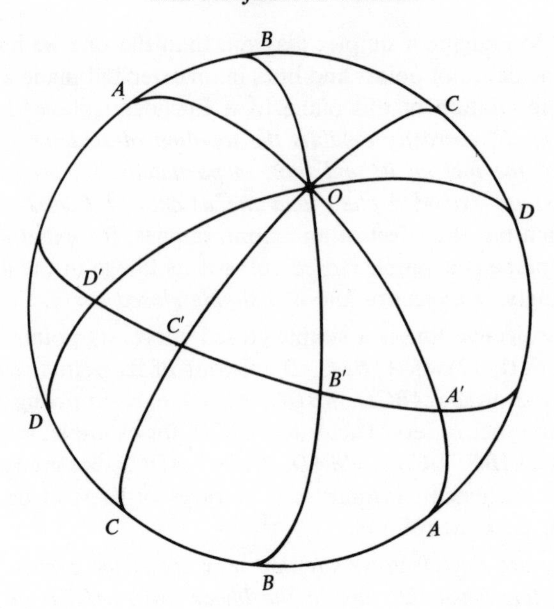

Fig. VIII, 8

Since cross-ratio is preserved by all projections, we then have:

10.2 *The separation or nonseparation of pairs of projective points is a projective property.*

This generalizes V, 9.7.

It is useful to analyze further the case in which $(AB, CD) > 0$. In Chapter V, working in the Euclidean plane, we saw that $0 < (AB, CD) < 1$ if these points are in one of the linear orders $ABDC$, $ACDB$, $DCAB$, and that $(AB, CD) > 1$ if they are in one of the orders $ABCD$, $ADCB$, $CDAB$ (V, §9, Ex. 9). By agreement (1) above, the latter orders correspond, respectively, to the cyclic orders $ABCD$, $ADCB$, $CDAB$ on the associated extended line. But the first of these implies the other two. Hence we can say that $(AB, CD) > 1$ if these points are Euclidean and in the cyclic order $ABCD$. In the same way we can show that $0 < (AB, CD) < 1$ if these points are Euclidean and in the cyclic order $ABDC$. By the same method used to prove 10.1 we could show that the preceding statements also hold when the points are not all Euclidean. This is left as an exercise. We thus have:

10.3 *If four points on a projective line are in the cyclic order ABCD, then* $(AB, CD) > 1$, *and if they are in the cyclic order ABDC, then* $0 < (AB, CD) < 1$.

Let us close this section by making a comparison. On a Euclidean line the relation of betweenness is fundamental and it goes hand in hand with linear order. Any statement involving one of these terms can be translated into a statement involving the other. The situation on a projective line is analogous for cyclic order and the relation of separation. Both terms are important and so we have used them both.

EXERCISES

1. Prove that $0 < (AB, CD) < 1$ if these points are in the cyclic order $ABDC$ and (a) all are Euclidean, (b) only D is ideal, (c) all are ideal.

2. Prove that $(AB, CD) > 1$ if these points are in the cyclic order $ABCD$ and (a) only D is ideal, (b) only A or B is ideal, (c) only C is ideal, (d) all are ideal.

3. Prove that $(AB, CD) < 0$ if these points are in the cyclic order $ACBD$.

4. Using the proposition of Ex. 3, prove the converse of each part of 10.3.

5. State 10.1 in terms of cyclic order.

6. Describe the nature of the locus in an extended plane consisting of (a) a parabola and the ideal point associated with its axis, (b) a hyperbola and the ideal points associated with its asymptotes. (*Hint*. See VII, §10.)

11. Figures in the Projective Plane. The discussions in this chapter thus far, even when they involved spatial considerations, have had one main objective: to help us build a more satisfactory two-dimensional projective geometry than was possible in the Euclidean plane. It should be reasonably clear by now that this can be done by working in the projective plane. We have in fact already learned much about the projective geometry of this plane, and to learn more need only continue what we have been doing, which is to study those figures in this plane, and those properties of these figures, which are preserved by projections. Having studied the points on a single line, their order and cross-ratio, the lines on a single point, and the incidence of points and lines, we shall now want to consider less simple figures. In the rest of this chapter, except where the context indicates otherwise, *point*, *line*, *plane* will mean *projective point*, *projective line*, *projective plane*, and all figures are in the same projective plane.

Triangles, quadrilaterals, other polygons, and conics naturally come to mind as figures to study next. In Chapter V, working in the Euclidean plane, we saw that a triangle sometimes projected into a triangle, and sometimes into a different kind of rectilinear figure. Since ideal points have been so helpful in removing exceptions, one might suppose that now a triangle always projects into a triangle. This would be reasonable except for the fact that "triangle" must now mean a figure in the projective plane, whereas in Chapter V it meant a Euclidean figure, three noncollinear points and the segments determined by them. If A, B, C are noncollinear points in a projective plane, and A is ideal, it makes no sense to talk of segments $\overline{AB}$ and $\overline{AC}$ since the term *segment* involves betweenness and has meaning only when applied to Euclidean points. If we wish to deal with triangles in the projective plane, then, we have to redefine them. This is very easy. By a **triangle** in a projective plane is meant a figure consisting of any three noncollinear points and the three (entire) lines determined by them. The points are called the *vertices* of the triangle, and the lines are called its *sides*. It follows immediately that a triangle always projects into a triangle, and hence is a suitable figure for study in the geometry of the projective plane. We are not concerned with angles in the geometry of the projective plane and so have not defined "angles of a triangle".

Quadrilaterals and other polygons, which presented the same kind of difficulty as did triangles when studied in Chapter V, must likewise be redefined for the projective plane. If A, B, C, D are distinct points, no three of which are collinear, the figure consisting of these points and the four lines AB, BC, CD, DA is called the **quadrilateral** $\boldsymbol{ABCD}$. The points and lines are called the *vertices* and *sides* of the quadrilateral. Pentagons, hexagons, and, in general, n-sided polygons are defined in similar fashion, so that the number of vertices and sides they possess is 5, 6, or n, respectively. These polygons are called **simple polygons.** (Presently we shall define another type of polygon.) As with the triangle, each simple polygon projects into another with the same number of vertices, and so *the property of being a simple polygon is projective*. According to our definition, the quadrilateral $ACBD$ has A, C, B, D as vertices, lines AC, CB, BD, DA as sides, and hence is not the same as the quadrilateral $ABCD$, whose sides are the lines AB, BC, CD, DA. A like remark can be made for any simple polygon when $n > 3$. Thus, it is not the vertices alone which determine a simple polygon, but also their order. If the definition of a triangle is dualized, it will be found that the resulting figure is also a triangle. Hence a triangle is said to be *self-dual*. Similarly, every simple polygon is self-dual.

Besides the simple polygons certain other polygons, called **complete polygons,** are also considered in projective geometry. Thus, a **complete quadrilateral** is the figure consisting of four lines, no three of which are concurrent, and their six points of intersection. The lines and points are called the *sides* and *vertices* of the quadrilateral. (In Fig. VIII, 9, lines q, r, s, t are the sides and A, B, C, D, E, F are the vertices.) Any two vertices not on the same side (such as A and C in Fig. VIII, 9) are called *opposite*. The three lines determined by the pairs of opposite vertices are the *diagonals* of the quadrilateral. In Fig. VIII, 9 the diagonals are AC, BD, EF.

The figure which is dual to a complete quadrilateral is called a **complete quadrangle**. It consists of four points, no three of which are collinear, and the six lines they determine. The points and lines are called the *vertices* and *sides* of the quadrangle. Any two sides not on the same vertex are called *opposite*. Fig. VIII, 10 shows a complete quadrangle with vertices A, B, C, D and pairs of opposite sides AB and CD, AC and BD, AD and BC. The three points of inter-

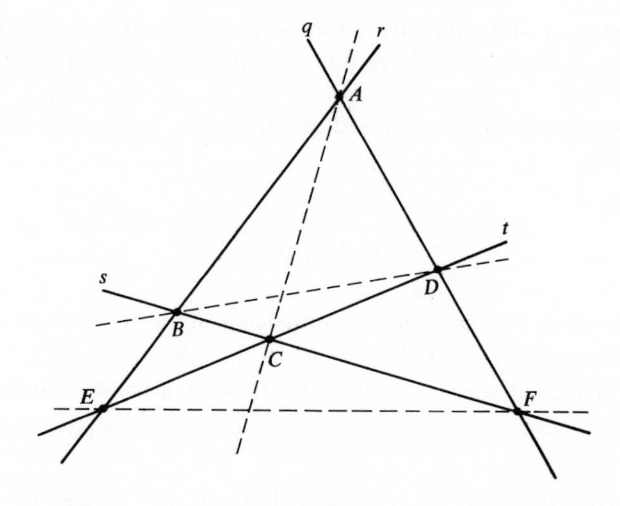

Fig. VIII, 9

section of the pairs of opposite sides are called the *diagonal points* of the quadrangle. In Fig. VIII, 10 they are E, F, G.

A complete quadrangle is also called a *complete* 4-*point*. When designated in this way, it is easily generalized. Thus a *complete* 5-*point* consists of five points, no three of which are collinear, and the ten lines they determine. Similarly, a complete quadrilateral is also called a *complete* 4-*line*, and thus readily generalized.

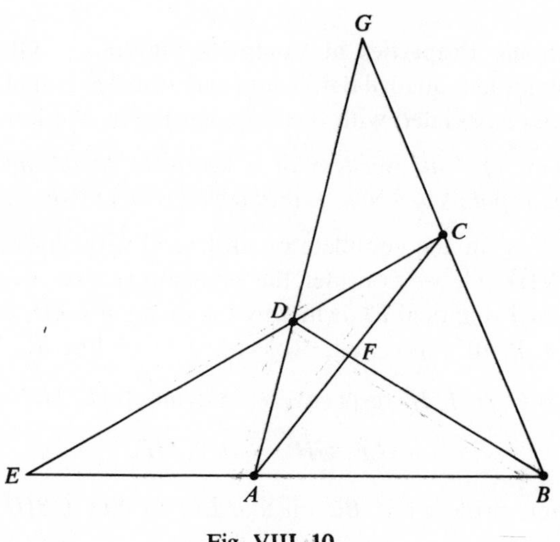

Fig. VIII, 10

As Figures VIII, 9 and VIII, 10 show, conventional drawings made with a ruler on a flat surface can be helpful in studying the geometry of the projective plane.

In this section we have discussed only polygons. Later in the chapter conics are considered.

EXERCISES

1. How many different simple quadrilaterals are there which have four given points, no three collinear, as vertices?

2. Define the following: (a) a simple hexagon; (b) a simple n-sided polygon.

3. Draw several different simple hexagons which have six given points, no three collinear, as vertices.

4. (a) Prove that a triangle is self-dual. (b) Do likewise for a simple quadrilateral and for a simple n-sided polygon.

5. (a) Prove that a triangle always projects into a triangle. (b) Do likewise for a simple quadrilateral and for a simple n-sided polygon.

6. How many different complete quadrilaterals are there with four given lines, no three concurrent, as sides, and how many simple quadrilaterals?

7. Define a complete 5-line.

8. Prove that complete quadrilaterals and complete quadrangles always project into figures of the same kind, respectively.

12. Harmonic Properties of Complete Figures. One reason for the importance of complete quadrilaterals and quadrangles is that many of the sets of points and lines associated with them are harmonic. We have, for example:

12.1 *Two opposite vertices of a complete quadrilateral are separated harmonically by the points in which their diagonal is met by the other two diagonals.*

To prove this for the complete quadrilateral with sides AB, AC, BD, CD shown in Fig. VIII, 11, we consider the opposite vertices E, F and the points G, H in which their diagonal EF is met by the diagonals AD, BC, and show that $(EF, GH) = -1$. With A as center, project line EF on line BC. Points E, F, G, H thus project into C, B, I, H, respectively; in symbols, $EFGH \overset{A}{\underset{\wedge}{=}} CBIH$. Hence

$$(EF, GH) = (CB, IH). \tag{1}$$

With D as center, project line BC on line EF, so that $CBIH \overset{D}{\underset{\wedge}{=}} FEGH$. Then

$$(CB, IH) = (FE, GH). \tag{2}$$

From (1) and (2) we obtain

(3) $$(EF, GH) = (FE, GH).$$

By 9.14,

$$(FE, GH) = 1/(EF, GH).$$

Substituting this in (3), we get

$$(EF, GH)^2 = 1.$$

Since the cross-ratio of four distinct points never has the value 1, we infer that $(EF, GH) = -1$. Thus, the points E, F, G, H, in that order, form a harmonic set, or G, H separate E, F harmonically. The proof is essentially the same for each of the other pairs of opposite vertices.

Corresponding to 12.1 we have the following *dual theorem* for complete quadrangles:

12.2 *Two opposite sides of a complete quadrangle are separated harmonically by the lines joining their diagonal point to the other two diagonal points.*

*Proof.** Consider the complete quadrangle in Fig. VIII, 11 with vertices A, B, C, D. As we saw in proving 12.1, lines AB, BC, CD, DA are the sides of a

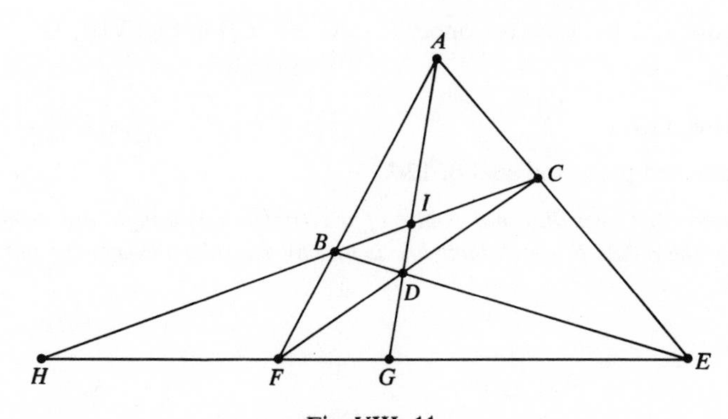

Fig. VIII, 11

complete quadrilateral; and by 12.1 the opposite vertices A, D are separated harmonically by G, I. Hence

$$(AD, GI) = -1.$$

Joining these points to E gives four lines EA, ED, EG, EI with the same cross-ratio by 9.12. Thus,

$$(EA\ ED, EG\ EI) = -1.$$

* Although 12.2 is an immediate consequence of 12.1 because of 8.1, the Principle of Duality, we are offering an additional proof since we did not prove 8.1.

Since lines EA, ED are opposite sides of the given quadrangle, E is their diagonal point, and F, I are the other diagonal points. Hence the theorem is proved for the opposite sides we have worked with. The proof is essentially the same for each of the other pairs of opposite sides.

Theorem 12.1 states a harmonic property involving two opposite vertices of a complete quadrilateral. The following two theorems state harmonic properties involving two sides and two diagonals, respectively, of the same figure. The proofs of these theorems, together with the statements and proofs of their duals, are covered in exercises.

12.3 *Two sides of a complete quadrilateral are separated harmonically by the diagonal through their point of intersection and the line joining this point to the point of intersection of the other two diagonals.*

12.4 *Two diagonals of a complete quadrilateral are separated harmonically by the lines joining their common point to the vertices on the third diagonal.*

EXERCISES

1. Prove 12.1 by using the opposite vertices A, D in Fig. VIII, 11.
2. Prove 12.2 by using the opposite sides AB, CD in Fig. VIII, 11.
3. Prove 12.3.
4. Prove 12.4.
5. State and prove the dual of 12.3.
6. Prove that *two diagonal points of a complete quadrangle are separated harmonically by the points in which their line is met by the sides through the third diagonal point.*

13. The Construction of Harmonic Conjugates. As we know from 9.15, if E, F, G are distinct collinear points, there is a unique point H which is the harmonic conjugate of G with respect to E, F. The preceding section gives us a way of constructing H, that is, of determining its position, when E, F, G are given. Take any two lines through E, and a line through F meeting them in A, B (Fig. VIII, 12). Let lines GA, EB meet in C, and lines FC, EA in D. Then E, F are opposite vertices in the complete quadrilateral with sides AB, BC, CD, DA, and diagonal EF is met by diagonal AC in G. By 12.1, the intersection of EF with the third diagonal BD is the desired position of H.

The construction may also be viewed as providing us with a complete quadrangle $ABCD$ in which E, F are two diagonal points and G is the point in

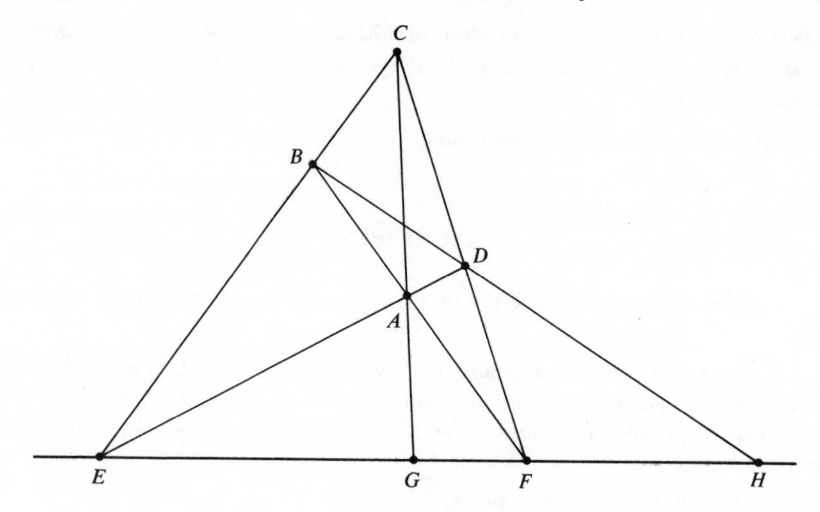

Fig. VIII, 12

which line EF is met by a side through the other diagonal point I. It then follows from §12, Ex. 6 that the other side through I meets line EF in the desired harmonic conjugate H.

According to 12.1, if four points are situated in a special way with regard to a given complete quadrilateral, they will form a harmonic set. Is the converse true, that is, if four given points form a harmonic set, is there a complete quadrilateral to which they are related in that special way? The answer is in the affirmative, for if it is given that $(EF, GH) = -1$, the preceding discussion shows that there is a complete quadrilateral for which E, F are opposite vertices and G, H are the intersections of the other two diagonals with the diagonal EF. We can therefore state the following proposition:

13.1 *Two pairs of points on a line form a harmonic set if and only if there exists a complete quadrilateral in which one pair are opposite vertices and the other pair are the intersections of the diagonal of these vertices with the other two diagonals.*

This theorem is noteworthy because it furnishes a criterion for a harmonic set of points which does not involve cross-ratio, and hence distance. When projective geometry is developed in a manner entirely free of metric notions, which we are not doing, the content of this theorem can be used as the definition of a harmonic set of points. Another criterion of this sort is stated in Exercise 2.

By means of the preceding section we can also construct the fourth harmonic line when the other three are given and prove the following purely projective criterion for a harmonic set of lines. The details, similar to those given above, are left to the student.

13.2 *Two pairs of lines through a point form a harmonic set if and only* *if there exists a complete quadrangle in which one pair are opposite sides and the other pair are the lines joining the diagonal point of these sides to the other two diagonal points.*

This theorem and 13.1 are duals.

EXERCISES

1. Show how to determine the fourth harmonic line to three given concurrent lines.

2. Prove that *two pairs of points on a line form a harmonic set if and only if there exists a complete quadrangle having one pair, E and F, as diagonal points and the other pair, G and H, as the intersections of line EF with the sides through the third diagonal point.*

3. Dualize the proposition of Ex. 2.

14. The Theorem of Desargues. Let us now turn from quadrangles and quadrilaterals to an even simpler figure, a triangle, and prove one of the basic theorems of projective geometry, the Theorem of Desargues. Our earlier statement of this theorem for the Euclidean plane (VI, 9.2) was unavoidably awkward because of the presence of exceptional cases. We did not mention "triangle", for example, because the property of being a triangle in that plane is not projective. This caused much of the awkwardness. By contrast, the following statement of the theorem for the projective plane is both simple and general:

14.1 Theorem of Desargues. *If two triangles are perspective from a point, they are perspective from a line.*

As was mentioned in VI, §9, *perspective from a point* means that the three lines joining corresponding vertices of the triangles are concurrent. More precisely, it means that the vertices of the triangles can be labeled A_1, B_1, C_1 and A_2, B_2, C_2, respectively, so that lines A_1A_2, B_1B_2, C_1C_2 are concurrent. For these triangles to be *perspective from a line* means that the intersections of corresponding sides, that is, A_1B_1 and A_2B_2, B_1C_1 and B_2C_2, and A_1C_1 and A_2C_2, are collinear.

To prove the theorem, let triangles $A_1B_1C_1$ and $A_2B_2C_2$ be given such that lines A_1A_2, B_1B_2, C_1C_2 meet in O (Fig. VIII, 13). The triangles are then perspective from O. We assume that no two corresponding vertices are the same and that O does not coincide with a vertex, for the proof is immediate in these cases. Let A_1B_1 and A_2B_2 meet in C_3, A_1C_1 and A_2C_2 in B_3, B_1C_1 and B_2C_2 in A_3. We shall prove A_3, B_3, C_3 collinear. Take any two points P_1, P_2 collinear with O

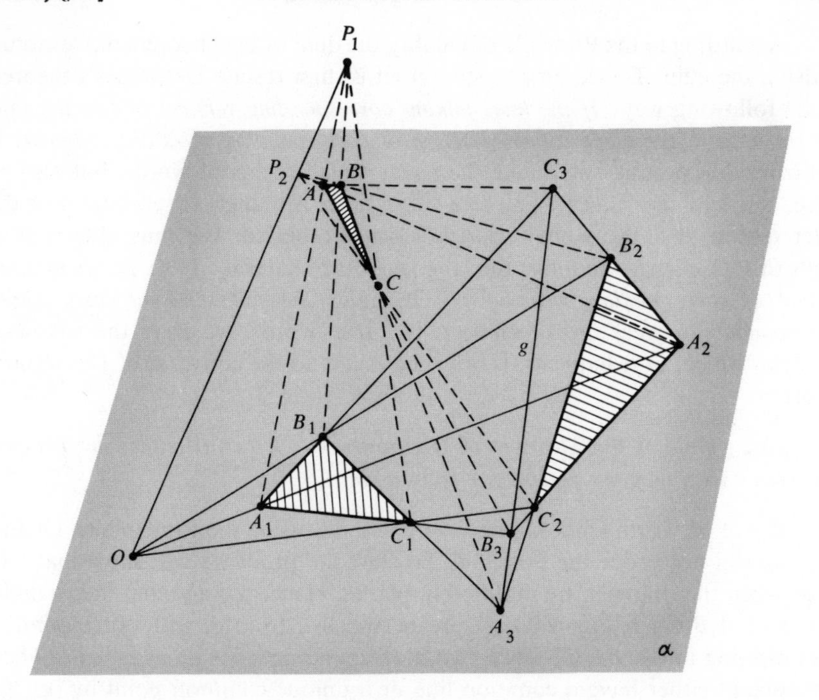

Fig. VIII, 13

but not in the plane α containing the triangles. Points O, A_1, P_1 are then noncollinear and hence determine a plane by (e), §6. This plane contains A_2, P_2 by (f), §6, and hence also lines A_1P_1, A_2P_2 by this same proposition. These lines, being coplanar, meet in a point A. In like manner, by working with B_1, B_2, and then with C_1, C_2, we find that lines B_1P_1, B_2P_2 meet in a point B, and lines C_1P_1, C_2P_2 in a point C. Noting that lines AA_1, BB_1, CC_1 meet in P_1, we infer that A, B, C are noncollinear, for only then could they be projected from P_1 into the noncollinear points A_1, B_1, C_1. Thus, A, B, C determine a plane β. Clearly none of these points is in α, so that $\beta \neq \alpha$. Line AB, being coplanar with line A_1B_1, meets the latter; for a like reason AB meets line A_2B_2. These intersections must be the same since line AB meets α in a unique point by (c), §6, and hence they coincide in C_3. Thus line AB meets α in C_3. Similarly, line AC meets α in B_3 and line BC meets it in A_3. Thus A_3, B_3, C_3 are common to α and β, and hence collinear. If g is the line of intersection of α and β, then the given triangles are perspective from g, as we wished to prove.

(Incidentally, we see that Desargues's theorem can be broadened to permit the two triangles to lie in different planes, for consider triangles ABC and $A_1B_1C_1$, which are perspective from P_1. Since lines AB and A_1B_1 meet in C_3, AC and A_1C_1 in B_3, and BC, B_1C_1 in A_3, the triangles are perspective from line $A_3B_3C_3$.)

According to the Principle of Duality the dual of the Theorem of Desargues is also a theorem. To see how to state it let us first restate Desargues's theorem in the following way: *If the lines joining corresponding vertices of two triangles are concurrent, the points of intersection of corresponding sides are collinear.* In dualizing this we not only make the usual replacements of words, but also replace "vertices" by "sides", and vice versa, since a triangle is self-dual. For this latter reason, too, the word "triangle" is not replaced. We thus obtain (after slight further changes to make the language more natural): *If the points of intersection of corresponding sides of two triangles are collinear, the lines joining corresponding vertices are concurrent.* In other words, we have the following theorem, which, as it happens, is both the dual and the converse of Desargues's theorem.

14.2 Dual of the Theorem of Desargues. *If two triangles are perspective from a line, they are perspective from a point.*

It will be worthwhile to prove 14.2 without using the Principle of Duality since we did not prove the principle. To start the proof we will show that 14.2 holds when the triangles are in different planes. Hence assume that the triangles ABC and $A_1B_1C_1$ in Fig. VIII, 13 are perspective from g, with corresponding sides meeting in A_3, B_3, C_3. Each two corresponding sides lie in a plane. These three planes either have a common line or a unique common point by (g), §6. Since they intersect by pairs in lines AA_1, BB_1, CC_1, they could have a common line only if these three lines coincided. This being impossible, the planes have a unique common point P_1. Hence lines AA_1, BB_1, CC_1 meet in P_1. The triangles are therefore perspective from P_1.

Again using Fig. VIII, 13, but otherwise starting afresh, let us now assume that triangles $A_1B_1C_1$ and $A_2B_2C_2$ are in the same plane α and are perspective from g, with corresponding sides meeting in A_3, B_3, C_3. Let β be a plane through g different from α. Take three lines in β through A_3, B_3, C_3, respectively, so that their intersections form a triangle ABC, as shown. Then triangles $A_1B_1C_1$ and ABC, being in different planes and perspective from g, are perspective from a point P_1. Similarly, triangles $A_2B_2C_2$ and ABC are perspective from a point P_2. Hence each of the planes AA_1A_2, BB_1B_2, CC_1C_2 contains P_1 and P_2, and therefore also line P_1P_2. The point O in which line P_1P_2 meets α must, of course, be on the line in which plane AA_1A_2 meets α. Thus O is on line A_1A_2. Similarly, using planes BB_1B_2 and CC_1C_2, we see that O is on lines B_1B_2 and C_1C_2. Thus, triangles $A_1B_1C_1$ and $A_2B_2C_2$ are perspective from O.

The proofs of this section depend on nothing more than the incidence properties of projective points, lines, and planes. Although the establishment of these properties required that we distinguish between Euclidean and ideal elements, it is clear from the above proofs and those of the preceding sections that this distinction is no longer necessary.

EXERCISES

1. Prove Desargues's theorem in each of the following cases:
(a) two corresponding vertices coincide;
(b) a vertex and the center of perspective coincide.

2. Show that two triangles are necessarily perspective from a point if two corresponding sides are the same, but that it is then meaningless to say that they are perspective from a line.

3. Prove that the property of being a figure consisting of two triangles perspective from a point is projective.

4. The statement of Desargues's Theorem for the Euclidean Plane given in VI, 9.2 was not proved. Show why we may now regard it as proved.

15. Other Perspective Figures. Desargues's theorem and its converse are important because of their usefulness in proving many theorems of projective geometry. In this section we apply them to perspective quadrangles and quadrilaterals. Other applications are given later.

We say that two complete quadrangles $ABCD$ and $A'B'C'D'$ are *perspective from a point* if their vertices can be paired (A with A', B with B', etc.) so that the lines determined by the four pairs are concurrent, and that they are *perspective from a line* if the intersections of corresponding sides are collinear. The following theorem then holds:

15.1 *If two complete quadrangles are perspective from a line, they are perspective from a point.*

Proof. Let the quadrangles be $ABCD$ and $A'B'C'D'$ (Fig. VIII, 14), with sides AB and $A'B'$, AC and $A'C'$, etc., meeting in points of line m. (This statement makes sense only if every two corresponding sides are distinct. We take this for granted.) Triangles ABC and $A'B'C'$ are then perspective from m. Hence lines AA', BB', CC' meet in a point P. Similarly, lines AA', BB', DD' meet in a point Q since triangles ABD and $A'B'D'$ are perspective from m. Lines AA' and BB' cannot coincide, for this would imply that lines AB and $A'B'$ coincide, which contradicts our agreement. Hence lines AA' and BB' meet in a point. Thus $P = Q$ and the proof is complete. Fig. VIII, 14 does not show P since the lines AA', BB', CC', DD' have approximately the same direction.

From familiarity with the triangle situation one might conjecture that the converse of 15.1 also holds. This is not so, as one can see with the aid of a diagram. The dual of 15.1, of course, does hold and is the following:

15.2 ***If two complete quadrilaterals are perspective from a point, they are also perspective from a line.***

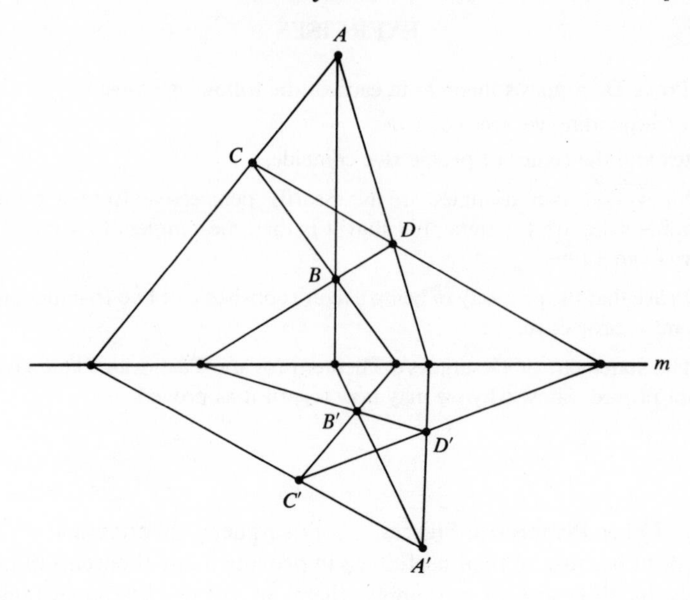

Fig. VIII, 14

The proof of this is left as an exercise. The converse of 15.2 is the dual of the converse of 15.1, and hence is not true according to the Principle of Duality. One can also see this with the aid of a diagram.

EXERCISES

1. Prove 15.2.

2. Show that the converses of 15.1 and 15.2 are duals.

3. Show by a diagram that two complete quadrangles can be perspective from a point without being perspective from a line.

4. Show by a diagram that two complete quadrilaterals can be perspective from a line without being perspective from a point.

5. Prove the theorem: *If five pairs of corresponding sides of two complete quadrangles meet in points of a line, then the sixth pair meet on this line and the quadrangles are perspective from it.*

6. (a) State the dual of the theorem of Ex. 5. (b) Prove it.

16. The Theorem of Pappus. This theorem was stated for the Euclidean plane in Chapter VI (see VI, 9.1), but not proved. Now, as in the case of

Desargues's theorem, we can state it more satisfactorily for the projective plane and give the proof. Some important preliminaries must be taken care of first.

We have seen that the projection, from a point O, of a line g on a coplanar line g_1 achieves a 1-1 mapping of the points of g on those of g_1. This mapping, besides being known as a projection, is also called a (one-dimensional) **perspectivity**, and O is the *center* of the perspectivity. The resultant of a finite sequence of one or more such perspectivities is called a (one-dimensional) projective transformation, or (one-dimensional) **projectivity**. Thus, if we project the above line g_1 on a line g_2, then g_2 on a line g_3, and so forth, ending up by projecting g_{n-1} on g_n, the resultant of these n perspectivities is a projectivity of g into g_n. This projectivity is clearly a 1-1 mapping of g on g_n. According to the preceding definitions, every perspectivity, of course, is a projectivity. A projectivity which maps g on g_n is often said to be a projectivity *between* the points of g and g_n.

If A, B, C are any points on a line g, and A', B', C' are any points on a line g', it is always possible to find a projectivity such that A, B, C go into A', B', C', respectively. To show this, suppose, first, that $g \neq g'$. If A and A' coincide, we project g on g' from O, the point of intersection of lines BB', CC' (Fig. VIII, 15a). This perspectivity, which is a projectivity, sends A, B, C into A', B', C'. The proof is similar if B and B' coincide, or C and C'. If $A \neq A'$, $B \neq B'$, $C \neq C'$, let g'' be a line through A' different from g and g' (Fig. VIII, 15b). Project g on g'' from O_1, which is any point on line AA' different from A and A'; then A, B, C go into A', B'', C''. Project g'' on g' from O_2, the point of intersection of lines $B'B''$, $C'C''$; then A', B'', C'' go into A', B', C'. Thus we have the two perspectivities

$$ABC \overset{O_1}{\underset{\wedge}{=}} A'B''C'' \overset{O_2}{\underset{\wedge}{=}} A'B'C',$$

and their resultant is a projectivity sending A, B, C into A', B', C'. If $g = g'$, we project g on a different line g'', so that A, B, C go into A'', B'', C''. Call this perspectivity T. Since $g' \neq g''$, there is a projectivity that sends g'' into g' so that

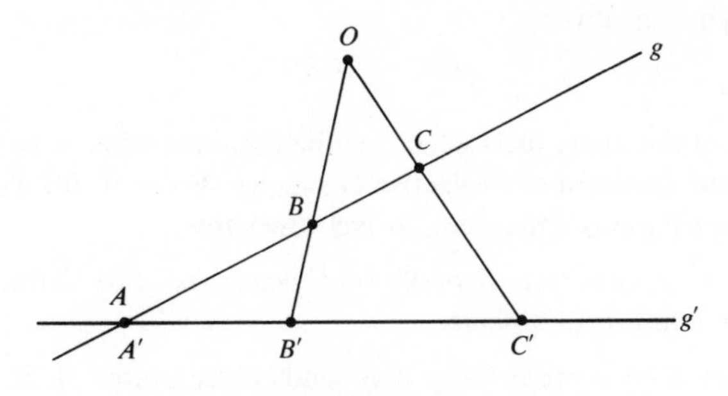

Fig. VIII, 15(a)

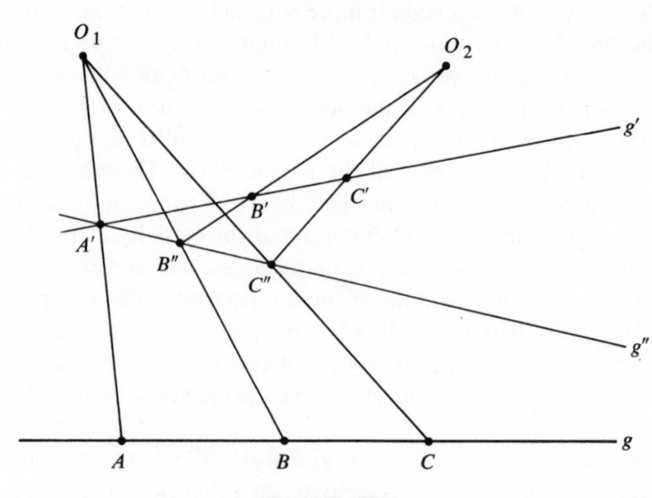

Fig. VIII, 15(b)

A'', B'', C'' go into A', B', C', as shown in the first part of the proof. The resultant of T and this projectivity is a projectivity that sends A, B, C into A', B', C'.

There cannot be more than one projectivity which sends A, B, C into A', B', C', respectively. For let $S(A, B, C) = T(A, B, C) = A', B', C'$, where S, T are projectivities. If D is an arbitrary point on g different from A, B, C, let $S(D) = D'$ and $T(D) = D''$. Also, let r denote the value of (AB, CD). Then, considering S, we have $r = (A'B', C'D')$; and considering T, we have $r = (A'B', C'D'')$. Hence $r = (A'B', C'D') = (A'B', C'D'')$. It follows from 9.15 that $D' = D''$. Thus S and T are the same projectivity.

We have now completely proved the following theorem:

16.1 *There is a unique projectivity that sends the points of one line into those of another (or the same) line so that three given points go into three given points, respectively.*

A projectivity in which points A, B, C go into points A', B', C', respectively, is expressed symbolically by

$$ABC \;\overline{\wedge}\; A'B'C'.$$

Because of the many facts which can be obtained with its aid, 16.1 is called the **Fundamental Theorem of Projective Geometry**. From it, for example, we can deduce 16.2 and Pappus's theorem, as we now show.

16.2 *A projectivity between the points of two distinct lines is a perspectivity if it has a fixed point.*

Proof. Let S be a projectivity that sends three points A, B, C of a line g, respectively, into three points A', B', C' of another line g', where $A = A'$. Thus the fixed point of S is the point of intersection of g and g' (Fig. VIII, 15a). Lines

BB' and CC' meet in a point O. If, with O as center, we project g on g', points A, B, C go into A', B', C', respectively. Denote this perspectivity by T. Since T is a projectivity, it must be identical with S by 16.1. Thus S is a perspectivity.

16.3 Theorem of Pappus. *If A, B, C are distinct points on a line and A', B', C' are distinct points on another line, the points of intersection of the lines AB' and $A'B$, BC' and $B'C$, AC' and $A'C$ are collinear.*

Proof. Let the points of intersection mentioned in the theorem be M, K, L, respectively (Fig. VIII, 16). Let lines ABC, $A'B'C'$ meet in O, lines AC', BA' in H, and lines BC', CA' in J. Projecting line $A'B$ on line $A'B'$ from A, and then line $A'B'$ on line BC' from C, we obtain

$$A'MHB \overset{A}{\underset{\wedge}{=}} A'B'C'O \overset{C}{\underset{\wedge}{=}} JKC'B.$$

The resultant of these two perspectivities is the projectivity

$$A'MHB \underset{\wedge}{-} JKC'B$$

of the points of line $A'B$ into those of line $C'B$. This projectivity has B as fixed point and hence is a perspectivity by 16.2. The lines joining corresponding points are therefore concurrent. Since lines $A'J$ and HC' meet in L, this must be the point of concurrence, that is, the center of the perspectivity. Hence line KM goes through L and the proof is complete.

The six given points in Pappus's theorem are the vertices of many simple hexagons. For the particular one, $AB'CA'BC'$, we note that the pairs of opposite

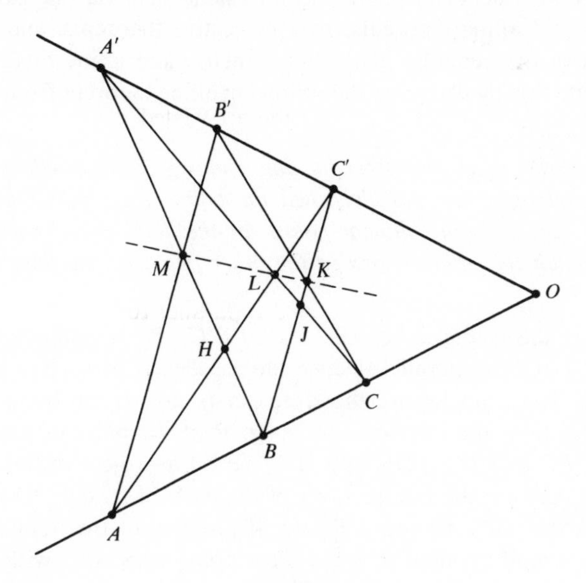

Fig. VIII, 16

sides are the pairs of lines mentioned in Pappus's theorem. Hence the latter may also be stated as follows:

16.4 Alternative Statement of Pappus's Theorem. *If the successive vertices of a simple hexagon lie alternately on two lines, the three pairs of opposite sides meet in collinear points.*

EXERCISES

1. (a) Show that the Theorem of Pappus holds if $A = A'$. (b) Show that the theorem holds if $A = B'$ provided that we define line AB' to mean any line through A.

2. State the dual of 16.3 and illustrate it in a diagram.

3. State the dual of 16.4 and illustrate it in a diagram.

4. We have seen that a perspectivity between the sets of points on two lines is a transformation of one set into the other in which the lines joining corresponding points are concurrent. (a) Dualize this so as to obtain the definition of a *perspectivity between two* (*coplanar*) *pencils of lines* (the line of intersection of corresponding lines of the pencils is called the *axis of the perspectivity*); (b) illustrate the perspectivity defined in (a) by a diagram; (c) define a *projectivity between two* (*coplanar*) *pencils of lines*; (d) state the dual of 16.1; (e) state and prove the dual of 16.2.

17. Connections with Euclidean Geometry. By taking advantage of the relation between a Euclidean plane and an extended plane one can sometimes deduce metric and affine theorems from projective theorems, and, conversely, obtain projective theorems by generalizing metric and affine theorems. Let us start to illustrate this by deducing the following affine theorem from Desargues's theorem:

17.1 (affine) *If the straight lines joining corresponding vertices of two Euclidean triangles are parallel, then the pairs of straight lines containing corresponding sides meet in collinear points, or the lines in one pair are parallel to the line through the intersections of the other pairs, or the lines in each pair are parallel.*

Proof. Let the triangles be ABC and $A'B'C'$. By hypothesis, the straight lines AA', BB', CC' are parallel. Hence the associated projective lines meet in an ideal point. The triangles are therefore perspective from this point and, by Desargues's theorem, the intersections P, Q, R of the pairs of projective lines AB and $A'B'$, BC and $B'C'$, AC and $A'C'$ lie on a projective line. If P, Q, R are Euclidean, they are the intersections of the pairs of straight lines mentioned in 17.1, that is, the pairs AB and $A'B'$, etc. If just two of the points P, Q, R are Euclidean, say P and Q, then R is the ideal point associated with the straight line PQ; hence straight lines AC, $A'C'$ are parallel to straight line PQ. The only remaining possibility is that P, Q, R are ideal, in which case the straight lines in each of the pairs mentioned in 17.1 are parallel. The drawing of diagrams to illustrate the three cases is left as an exercise.

If "parallel" is replaced by "concurrent" in the hypothesis of 17.1, the resulting statement is another affine theorem. We shall leave its proof and illustration as an exercise. Despite the similarity of the two theorems, the figures they deal with are essentially different in affine geometry, for the figure consisting of two triangles such that the lines joining corresponding vertices are parallel is not affinely equivalent to the figure consisting of two triangles such that the lines joining corresponding vertices are concurrent.

Analogous to 17.1 we have the following affine theorem which can be deduced from Pappus's theorem:

17.2 (affine) *If the successive vertices of a Euclidean hexagon lie alternately on two straight lines, the pairs of straight lines containing opposite sides meet in collinear points, or the lines in one pair are parallel to the line through the intersections of the other pairs, or the lines in each pair are parallel.*

Now, reversing the procedure just used, let us obtain a projective theorem by generalizing the affine theorem: *the medians of a triangle are concurrent.* To this end we recall from 9.2 that the midpoints of the sides of a Euclidean triangle are the harmonic conjugates, with respect to the vertices on their sides, of the ideal points on the projective lines containing the sides. Since these midpoints are noncollinear and the ideal points are collinear, we are led to the following two generalizations of the given theorem: (1) if three points, one on each side of a projective triangle, are collinear, the lines joining their harmonic conjugates, with respect to the vertices, to the opposite vertices are concurrent; (2) if the lines joining three noncollinear points, one on each side of a projective triangle, to the opposite vertices are concurrent, the harmonic conjugates of the three points with respect to the vertices are collinear. These statements, which are converses, are both correct, as we now prove.

17.3 (projective) *If three points, one on each side of a triangle, are collinear, the lines joining their harmonic conjugates, with respect to the vertices, to the opposite vertices are concurrent.*

Proof. Let ABC be the given triangle, and A', B', C' the collinear points (Fig. VIII, 17). If the specified harmonic conjugates A'', B'', C'' are carefully constructed, it will appear that they form a triangle whose sides meet the like-labeled sides of the given triangle in A', B', C'. If this is so, the lines AA'', BB'', CC'', joining corresponding vertices, are concurrent by the converse of Desargues's theorem, and the proof is complete. The suggestion is, indeed, correct. To show that lines BC and $B''C''$ meet in A' we project line AB on line AC, using A' as center, and obtain

$$ABC' \overset{A'}{\underset{\wedge}{=}} ACB'.$$

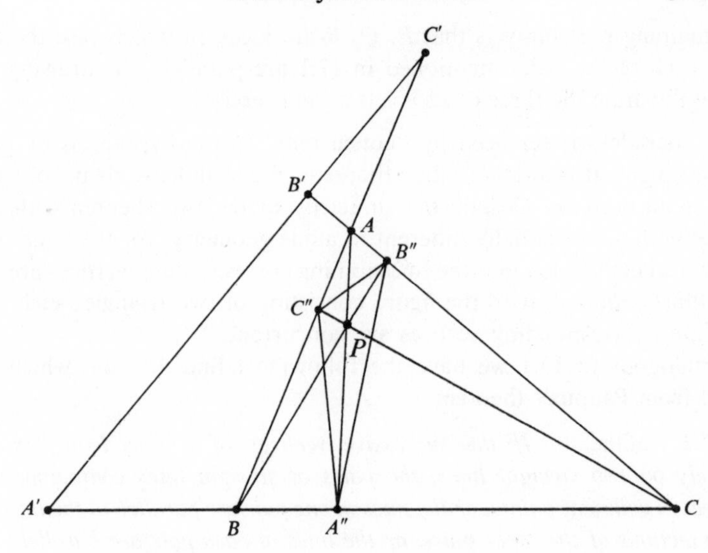

Fig. VIII, 17

Since $(AB, C'C'') = -1$, C'' projects into the harmonic conjugate of B' with respect to A, C, that is, into B''. Hence A', B'', C'' are collinear, that is, lines BC, $B''C''$ meet in A'. Similarly, by using projections from B' and C', respectively, we can show that lines AC, $A''C''$ meet in B', and lines AB, $A''B''$ in C'. (It is to be noted that the collinearity of A', B'', C'' allows us to infer that A'', B'', C'' are noncollinear, for in the contrary case A'' would coincide with A', which is impossible.)

17.4 (projective) *If the lines joining three noncollinear points, one on each side of a triangle, to the opposite vertices are concurrent, the harmonic conjugates of the three points, with respect to the vertices, are collinear.*

Proof. We again use Fig. VIII, 17, where now A'', B'', C'' are given and their harmonic conjugates A', B', C' are to be proved collinear. Since lines AA'', BB'', CC'' meet in a point P by hypothesis, this collinearity will follow if A', B', C' are shown to be the intersections of corresponding sides of triangles ABC and $A''B''C''$. To prove that $B''C''$ and BC meet in A', we note that two sides of the complete quadrangle $AB''PC''$ go through B, two sides go through C, and one side goes through A''. Since $(BC, A''A') = -1$, the remaining side must go through A' (see §13, Ex. 2). Similarly, by considering quadrangles $CA''PB''$ and $BA''PC''$, respectively, we can show that lines AB, $A''B''$ meet in C', and lines AC, $A''C''$ in B'.

If we now state 17.5, which is the dual of 17.3, leaving its proof as an exercise, we find that the metric theorem 17.6 can be deduced from it.

17.5 (projective) *If three lines, one through each vertex of a triangle,* *are concurrent, the three points in which their harmonic conjugates, with respect to the sides of the triangle, meet the opposite sides are collinear.*

To obtain the metric theorem let us consider the special case of 17.5 in which the triangle is Euclidean and the three lines are the bisectors of its angles. The harmonic conjugates in question are then the bisectors of the exterior angles of the triangle (see V, §12). It is easily verified that each such bisector will meet the opposite side (extended) only if the triangle is scalene. Hence we have:

17.6 (metric) *If a triangle is scalene, the bisectors of its exterior angles meet the opposite sides (extended) in three collinear points.*

Let us now return to 17.3, which was obtained by generalizing an affine theorem, and note that it is a special case of the following more general proposition:

17.7 (projective) *If A', B', C' are collinear points, one on each side of a triangle ABC, and A'', B'', C'' are three other points, one on each side of the triangle (A' and A'' on BC, B' and B'' on AC, etc.), then lines AA'', BB'', CC'' are concurrent if and only if*

$$(AB, C''C') \cdot (BC, A''A') \cdot (CA, B''B') = -1.$$

To deduce 17.3 from 17.7 we choose A'' to be the harmonic conjugate of A' with respect to B and C, B'' the harmonic conjugate of B' with respect to A and C, and so forth. The above equation is then satisfied, for each of the specified cross-ratios has the value -1. Hence the lines AA'', BB'', CC'' are concurrent.

We shall not prove 17.7.*

EXERCISES

1. Illustrate by diagrams the various conclusions in 17.1.

2. State and illustrate by a diagram the affine theorem resulting when "parallel" is replaced by "concurrent" in the hypothesis of 17.1.

3. Prove 17.2 and illustrate it with a diagram.

4. Prove 17.5 by using 17.4, and hence show that the converse and dual of 17.3 are essentially the same.

5. Deduce from 17.5 a metric theorem concerning the bisectors of the exterior angles of (a) an isosceles triangle, (b) an equilateral triangle.

6. Show that the affine Theorem of Ceva (IV, 7.2) can be deduced from 17.7. (*Hint*. Let A', B', C' be ideal points.)

7. Show that 17.4 is a special case of the following projective theorem: *If A', B', C' are collinear points, one on each side of a triangle ABC, and A'', B'', C'' are three*

---

* See W. C. Graustein, *Introduction to Higher Geometry*, p. 80.

other points, one on each side of the triangle (A' and A'' on BC, B' and B'' on AC, etc.), then A'', B'', C'' are collinear if and only if

$$(AB, C''C') \cdot (BC, A''A') \cdot (CA, B''B') = 1.$$

8. Deduce the affine Theorem of Menelaus (IV, 7.1) from the projective theorem of Ex. 7.

18. Projective Conics. In our discussions of the figures which are of interest in the geometry of the projective plane nothing has thus far been said about the conic sections. When studying the projections of one Euclidean plane α on another, α', in Chapter V, we saw that each conic c goes into a conic c', but that there could be exceptional points. One or two points of c, for example, might have no images, and one or two of c' no originals, these exceptional points being the intersections of c and c' with the vanishing lines in α and α', respectively. With the addition of ideal points to each plane and the broadening of projections to include them, these exceptional points cease to be exceptional, as we know. Thus, if c has one exceptional point V, then among the ideal points added to α' there is one, V', which is its image. Since it is reasonable to regard V' as added to c', we are led to the concept of an *extended conic* $\bar{c}'$, consisting of c' and V'. Similarly, if c' has two exceptional points W', X', there are two ideal points W, X among those added to α which are the originals of W', X'. The extended conic $\bar{c}$ then consists of c, W, and X. Instead of the imperfect projection of c on c' we now have a projection of $\bar{c}$ on $\bar{c}'$ which holds without exception.

In putting the preceding discussion on a more precise basis it will be helpful if we make use of the circular models of a Euclidean plane and the associated extended plane. In each of these models, as we saw in the preceding chapter (VII, §10), the curve k corresponding to a parabola q is a simple arc lacking one point of being closed, namely, a certain point of the boundary γ (Fig. VII, 19). This point, which we have denoted by A, is meaningless in the Euclidean model since it corresponds to nothing in the Euclidean plane, but in the model of the extended plane it corresponds to the ideal point I associated with the axis of the parabola, as is clear from VII, §10. Let $\bar{k}$ denote the locus in the model of the extended plane consisting of k and A, and $\bar{q}$ the corresponding locus in the extended plane. (An extended plane and its model, we again emphasize, are not the same thing.) Thus, $\bar{q}$ consists of q and I. We call $\bar{q}$ an *extended parabola*, or, more exactly, *the extended parabola associated with* q. Since $\bar{k}$ is a simple closed curve,* the same is true of $\bar{q}$ because of previous agreements concerning the nature of loci in an extended plane (see VIII, §10). Thus we have the following definition and theorem:

*** We offer no separate diagram of $\bar{k}$ since $\bar{k}$ differs from the curve of Fig. VII, 19 only in containing A.**

18.1 Definition. *An extended parabola is a locus that consists of a parabola and the ideal point associated with its axis.*

18.2 *An extended parabola is a simple closed curve.*

We follow a similar procedure in extending hyperbolas. As we saw in VII, §10, the curve c in the Euclidean model corresponding to a hyperbola h comes arbitrarily close to four points of γ and consists of two nonintersecting simple arcs; each arc comes arbitrarily close to a pair of these points, the points in one pair being diametrically opposite those in the other. This is illustrated in Fig. VII, 20, where the pairs are A, B and A', B'. These points are meaningless in this model since they correspond to nothing in the Euclidean plane; but in the model of the extended plane, where they constitute only two points since $A = A'$, $B = B'$, they correspond to the ideal points I_1, I_2 associated with the asymptotes of h (VII, §10). Let $\bar{c}$ denote the locus in the model of the extended plane consisting of c, A, B, and hence also A', B', and let $\bar{h}$ be the corresponding locus in the extended plane, that is, $\bar{h}$ consists of h, I_1, I_2. We call $\bar{h}$ an *extended hyperbola*, or, more exactly, *the extended hyperbola associated with h.* Careful examination of $\bar{c}$ (Fig. VIII, 18) shows it to be a simple closed curve. Hence, according to the agreement made in §10, the same is true of $\bar{h}$. Thus we have the following definition and theorem:

18.3 Definition. *An extended hyperbola is a locus that consists of a hyperbola and the two ideal points associated with its asymptotes.*

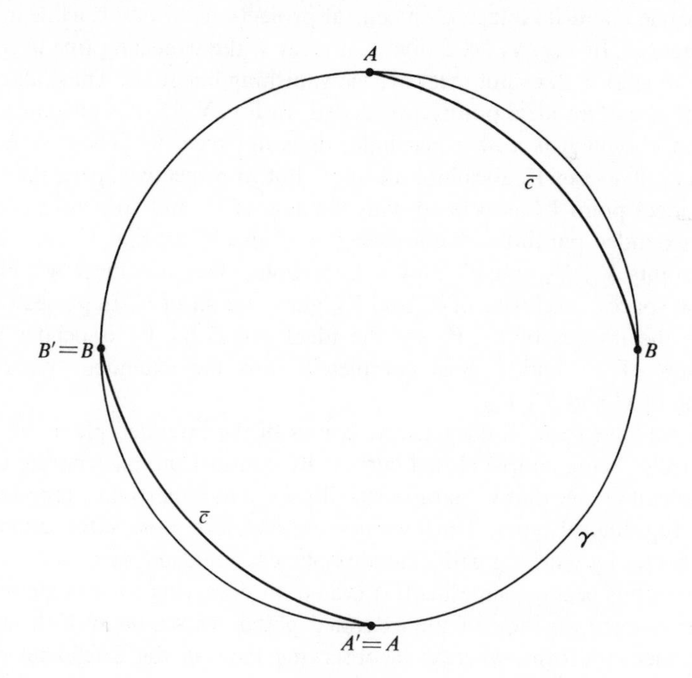

Fig. VIII, 18

18.4 *An extended hyperbola is a simple closed curve.*

Extended hyperbolas and extended parabolas, like extended lines, are abstractions and cannot be visualized directly. Some of their properties, however, can be illustrated by models such as the one we have been using. Actually, the same is true of all mathematical concepts. We never really *see* a Euclidean parabola, for example, except in terms of a physical model, such as a pencil drawing.

Turning to ellipses, we recall that the curves in the Euclidean model which correspond to them are closed and do not come arbitrarily close to any point of γ. This suggests that no ideal points need be added to ellipses in order to perfect any projections involving them. To check this, let c be an ellipse in a Euclidean plane α, and c' its image, by projection, in a Euclidean plane α'. Assume V' is a point of c' which has no original on c. Then V' is on the vanishing line in α'. As a variable point of c' approaches V', its original on c moves off indefinitely far. This is impossible since c is of limited extent. Hence every point of c' has an original, and no ideal point need be added to c to perfect the projection. Thus, *there are no extended ellipses.*

We shall call extended parabolas, extended hyperbolas, and ordinary ellipses **nondegenerate conics of the extended plane**. As was suggested by the discussion at the beginning of this section, these conics always go completely into one another in projections involving extended planes. We can now illustrate this in a precise way by returning to Chapter V, §7, where Figs. V, 14–16 show a circle c and its image c' in central projections of one Euclidean plane α on another, α'. In Fig. V, 14, c does not meet v, the vanishing line in α, and c', which is an ellipse, does not meet w', the vanishing line in α'. Thus, all of c goes into all of c' and no ideal points are needed. In Fig. V, 15, c is tangent to v at V, and again c', which is now a parabola, does not meet w'. Thus, in Euclidean space, c, exclusive of V, goes into all of c'. But in projective space the image of V is the ideal point V' associated with the axis of c', and so c goes completely into the extended parabola $\bar{c}'$, consisting of c' and V'. In Fig. V, 16, c crosses v at two points V_1, V_2 and c', now a hyperbola, does not meet w'. Hence, in Euclidean space c, exclusive of V_1 and V_2, goes into all of c'. In projective space, however, the images of V_1, V_2 are the ideal points V_1', V_2' associated with the asymptotes of c', and c goes completely into the extended hyperbola $\bar{c}'$, consisting of c' and V_1', V_2'.

As we have seen, nondegenerate conics of the extended plane are all alike topologically, being simple closed curves. By contrast, nondegenerate conics of the Euclidean plane, that is, parabolas, ellipses, and hyperbolas, represent three different topological types. Thus, we not only achieve more satisfactory projections of conics by working with extended planes, but also find that the nature of these curves has become simplified. It is as if, in observing nondegenerate conics from the vantage position of the extended plane, we see them in their entirety and find them uniform, whereas in observing them in the Euclidean plane we get only partial and differing views.

A degenerate conic in the Euclidean plane is either a single point or a pair of straight lines (which may coincide). Similarly, we define a **degenerate conic in an extended plane** to be either a single point or a pair of projective lines (which may coincide). Such conics clearly go into one another in projections involving extended planes.

Conics in the ideal plane can also be defined, but we shall not go into this. Conics in the ideal plane and in extended planes are known collectively as **conics of the projective plane**, or **projective conics**, or, when there is no possibility of confusion, simply **conics**. In all our discussions, however, these terms will mean only conics in extended planes, and the projective planes in those discussions will be understood to be extended planes.

We can state the following in summary of our discussion:

18.5 *The property of being a degenerate or a nondegenerate conic of the projective plane is projective, that is, each such conic goes into another in projections involving such planes. Nondegenerate projective conics are simple closed curves.*

It is shown in analytic geometry that there is a unique Euclidean conic passing through five coplanar points, no three of which are collinear. The same is true of projective conics, but we shall not prove this.

18.6 *Through five coplanar points, no three of which are collinear, there passes a unique projective conic and it is nondegenerate.*

In Euclidean geometry it is customary to define a tangent to a curve by means of a limiting process that involves metric concepts. When this process is applied to parabolas, ellipses, and hyperbolas, it turns out that (1) these curves have a unique tangent at each of their points, and (2) each tangent meets the curve to which it is tangent in no point other than the point of tangency. Only in the case of the ellipse is it true, conversely, that any line meeting the ellipse in just one point is a tangent. The axis of a parabola and each line parallel to it meet the parabola in just one point, and each line parallel to an asymptote of a hyperbola meets the hyperbola in just one point, but none of these lines is a tangent.

Let us now try to define a tangent to a nondegenerate projective conic. If g is the tangent to a parabola, ellipse, or hyperbola at a point P, we shall want to regard the extended line $\bar{g}$ as being tangent to the associated projective conic at P. This would take care of the Euclidean points of these conics. Since an extended parabola is represented in the circular model of an extended plane by a curve tangent to γ, we shall wish to regard the ideal line as the tangent to this conic at its ideal point. In further support of this view is the fact that the ideal line meets this conic in but one point. An ordinary graph of a hyperbola and its asymptotes shows how satisfying it would be, intuitively, if we could say that the extended asymptotes of the extended hyperbola are tangent to it at its ideal points.

All of the above can be achieved if we adopt the following definition, whose nonmetric character is to be noted:

18.7 Definition. *A projective line is said to be* **tangent** *to a nondegenerate projective conic (in the same plane) if and only if it meets the conic in just one point.*

Using this definition we can then prove:

18.8 *There is a unique tangent to a nondegenerate projective conic at each of its points.*

The details are left as an exercise. We shall likewise leave the following to be verified:

18.9 *The property of being tangent to a nondegenerate projective conic is projective.*

We study projective conics algebraically in the next chapter, and in Chapter X give considerable attention to them in a purely geometrical way, proving among other theorems those of Pascal and Brianchon. Now we wish to do no more with projective conics than to state these two theorems so that the reader may note their simplicity and generality when compared with the same-named theorems for the Euclidean plane (VI, §9).

18.10 Theorem of Pascal. *If a simple hexagon is inscribed in a nondegenerate projective conic, the three intersections of the pairs of opposite sides are collinear.*

18.11 Theorem of Brianchon. *If a simple hexagon is circumscribed about a nondegenerate projective conic, the three lines joining the pairs of opposite vertices are concurrent.*

The term *inscribed* here means, as usual, that the vertices of the hexagon are points of the conic, and *circumscribed* means that the sides of the hexagon are tangent to the conic.

In our statement of these theorems for the Euclidean plane in Chapter VI we avoided the term *hexagon* since the property of being a hexagon is not projective for that plane, that is, hexagons do not always project into hexagons. This made for much awkwardness in those statements. The latter were also marred by their provisional character: there was no certainty that the specified intersections exist. In the projective plane, however, these intersections always exist, and the property of being a hexagon is projective, so that much improved statements of the theorems are possible.

EXERCISES

1. We saw that certain straight lines meet a parabola in just one point without being tangent. Are the associated extended lines tangent to the associated extended parabola? Justify your answer.

2. Solve Ex. 1, with "parabola" replaced by "hyperbola."

3. Prove: (a) the ideal line is tangent to each extended parabola; (b) the extended asymptotes, which are the extended lines associated with the asymptotes of a hyperbola, are tangent to the associated extended hyperbola.

4. Prove 18.8.

5. Prove 18.9.

6. Show that a projective line and a coplanar nondegenerate projective conic meet in at most two points. (Use the fact that a straight line and a nondegenerate conic of the same Euclidean plane meet in at most two points.)

7. In Pascal's theorem for the Euclidean plane (VI, 9.3), one or more of the specified intersections might not exist because of parallelism. Show that the following more precise statements concerning these exceptional cases can be deduced from 18.10: (a) if just two intersections exist, the line joining them is parallel to the lines in the third pair; (b) it is not possible for there to be just one intersection, that is, if the lines in each of two of the specified pairs are parallel, so are the lines of the third pair.

19. Transformations of a Projective Plane Into Itself. The many things done in this chapter have had as their objective the creation of a more satisfactory system of two-dimensional projective geometry than was possible in the Euclidean plane. The new system is called **projective geometry of the projective plane**. In view of the procedure we followed in obtaining it, which involved a considerable use of three-dimensional space, one could describe this system as being concerned with those properties of geometric figures which are preserved in projections of one projective plane on another, such as collinearity, concurrence, cross-ratio, etc.

There is another way of describing projective geometry of the projective plane which makes no use of three-dimensional space and hence is more in keeping with the method used in earlier chapters to characterize the various geometries of the Euclidean plane. It was by means of groups of transformations of this plane into itself that we characterized these geometries, and we can now follow an analogous procedure in obtaining the alternative description of projective geometry of the projective plane. First, we make the following definition:

19.1 Definition. *By a projective transformation of, or a projectivity in, a projective plane, we mean a* 1-1 *mapping of that plane onto itself which preserves collinearity and the cross-ratio of points.**

This is analogous to the definition of a projective transformation of a

* Ordinarily, the term "projective transformation of a projective plane" has a broader meaning than this, and includes also certain transformations of a projective plane into itself which send points into *lines*. Projectivities of the latter type are known as *correlations*, and those described in 19.1 are called *collineations*.

Euclidean plane given in Chapter VI, but is simpler since there are no vanishing points to contend with.

A projective transformation of a projective plane necessarily preserves *noncollinearity*. To prove this, let us assume that noncollinear points A, B, C go into collinear points A', B', C'. Since collinearity is preserved, it then follows that all points of lines AB, AC, BC go into points on line $A'B'C'$. If P is a point not on any of the lines AB, AC, BC, it is collinear with at least two points on these lines and hence is sent into a point of line $A'B'C'$. Thus the plane goes into this line, which contradicts that the plane goes into itself.

The following can now be proved. The details are left to the student.

19.2 *The set of all projectivities in a projective plane is a group.* (We call this group the **projective group** in that plane.)

It has not yet been shown that the transformations we have been discussing actually exist. This is easily done. Let α be any projective plane. Project it on another such plane; then project the latter back on α. The resultant of these projections is a mapping of α onto itself which has the same properties as projections and hence, in particular, is 1-1, preserves collinearity, and preserves the cross-ratio of points. Instead of using just two projections, we can use any number, provided that we always end up with a projection on the same plane α with which we start.

Thus, projectivities in any projective plane do exist. In fact, it can be shown that there is always at least one such transformation which sends four given points, no three of which are collinear, into four given points, no three of which are collinear.* Using this fact we shall now prove:

19.3 *There is a unique projectivity in a projective plane that sends four given points into four given points, respectively, if no three points in either set are collinear.*

Proof. Let A, B, C, D and A', B', C', D' be two such sets of points in a projective plane α. Then, by the above-mentioned fact, there exists a projectivity, S, in α such that $S(A, B, C, D) = A', B', C', D'$. Suppose that T is also a projectivity in α with this same property. We shall prove that $S = T$. Denote the intersection of lines AB, CD by E, and that of lines $A'B'$, $C'D'$ by E' (Fig. VIII, 19). Since S, T both send lines AB, CD into $A'B'$, $C'D'$, they must both send E into E'. Let P be any point on line AB different from A, B, E. If r is the value of (AB, EP), there is a unique point P' on line $A'B'$ such that $(A'B', E'P') = r$ (9.15). Since S, T both send A, B, E into A', B', E' and preserve cross-ratio, they must both send P into P'. Thus S and T are identical as far as the points of line AB are concerned, and the same can be shown for the points of lines AC and BC. Finally, let Q be a point not on any of the lines AB, AC, BC. A line g through Q can be chosen which meets the foregoing lines in three points P_1, P_2, P_3. It then follows from what was shown above that S and T send P_1, P_2, P_3 into the

---

* See O. Veblen and J. W. Young, *Projective Geometry*, vol. 1, §28, Theorem 10.

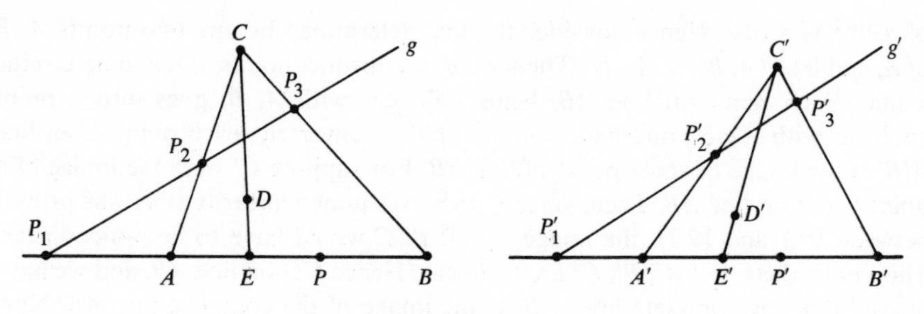

Fig. VIII, 19

same collinear points P_1', P_2', P_3'. Since cross-ratio is preserved, they must therefore both send Q into the same point Q'. Thus $S = T$ throughout α.

Prior to stating 19.3 we showed how projectivities in a projective plane can be generated by means of projections. The following converse of this is also true, although we shall not prove it:*

19.4 *Any given projectivity in a projective plane can be obtained as the resultant of a sequence of projections of one projective plane on another.*

In the light of 19.4 we see that projectivities in a projective plane not only possess the properties stated in Definition 19.1, but also all the other properties preserved by projections.

19.5 *The properties preserved by the projectivities in a projective plane are the same as the properties preserved in projections of one projective plane on another.*

Therefore, instead of characterizing two-dimensional projective geometry as being concerned with the properties preserved by all projections of one projective plane on another and calling those properties projective, which is what we have been doing, we could as well say that it is concerned with the properties preserved by all projectivities in a *single* projective plane and call those properties projective. This description, which makes no use of three-dimensional space, is the alternative description we were seeking.

It might seem, because of 19.4, that this description is a mere verbal achievement and that projections of one plane on another are really indispensable in developing two-dimensional projective geometry. This is not the case. Two-dimensional projective geometry can be completely developed by working in a single projective plane and using, when necessary, the group of projectivities in that plane.†

To illustrate this let us prove that any projectivity T in a projective plane α preserves the concurrence of lines. To do this we must first show that the image

---

* A proof can be obtained by combining 19.3 with the discussion given in §28 of *Projective Geometry*, vol. 1, by O. Veblen and J. W. Young.

† This is done, for example, in H. S. M. Coxeter, *The Real Projective Plane.*

of a line is a line. Hence consider the line determined by any two points A, B of α, and let $T(A, B) = A', B'$. Then A', B' are distinct points of α and determine a line. Each point on line AB, being collinear with A, B, goes into a point collinear with A', B', and hence on line $A'B'$. Conversely, each point C' of line $A'B'$ is the image of some point of line AB. For suppose C' were the image of a point C not on line AB. Then, since T preserves noncollinearity (this was proved between 19.1 and 19.2), the images of A, B, C would have to be noncollinear. This contradicts that A', B', C' are collinear. Hence C is on line AB, and we have proved that the complete line $A'B'$ is the image of the complete line AB. Now let a, b be two lines of α and let O be their common point. If $T(a, b) = a', b'$, then a', b' are lines of α, and their common point O' is necessarily the image of O. Similarly, any third line through O must go into a line through the image of O. Hence, the concurrent lines a, b, c go into the concurrent lines a', b', c'.

What we have just proved without using projections was established earlier in the chapter by means of projections. Our discussion of projective conics in §18, also, was largely based on projections. Later, in Chapters IX and X, we discuss conics again, without using projections at all.

19.6 Definition. *Two geometric figures in a projective plane are called projectively equivalent if there is a projectivity in the plane which maps one figure on the other.*

As an immediate consequence of 19.3, we then have:

19.7 *Any two lines, triangles, simple quadrilaterals, complete quadrilaterals, or complete quadrangles in the same plane are projectively equivalent.*

In the following chapter we shall also see that *any two nondegenerate projective conics in the same plane are projectively equivalent* (IX, 8.3).

Let us close this section by noting that not all 1-1 mappings of a projective plane onto itself are projectivities. To show this let us recall from our study of topological transformations that $x' = x$, $y' = y^3$ is a 1-1 mapping of a Euclidean plane onto itself which does not preserve collinearity (VII, §3). Let α be the extended plane associated with this Euclidean plane. Then the mapping of α onto itself, in which each ideal point remains fixed and each Euclidean point (x, y) goes into the Euclidean point (x', y') in accordance with the above equations, is a 1-1 mapping of a projective plane onto itself which does not preserve collinearity, and hence is not a projectivity.

EXERCISES

1. Prove 19.2.

2. Without using projections, prove that a projectivity in a projective plane
(a) establishes a 1-1 correspondence among the lines of the plane,
(b) preserves the cross-ratio of lines,

(c) sends triangles, complete quadrangles, and complete quadrilaterals into figures of the same type, respectively,
(d) sends two triangles which are perspective from a point into a figure of the same kind.

3. Show that there is a unique projectivity in a projective plane which leaves fixed each of four points, no three collinear, and that it is the identity.

4. Given the projectivity in an extended plane such that $(1, 1) \to (-1, 1)$, $(-1, 1) \to (-1, -1)$, $(-1, -1) \to (1, -1)$, $(1, -1) \to (1, 1)$, where the arrow means "goes into," show that it leaves the origin fixed and sends the ideal point associated with the line $y = x$ into the ideal point associated with the line $y = -x$.

5. Prove 19.7.

20. Other Methods of Developing Projective Geometry. In the method we have been using to develop projective geometry the subject appears very closely related to Euclidean geometry, for in that method a projective plane can be obtained by extending a Euclidean plane. Besides this method, whose use brought projective geometry to a high state of development historically, there are two others, in which the subject is developed independently of Euclidean geometry. In one of these, projective geometry is based on axioms of its own, which, as might be expected, differ greatly from those of Euclidean geometry. There are, for example, no axioms on congruence, betweenness, or parallelism. In the other method projective geometry is developed algebraically. A point, for example, is defined as an ordered number-triple, and a projective line as the set of all such triples that satisfy a first-degree equation in three unknowns. For discussions of these important methods the student can consult the references at the end of the book.

CHAPTER IX

Analytic Projective Geometry

1. Introduction. In an earlier chapter we obtained the equations of the projective transformations of the Euclidean plane (VI, §4). Although these equations still apply to the Euclidean points of the associated projective plane, they do not, of course, apply to the ideal points, for the variables in the equations represent rectangular coordinates and only Euclidean points have such coordinates. Since it is reasonable to want to have equations representing the projective transformations of the *entire* projective plane, our first task is to assign coordinates of some sort to its ideal points. After doing this suitably we are able not only to obtain the desired equations, but also to develop the projective geometry of this plane from an analytic, or algebraic, viewpoint. However, we do not go so far with this development as to duplicate things done in Chapter VIII. Algebraic proofs of the Theorems of Desargues and Pappus, for example, are not given.

As the foregoing remarks indicate, the entire discussion is confined to the figures in a single extended plane, which is referred to as "the plane" or "the projective plane." Except where the contrary is indicated, "line" means "projective line."

2. Homogeneous Coordinates of Points. Rather than retain the rectangular coordinates (x, y) for Euclidean points and use a different type for ideal points, we introduce a new system of coordinates applicable to all the points of the plane. These coordinates are called *homogeneous coordinates.* The rectangular coordinates (x, y) are often called *nonhomogeneous coordinates.* One novelty of the new system, as we shall see, is that each point has three coordinates instead of the usual two. We first define the new coordinates for Euclidean points and later extend them to ideal points.

2.1 Definition. *Three numbers x_1, x_2, x_3, in this order, are called homogeneous coordinates of the Eucliden point (x, y) if and only if*

$$\frac{x_1}{x_3} = x \qquad \text{and} \qquad \frac{x_2}{x_3} = y.$$

We write these homogeneous coordinates in the form (x_1, x_2, x_3).

Thus (6, 8, 2) are homogeneous coordinates of the point (3, 4), as are (9, 12, 3), (3, 4, 1), and $(3k, 4k, k)$, where k is any number not zero. As this example suggests, every Euclidean point has infinitely many sets of homogeneous coordinates, any two sets being proportional, and x_3 in each set being different from zero. If the point is (a, b), then one set of homogeneous coordinates is $(a, b, 1)$, the other sets being (ka, kb, k), where k is any number not zero. Conversely, any three given numbers x_1, x_2, x_3 taken in this order are homogeneous coordinates of a Euclidean point if $x_3 \neq 0$.

Number triples (x_1, x_2, x_3) in which $x_3 = 0$ are still at our disposal and we shall assign them to the ideal points. Thus, (1, 2, 0) and (−3, 4, 0), for example, will represent ideal points. Since we do not yet know on which lines these points are situated, we must make the assignment of coordinates to ideal points more precise.

Consider, for example, the family of straight lines with slope 2. The equation of the family is $y = 2x + b$, where b is arbitrary. The lines of this family, being parallel, are all associated with the same ideal point I. To decide what coordinates I is to have, we shall consider a Euclidean point P on a line of the family and let it recede indefinitely far along the line (Fig. IX, 1a). The point P' in the circular model corresponding to P will then approach the point I' on γ corresponding to I (Fig. IX, lb). If under these circumstances the coordinates of P approach definite limits, the third coordinate in particular approaching zero, it is reasonable to take these limits as the coordi-

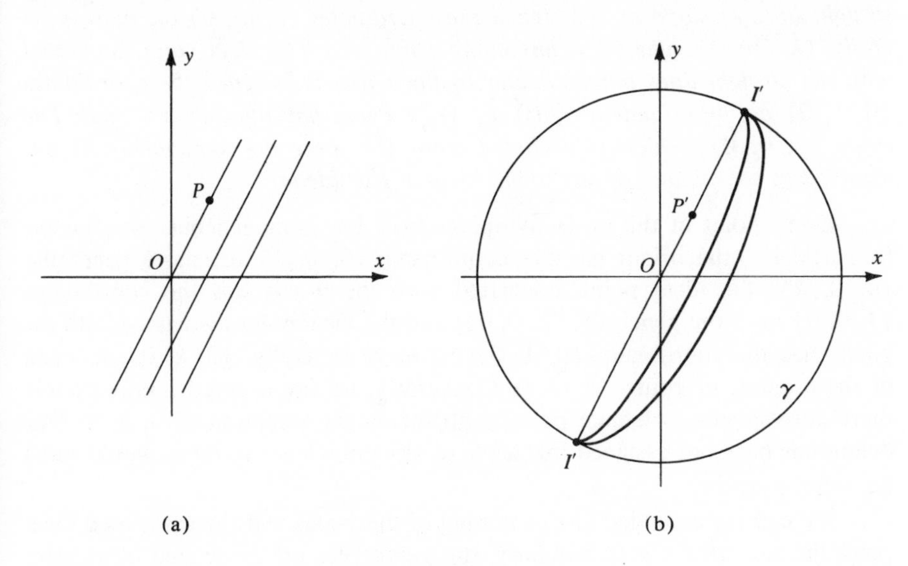

Fig. IX, 1

nates of I. If (ξ, η) are the rectangular coordinates of P, then $(\xi, \eta, 1)$ are homogeneous coordinates of P, and so are $(1, \eta/\xi, 1/\xi)$ if $\xi \neq 0$, that is, if P is not on the y-axis. Since $\eta = 2\xi + b$, these last coordinates can be written

$$(1, \frac{2\xi + b}{\xi}, \frac{1}{\xi}), \quad \text{or} \quad (1, 2 + \frac{b}{\xi}, \frac{1}{\xi}).$$

When P recedes without bound in either direction along the line, $|\xi|$ becomes arbitrarily great and b remains constant. Hence b/ξ and $1/\xi$ approach zero, and the homogeneous coordinates of P approach 1, 2, 0, respectively. It is therefore reasonable to take these numbers as coordinates for I.

Had we dealt with the family of slope 3 we would have been led to $(1, 3, 0)$ as coordinates for the ideal point associated with this family, and, more generally, we would have been led to $(1, m, 0)$ as coordinates of the ideal point associated with the family of slope m. One family remains to be considered, that of lines perpendicular to the x-axis, for these lines have no slope. By a procedure like that used above we would be led to $(0, 1, 0)$ as coordinates for the ideal point associated with this family. The justification of this is left as an exercise. Finally, we want every ideal point, like every Euclidean point, to have infinitely many sets of coordinates, each two sets being proportional. Thus, $(1, 2, 0)$, $(2, 4, 0)$, and $(k, 2k, 0)$, where k is any number not zero, should represent the same ideal point.

Guided by the preceding results, we now make the following definition:

2.2 Definition. *The ideal point associated with the family of straight lines of slope m is assigned the coordinates $(1, m, 0)$ or, more generally, $(k, km, 0)$, where k is any number not zero. The ideal point associated with the straight lines perpendicular to the x-axis is assigned the coordinates $(0, 1, 0)$ or, more generally, $(0, k, 0)$, k being any number not zero. The triple $(0, 0, 0)$, which is excluded from the foregoing agreements, is not assigned as coordinates of any point, ideal or Euclidean.*

Every point in the projective plane now has homogeneous coordinates. In particular, the origin has the coordinates $(0, 0, 1)$ or, more generally, $(0, 0, k)$; the ideal point associated with the x-axis has the coordinates $(1, 0, 0)$ or, more generally, $(k, 0, 0)$; and the ideal point associated with the y-axis has the coordinates $(0, 1, 0)$ or, more generally, $(0, k, 0)$. In each of these cases, of course, $k \neq 0$. Conversely, to any ordered number triple there corresponds a definite point, with the single exception of 0, 0, 0. Our definitions have not assigned this triple to any point since to do so would serve no useful purpose.

We call the extended line consisting of the x-axis and the associated ideal point the *extended x-axis*. Similarly, the y-axis plus the associated ideal point will be called the *extended y-axis*. These two ideal points are exhibited in

Fig. IX, 2, which utilizes the circular model of the projective plane. Also shown are the ideal points associated with the lines $y = \pm x$.*

It should be clear from the discussion that two given triples (a_1, a_2, a_3), (b_1, b_2, b_3)† represent the same point only if they are in proportion, that is, only if the numbers in each triple equal those in the other multiplied by a

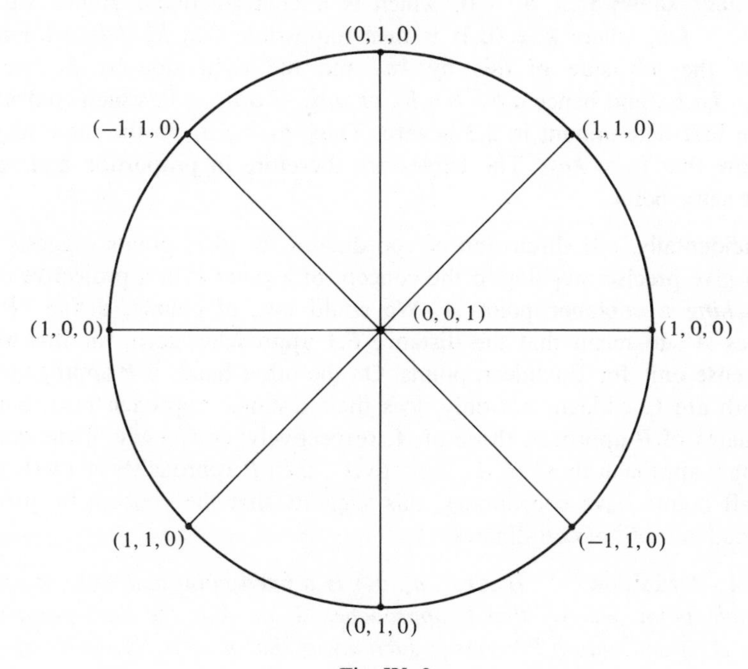

Fig. IX, 2

nonzero constant. (This is not exactly the same as saying that $a_1/b_1 = a_2/b_2 = a_3/b_3$.) Another useful way of expressing this proportionality is the following:

2.3 *Two triples (a_1, a_2, a_3), (b_1, b_2, b_3) represent the same point if and only if the determinants*

$$\begin{vmatrix} a_1 & a_2 \\ b_1 & b_2 \end{vmatrix}, \quad \begin{vmatrix} a_2 & a_3 \\ b_2 & b_3 \end{vmatrix}, \quad \begin{vmatrix} a_3 & a_1 \\ b_3 & b_1 \end{vmatrix}$$

are all zero.

---

* Strictly, the model does not show these ideal points and the extended axes, but only their representatives or images. As is customary in map-making, however, we label the representatives with the names of the originals.

† We shall always assume that a given triple expressed in letters is not (0, 0, 0).

Proof. Suppose, first, that the triples represent the same point. They are then proportional, so that $b_1 = ka_1$, $b_2 = ka_2$, $b_3 = ka_3$, where $k \neq 0$. Substituting these numbers for b_1, b_2, b_3 in the above determinants, we immediately find that the latter are all zero. Conversely, suppose all the determinants are zero. Then, a_1, b_1 are both zero or both nonzero. (For suppose $a_1 = 0$, $b_1 \neq 0$. The first detriminant then shows that $a_2 = 0$. Since $a_3 \neq 0$, the third determinant shows that $b_1 = 0$, which is a contradiction.) Hence we can write $b_1 = ka_1$, where $k \neq 0$. It is then impossible that $b_2 \neq ka_2$, for if we multiply the left side of this by ka_1 and the right side by b_1, we get $ka_1b_2 \neq ka_2b_1$, and hence $a_1b_2 \neq a_2b_1$, or $a_1b_2 - a_2b_1 \neq 0$, which contradicts that the first determinant in 2.3 is zero. Thus, $b_2 = ka_2$. In the same way we can show that $b_3 = ka_3$. The triples are therefore in proportion and represent the same point.

Incidentally, our discussion of coordinates for ideal points suggests how we can give precise meaning to the concept of a point P in a projective plane *approaching* a coplanar point A. We could not, of course, define "P approaches A" to mean that the distance PA approaches zero, for this would make sense only for Euclidean points. On the other hand, if P approaches A, and both are Euclidean, not only does their distance approach zero, but the coordinates of P approach those of A, respectively; conversely, if the coordinates of P approach those of A, respectively, then P approaches A (VII, § 1). Since all points have coordinates, this suggests that the concept in question be defined in terms of coordinates.

2.4 Definition. *If $A(a_1, a_2, a_3)$ is a fixed point and $P(x_1, x_2, x_3)$ is a variable point, we say that P approaches A, or A is the limit point of P, or $P \to A$, if and only if P varies in such a way that $x_1 \to a_1$, $x_2 \to a_2$, $x_3 \to a_3$.*

Returning to our earlier discussion, we can now say the ideal point $I(1, 2, 0)$ is approached by the variable Euclidean point $P(1, 2 + b/\xi, 1/\xi)$ as P varies so that ξ becomes infinite. In this application of 2.4 we have $a_1 = 1$, $a_2 = 2$, $a_3 = 0$ and $x_1 = 1$, $x_2 = 2 + b/\xi$, $x_3 = 1/\xi$, so that $x_1 \to 1$, $x_2 \to 2$, $x_3 \to 0$ when P varies in such a way that ξ becomes infinite. More generally, we have the following theorem:

2.5 *The ideal point* $(1, m, 0)$ *is approached by the variable Euclidean point* $(1, m + b/x, 1/x)$ *on the line* $y = mx + b$ *when* x *becomes infinite, and* $(0, 1, 0)$ *is approached by the variable point* $(a, y, 1)$, *or* $(a/y, 1, 1/y)$, *on the line* $x = a$ *when* y *becomes infinite.*

A conventional Cartesian diagram is of no value in visualizing these limits, but our circular model is quite useful here. The approach of the above point $P(1, 2 + b/\xi, 1/\xi)$ to the point $I(1, 2, 0)$, for example, can be visualized in terms of their representatives P' and I' in the circular model, for the distance $P'I'$ actually approaches zero when $P \to I$ (see Fig. IX, 1b).

EXERCISES

1. Give two sets of homogenous coordinates for each of the following points: (a) $(\frac{1}{2}, \frac{1}{3})$; (b) $(1, 0)$; (c) $(0, 0)$; (d) the ideal point associated with the line $y = -\frac{2}{3}x + 1$; (e) the ideal point on the extended x-axis; (f) the ideal point on the extended y-axis.

2. Show that the ideal point associated with the line $ax + by + c = 0$ is $(-b, a, 0)$ or, more generally, $(-kb, ka, 0)$, $k \neq 0$.

3. Describe the location of each of the following points:
(a) $(1, 2, 3)$; (b) $(2, 4, 0)$; (c) $(k, 0, 0)$; (d) $(0, 0, k)$.

4. Find the equation of the straight line through $(3, 2, 1)$ and $(2, 1, -1)$.

5. Show why we take $(0, 1, 0)$ as coordinates of the ideal point associated with the family of straight lines perpendicular to the x-axis.

6. Verify that $(x_1, x_2, 0)$, where $x_1 \neq 0$, is the ideal point associated with the family of straight lines of slope x_2/x_1.

7. Show that (a_1, a_2, a_3) and (b_1, b_2, b_3) are the same point if $a_1/b_1 = a_2/b_2 = a_3/b_3$. Is the converse true?

8. Find the ideal points associated with the axes of the following parabolas:
(a) $y = ax^2$; (b) $x = ay^2$; (c) $y - k = a(x - h)^2$; (d) $x - h = a(y - k)^2$.

9. Find the ideal points associated with the asymptotes of the following hyperbolas:
(a) $xy = a$; (b) $x^2 - y^2 = 1$; (c) $b^2x^2 - a^2y^2 = 1$; (d) $b^2y^2 - a^2x^2 = 1$.

10. When it is not vertical, that is, when B, C are not both zero, the axis of the general parabola

$$Ax^2 + Bxy + Cy^2 + Dx + Ey + F = 0,$$

where $B^2 - 4AC = 0$, has the equation

$$(A + C)(2Bx + 4Cy) + BD + 2EC = 0.*$$

Show that the ideal point associated with the axis is then $(-2C, B, 0)$, where $C \neq 0$.

11. The asymptotes of the general hyperbola.

$$Ax^2 + Bxy + Cy^2 + Dx + Ey + F = 0,$$

where $B^2 - 4AC > 0$, are either parallel to, or coincident with, the lines through the origin with equation $Ax^2 + Bxy + Cy^2 = 0$, that is, the lines $y = Kx/2C$ if $C \neq 0$, $x = Ky/2A$ if $A \neq 0$, and $xy = 0$ if $A = C = 0$, where $K = -B \pm \sqrt{B^2 - 4AC}$. Show that the ideal points associated with the asymptotes are $(2C, -B \pm \sqrt{B^2 - 4AC}, 0)$ when $C \neq 0$. Also, find the ideal points in the other cases.

12. Find the point A which is approached by the variable point $P(a, a^2, 2)$ when $a \to 3$ and show that the distance PA approaches zero.

13. Find the limit point in Ex. 12 when $a \to 0$; when $a \to \infty$.

---

* See L. P. Eisenhart, *Coordinate Geometry*, p. 224.

14. Prove 2.5.

15. Show that the extended x-axis is a simple closed curve by mapping it topologically on the circle $x^2 + y^2 - 2y = 0$.

3. Equations of Projective Lines. In algebra, a polynomial equation is called *homogeneous* if all of its nonzero terms are of the same degree. Thus $x^3 + y^3 - z^3 = 0$ is a homogeneous equation of the third degree, while $x^3 + y^3 - z^3 = 5$ is a nonhomogeneous equation of the same degree. In plane analytic geometry most polynomial equations of curves expressed in rectangular coordinates (x, y) are nonhomogeneous. Such nonhomogeneous equations in (x, y) can be made homogeneous in (x_1, x_2, x_3) by substituting $x = x_1/x_3$, $y = x_2/x_3$ and eliminating fractions. In this way, for example, $4x - y = 5$ is changed to $4x_1 - x_2 = 5x_3$. We see here a good reason for calling (x_1, x_2, x_3) homogeneous coordinates, and (x, y) nonhomogeneous coordinates.

Equations in (x_1, x_2, x_3) have the obvious advantage over equations in (x, y) that they can apply to ideal as well as Euclidean points. This is illustrated in the present section, where we shall see that each projective line in its entirety can be represented by a homogeneous equation of the first degree in x_1, x_2, x_3.

It is helpful in showing this if we first consider straight lines. Suppose, for example, that g is the straight line with equation

$$3x - 2y + 4 = 0.$$

If we replace x and y in (1) by x_1/x_3 and x_2/x_3, respectively, we obtain

$$3\frac{x_1}{x_3} - 2\frac{x_2}{x_3} + 4 = 0, \tag{2}$$

or

$$3x_1 - 2x_2 + 4x_3 = 0. \tag{3}$$

This is an equation of g in homogeneous coordinates. Equations (1) and (3) represent exactly the same locus of Euclidean points. For each point with coordinates (x, y) satisfying (1) has homogeneous coordinates (x_1, x_2, x_3) satisfying (2), and hence (3); conversely, each Euclidean point with coordinates (x_1, x_2, x_3) satisfying (3), and hence (2), has coordinates satisfying (1).

It remains to determine whether the coordinates of any ideal points satisfy (3). Substituting 0 for x_3 in (3), we obtain $3x_1 - 2x_2 = 0$, or $x_1/x_2 = 2/3$. Hence, any ideal point for which the ratio of x_1 to x_2 is $2/3$ will have coordinates that satisfy (3). There is just one such ideal point, $(2k, 3k, 0)$, where $k \neq 0$. This is the ideal point associated with straight lines of slope

$3/2$, and hence with g. Thus, in addition to being satisfied by the coordinates of no Euclidean points but those of g, (3) is satisfied by the coordinates of no ideal point but the one associated with g. Hence (3) is the equation of the projective line $\bar{g}$ associated with g.

Apparently (3) is the equation of both g and $\bar{g}$. There is nothing contradictory in this. When we speak of (3) as the equation of g we are thinking only of Euclidean points; when we regard (3) as the equation of $\bar{g}$ we are thinking of projective points.

The preceding discussion can be generalized. Let g be any straight line

$$ax + by + c = 0, \tag{4}$$

where a, b are not both zero. The equation of g in homogeneous coordinates is then

$$a\frac{x_1}{x_3} + b\frac{x_2}{x_3} + c = 0,$$

or

$$ax_1 + bx_2 + cx_3 = 0. \tag{5}$$

Every Euclidean point with coordinates (x, y) satisfying (4) has homogeneous coordinates (x_1, x_2, x_3) satisfying (5), and every Euclidean point with coordinates (x_1, x_2, x_3) satisfying (5) has nonhomogeneous coordinates (x, y) satisfying (4). Hence, as far as Euclidean points are concerned, the locus of (5) is identical with that of (4).

Putting 0 for x_3 in (5), we obtain $ax_1 + bx_2 = 0$, so that $x_1/x_2 = -b/a$ if $a \neq 0$, and $x_2/x_1 = -a/b$ if $b \neq 0$. Hence (5) will be satisfied by the coordinates of any ideal point for which x_1/x_2 equals $-b/a$, or for which x_2/x_1 equals $-a/b$. There is just one such point, namely, $(-b, a, 0)$ or, more generally, $(-kb, ka, 0)$, where $k \neq 0$. This is the ideal point associated with g (see §2, Ex. 2). Hence, (5) is the equation of the projective line associated with g.

It remains to find an equation for the ideal line, which is the only projective line not associated with a straight line. The ideal line, being the locus of all the ideal points in the plane, has the simple equation $x_3 = 0$, or, more generally, $cx_3 = 0$, where $c \neq 0$. If this is written in form $0{\cdot}x_1 + 0{\cdot}x_2 + cx_3 = 0$, we see that it is the special case of (5) in which a, b are zero.

We have now proved that every projective line has an equation of the form (5), where a, b, c are not all zero. The converse is also true. For let (5) be given, with a, b, c not all zero. If a, b are zero, then $c \neq 0$, so that (5) represents the ideal line. If a, b are not both zero, then (5) represents the projective line associated with the Euclidean line $ax + by + c = 0$, as the reader can show by reversing the steps used earlier in this proof. Thus we have the theorem:

3.1 *Every projective line has an equation of the form*

$$ax_1 + bx_2 + cx_3 = 0, \tag{6}$$

where a, b, c are not all zero. Conversely, the locus of every such equation, with a, b, c not all zero, is a projective line.

It is to be noted that (6) is a homogeneous equation of the first degree in x_1, x_2, x_3.

Figure IX, 3 uses the circular model of the projective plane to exhibit the extended y-axis $x_1 = 0$, the extended x-axis $x_2 = 0$, the ideal line $x_3 = 0$, and the projective lines $x_1 \pm x_2 = 0$, which are associated with the lines $x \pm y = 0$.

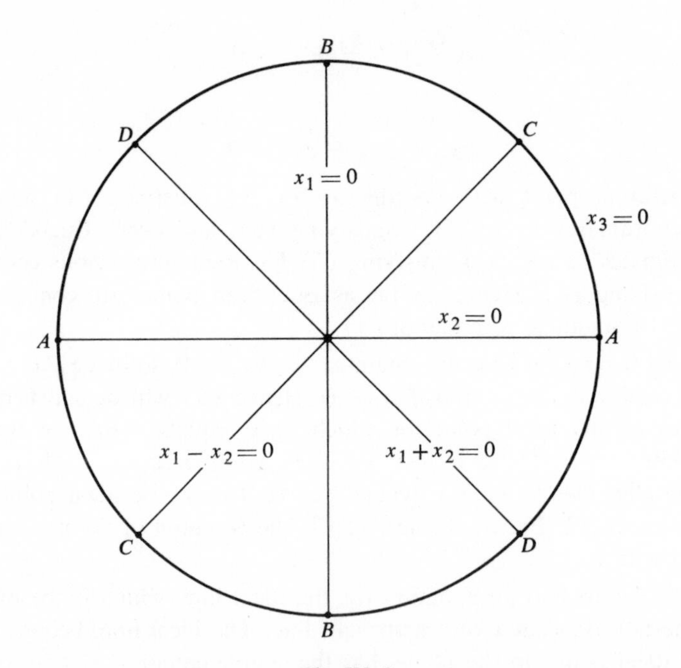

Fig. IX, 3

Two equations of the form (6) represent the same line if and only if their coefficients are in proportion, that is, if the coefficients in one equation are k times those in the other, where $k \neq 0$. (The proof of this, being like the proof of the corresponding fact for two equations of the form (4), is left to the reader.) This proportionality can also be expressed by means of determinants, as follows:

3.2 *Two first-degree equations* $A_1x_1 + A_2x_2 + A_3x_3 = 0$ and

$B_1x_1 + B_2x_2 + B_3x_3 = 0$ *represent the same projective line if and only if the determinants*

$$\begin{vmatrix} A_1 & A_2 \\ B_1 & B_2 \end{vmatrix}, \quad \begin{vmatrix} A_1 & A_3 \\ B_1 & B_3 \end{vmatrix}, \quad \begin{vmatrix} A_2 & A_3 \\ B_2 & B_3 \end{vmatrix}$$

are all zero.

This is analogous to 2.3 and can be proved in the same way.

In the preceding chapter (VIII, §5) we called attention to a model of the projective plane in which the straight lines and Euclidean planes through any point O in Euclidean space play the roles of projective points and lines, but left until now the proof that these Euclidean lines and planes are in 1-1 correspondence, respectively, with the projective points and lines. Take a three-dimensional rectangular coordinate system with O as origin, and with x-, y-, z-axes. Then each straight line through O has direction numbers a_1, a_2, a_3, not all zero, and no two of these lines have the same direction numbers; conversely, any given numbers a_1, a_2, a_3, not all zero, are the direction numbers of one and only one of these lines. It follows that a 1-1 correspondence between the lines through O and the points of the projective plane can be established by pairing with any given point (x_1, x_2, x_3) the line which has direction numbers equal to x_1, x_2, x_3. Consider, next, the Euclidean planes through O. Each has an equation of the form $Ax + By + Cz = 0$, where A, B, C are not all zero, and distinct planes have different equations; conversely, any given equation of this form, with A, B, C not all zero, is the equation of one such plane. It follows that a 1-1 correspondence between these planes and the lines of the projective plane can be established by pairing with any given line $ax_1 + bx_2 + cx_3 = 0$ the plane $Ax + By + Cz = 0$ such that $a = A$, $b = B$, $c = C$.

EXERCISES

1. Give homogeneous equations of the following straight lines: (a) $y = 2x + 3$; (b) $\frac{1}{2}x + \frac{1}{3}y = 1$; (c) $x = 5$; (d) $y = 3$; (e) $x + y = 0$; (f) $4x - 3y = 0$.

2. Whenever possible, find nonhomogeneous equations of the following: (a) $2x_1 - 3x_2 + x_3 = 0$; (b) $2x_1 - 3x_2 = 0$; (c) $2x_1 + x_3 = 0$; (d) $-3x_2 + x_3 = 0$; (e) $2x_1 = 0$; (f) $-3x_2 = 0$; (g) $5x_3 = 0$; (h) $2x_1 + 3x_2 = 0$.

3. Show how the projective lines in Exs. 2(a),(b) look in the circular model.

4. Show how the following families of projective lines look in the circular model: (a) $ax_1 + bx_2 = 0$; (b) $ax_1 + cx_3 = 0$; (c) $bx_2 + cx_3 = 0$.

5. Find the coordinates of the ideal points on the lines in Exs. 2 (a), (b), (c).

6. Find the equations of the projective lines determined by the following pairs of points:

(a) (0, 0), (2, 1); (b) (1, 2), (− 3, 4); (c) (0, 0, −3), (4, 2, 2);
(d) (0, 0, 1), (1, 0, 0); (e) (0, 0, 1), (0, 1, 0); (f) (3, 2. −1), (− 4, − 2, 0);
(g) (5, − 4, 0), (− 7, 6, 0); (h) (5, 0, −4), (− 7, 0, 6).

7. Find homogeneous coordinates for the points of intersection of the following pairs of projective lines:

(a) $x_1 = 0, x_2 = 0$; (b) $x_1 = 0, x_3 = 0$; (c) $x_2 = 0, x_3 = 0$;
(d) $x_1 = 0, x_1 = 2x_3$; (e) $x_2 = 0, x_2 = 5x_3$; (f) $x_1 + x_2 = 0, x_1 - x_2 = 0$;
(g) $x_1 + 3x_2 - x_3 = 0, 2x_1 - x_2 + x_3 = 0$;
(h) $x_1 + 3x_2 - x_3 = 0, x_1 + 3x_2 + x_3 = 0$.

8. Verify that *the equation of the line through two points* (b_1, b_2, b_3) *and* (c_1, c_2, c_3) *is*

$$\begin{vmatrix} x_1 & x_2 & x_3 \\ b_1 & b_2 & b_3 \\ c_1 & c_2 & c_3 \end{vmatrix} = 0,$$

or

$$\begin{vmatrix} b_2 & b_3 \\ c_2 & c_3 \end{vmatrix} x_1 + \begin{vmatrix} b_3 & b_1 \\ c_3 & c_1 \end{vmatrix} x_2 + \begin{vmatrix} b_1 & b_2 \\ c_1 & c_2 \end{vmatrix} x_3 = 0.$$

9. Solve Exs. 6 (c), (f) by using the fact stated in Ex. 8.

10. Verify that *the point of intersection of two lines* $A_1x_1 + A_2x_2 + A_3x_3 = 0$ *and* $B_1x_1 + B_2x_2 + B_3x_3 = 0$ *has the coordinates*

$$\begin{vmatrix} A_2 & A_3 \\ B_2 & B_3 \end{vmatrix}, \quad \begin{vmatrix} A_3 & A_1 \\ B_3 & B_1 \end{vmatrix}, \quad \begin{vmatrix} A_1 & A_2 \\ B_1 & B_2 \end{vmatrix}.$$

11. Solve Exs. 7 (g), (h) by using the fact stated in Ex. 10.

12. Using the fact stated in Ex. 8 show that *three points* (a_1, a_2, a_3), (b_1, b_2, b_3), (c_1, c_2, c_3) *are collinear if and only if*

$$\begin{vmatrix} a_1 & a_2 & a_3 \\ b_1 & b_2 & b_3 \\ c_1 & c_2 & c_3 \end{vmatrix} = 0.$$

13. Using the fact stated in Ex. 10 show that *three lines*
$A_1x_1 + A_2x_2 + A_3x_3 = 0$, $B_1x_1 + B_2x_2 + B_3x_3 = 0$, $C_1x_1 + C_2x_2 + C_3x_3 = 0$ *are concurrent if and only if*

$$\begin{vmatrix} A_1 & A_2 & A_3 \\ B_1 & B_2 & B_3 \\ C_1 & C_2 & C_3 \end{vmatrix} = 0.$$

14. Prove that two homogeneous first-degree equations represent the same line if and only if their coefficients are in proportion.

15. A variable line $px_1 + qx_2 + rx_3 = 0$ is said to *approach* a fixed line $ax_1 + bx_2 + cx_3 = 0$, or to have this line as a limit line, if and only if it varies so that $p \to ka$, $q \to kb$, $r \to kc$, where $k \neq 0$. If A, B are distinct points and P is a point such that $P \to B$, prove that line $AP \to$ line AB. (*Hint.* Use Ex. 8.)

16. What is the locus of equation (6) in Theorem 3.1 when a, b, c are zero?

17. Describe the locus of the equation $x_1 = 2$.

18. Verify that the locus of $x_1 + x_2 = 5$ is not a line.

4. Linear Combination of Points. If h and k are not both zero, the number triple $(ha_1 + kb_1, ha_2 + kb_2, ha_3 + kb_3)$ is said to be a **linear combination** of the triples (a_1, a_2, a_3) and (b_1, b_2, b_3). For example, if $h = 2$, $k = 3$, then $(2, -3, 4)$ is a linear combination of $(1, 0, 2)$ and $(0, -1, 0)$.

The concept of linear combination is useful because it permits one to deal with collinear points in terms of their coordinates instead of their equations, as we shall see. The case in which $h = 1$ and $k \neq 0$ is the only one needed for our purposes.

4.1 *If $P(a_1, a_2, a_3)$ and $Q(b_1, b_2, b_3)$ are distinct points, then $(a_1 + kb_1, a_2 + kb_2, a_3 + kb_3)$, where $k \neq 0$, represents a point other than P or Q.*

Proof. If $(a_1 + kb_1, a_2 + kb_2, a_3 + kb_3)$ is not a point, it must be the triple $(0, 0, 0)$, in which case $a_1 + kb_1 = 0$, $a_2 + kb_2 = 0$, $a_3 + kb_3 = 0$, or $a_1 = -kb_1$, $a_2 = -kb_2$, $a_3 = -kb_3$. Since $k \neq 0$, it follows that P, Q are the same point, contradicting the hypothesis. Thus $(a_1 + kb_1, a_2 + kb_2, a_3 + kb_3)$ is a point. Denote it by R. Then $R \neq P$. For if $R = P$, then $a_1 + kb_1 = ja_1$, $a_2 + kb_2 = ja_2$, $a_3 + kb_3 = ja_3$, where j is any number not zero, and hence $kb_1 = (j - 1)a_1$, $kb_2 = (j - 1)a_2$, $kb_3 = (j - 1)a_3$. This implies that Q is the triple $(0, 0, 0)$ when $j = 1$, and that $Q = P$ when $j \neq 1$, both of which are contradictions. Similarly, $R \neq Q$.

4.2 *If $P(a_1, a_2, a_3)$ and $Q(b_1, b_2, b_3)$ are distinct points, then* $(a_1 + kb_1, a_2 + kb_2, a_3 + kb_3)$, where $k \neq 0$, *represents another point on their line. Conversely, any other point on their line has coordinates expressible as $(a_1 + kb_1, a_2 + kb_2, a_3 + kb_3)$, where $k \neq 0$.*

To prove the first part of 4.2, let the equation of line PQ be

$$Ax_1 + Bx_2 + Cx_3 = 0. \tag{1}$$

Then, since P and Q are on this line,

$$Aa_1 + Ba_2 + Ca_3 = 0$$
$$Ab_1 + Bb_2 + Cb_3 = 0.$$

Hence, if k is any number,

$$(Aa_1 + Ba_2 + Ca_3) + k(Ab_1 + Bb_2 + Cb_3) = 0$$

or

$$A(a_1 + kb_1) + B(a_2 + kb_2) + C(a_3 + kb_3) = 0,$$

which shows that the triple $(a_1 + kb_1,\ a_2 + kb_2,\ a_3 + kb_3)$ satisfies (1). By 4.1 this triple is a point distinct from P and Q when $k \neq 0$.

To prove the second part of 4.2, let $R(c_1, c_2, c_3)$ be a point of line PQ distinct from P and Q. Since (jc_1, jc_2, jc_3), where j is any nonzero number, are also coordinates of R, we must find nonzero constants j, k so that

$$\text{(2)}\qquad \begin{cases} a_1 + kb_1 = jc_1 \\ a_2 + kb_2 = jc_2 \\ a_3 + kb_3 = jc_3. \end{cases}$$

To solve these three equations in the two unknowns j, k, we write them in the form

$$jc_1 - kb_1 = a_1$$
$$jc_2 - kb_2 = a_2$$
$$jc_3 - kb_3 = a_3.$$

Since R is different from Q, at least one of the determinants

$\begin{vmatrix} b_1 & b_2 \\ c_1 & c_2 \end{vmatrix}$, $\begin{vmatrix} b_1 & b_3 \\ c_1 & c_3 \end{vmatrix}$, $\begin{vmatrix} b_2 & b_3 \\ c_2 & c_3 \end{vmatrix}$ is not zero according to 2.3. Suppose that $\begin{vmatrix} b_1 & b_2 \\ c_1 & c_2 \end{vmatrix} \neq 0$. Then $\begin{vmatrix} c_1 & -b_1 \\ c_2 & -b_2 \end{vmatrix} \neq 0$ and we can solve the first two equations, obtaining

$$\text{(3)}\qquad j = \frac{\begin{vmatrix} a_1 & -b_1 \\ a_2 & -b_2 \end{vmatrix}}{\begin{vmatrix} c_1 & -b_1 \\ c_2 & -b_2 \end{vmatrix}}, \quad k = \frac{\begin{vmatrix} c_1 & a_1 \\ c_2 & a_2 \end{vmatrix}}{\begin{vmatrix} c_1 & -b_1 \\ c_2 & -b_2 \end{vmatrix}}.$$

These values also satisfy the third equation, as substitution will show. Finally, neither value is zero. For if we put 0 for j in the three equations, a contradiction of the sort encountered in the first part of the proof results, and the same is true if we put 0 for k.

It may be helpful to illustrate the second part of the proof with an example. Let P, Q, R be the distinct points (1, 1, 2), (3, 2, 5), (4, −1, 3) on the line $x_1 + x_2 - x_3 = 0$. Substituting in (3) gives $j = -\frac{1}{11}$, $k = -\frac{5}{11}$. Equations (2) are satisfied by these values:

$$1 + (-\tfrac{5}{11})3 = (-\tfrac{1}{11})4$$
$$1 + (-\tfrac{5}{11})2 = (-\tfrac{1}{11})(-1)$$
$$2 + (-\tfrac{5}{11})5 = (-\tfrac{1}{11})3.$$

Thus, if the given coordinates of R are multiplied by $-\frac{1}{11}$, we obtain $(-\frac{4}{11}, \frac{1}{11}, -\frac{3}{11})$, and these coordinates of R are a linear combination of the coordinates of P and Q in which $k = -\frac{5}{11}$.

Sets of homogeneous coordinates, being vectors with three components, can be expressed more concisely if vector notation (II, §6) is used. Thus, (a_1, a_2, a_3) can be denoted by $\mathbf{a}$, (b_1, b_2, b_3) by $\mathbf{b}$, and so forth. The definitions and theorems of this section, as well as other statements involving linear combinations, likewise take a more concise form when expressed in vector notation. To do this one must first define *multiplication of a vector by a scalar*, that is, by a number. By the **product** $h(a_1, a_2, a_3)$, or $h\mathbf{a}$, of the vector (a_1, a_2, a_3) by the number h is meant the vector (ha_1, ha_2, ha_3).

Using this definition we can now denote $(ha_1 + kb_1, ha_2 + kb_2, ha_3 + kb_3)$, the general linear combination of (a_1, a_2, a_3) and (b_1, b_2, b_3), *by* $h\mathbf{a} + h\mathbf{b}$. Also, we can state 4.1 and 4.2, combined, as follows: *If* $\mathbf{a}$ *and* $\mathbf{b}$ *represent distinct points, then* $\mathbf{a} + k\mathbf{b}$, *where* $k \neq 0$, *represents another point on their line; conversely, any other point on their line can be so represented.*

EXERCISES

1. If $\mathbf{a}$ is $(0, 1, -1)$ and $\mathbf{b}$ is $(2, -3, 0)$, find
(a) $2\mathbf{a} + 3\mathbf{b}$, (b) $-2\mathbf{a} + 3\mathbf{b}$.

2. If $\mathbf{a}$ is $(3, -1, 2)$ and $\mathbf{b}$ is $(4, 0, 5)$, give the coordinates of the following points and verify that they satisfy the equation of the line through $\mathbf{a}$ and $\mathbf{b}$:
(a) $\mathbf{a} + \mathbf{b}$; (b) $\mathbf{a} - \mathbf{b}$; (c) $\mathbf{a} + 2\mathbf{b}$; (d) $\mathbf{a} - 2\mathbf{b}$.

3. The following points are on the line of Ex. 2, and hence have coordinates of the form $\mathbf{a} + k\mathbf{b}$. Find k in each case:
(a) $(1, 1, 3)$; (b) $(-2, 2, 1)$; (c) $(-7, 5, 0)$.

4. If $\mathbf{a}$ and $\mathbf{b}$ are the coordinates of distinct points, prove that $h\mathbf{a} + k\mathbf{b}$, where h, k are not both zero, are the coordinates of a point on their line and that, conversely, every point on their line has coordinates expressible in this way.

5. If $\mathbf{a}$, $\mathbf{b}$ are the coordinates of distinct points, show that $\mathbf{a} + k\mathbf{b}$ and

$\mathbf{a} + k'\mathbf{b}$, where $k \neq 0$, $k' \neq 0$, are the coordinates of distinct points if and only if $k \neq k'$.

6. If A, B, C, D are distinct collinear points with coordinates $\mathbf{a}, \mathbf{b}, \mathbf{c}, \mathbf{d}$, respectively, and $\mathbf{c} = \mathbf{a} + k\mathbf{b}$,* $\mathbf{d} = \mathbf{a} + k'\mathbf{b}$, where $k \neq k'$, (a) show that $\mathbf{c} - \frac{k}{k'}\mathbf{d}$ and $\mathbf{c} - \mathbf{d}$ are coordinates of A and B, respectively; (b) find the linear combination of $\mathbf{c}$ and $\mathbf{d}$ which equals $\mathbf{a}$; (c) find the linear combination of $\mathbf{c}$ and $\mathbf{d}$ which equals $\mathbf{b}$.

5. Linear Combination and Cross-Ratio. If A, B, C, D are distinct collinear points and the coordinates of C, D are expressed as linear combinations of those of A, B, a very simple formula can be obtained for the cross-ratio (AB, CD). To determine it, let the coordinates of A, B, C, D be $\mathbf{a}, \mathbf{b}$, $\mathbf{a} + k\mathbf{b}$, $\mathbf{a} + k'\mathbf{b}$, respectively, where $k, k' \neq 0$ and $k \neq k'$.

We shall suppose, first, that the points are Euclidean. The rectangular coordinates of A, B, C are then

$$A\left(\frac{a_1}{a_3}, \frac{a_2}{a_3}\right),\ B\left(\frac{b_1}{b_3}, \frac{b_2}{b_3}\right),\ C\left(\frac{a_1 + kb_1}{a_3 + kb_3}, \frac{a_2 + kb_2}{a_3 + kb_3}\right).$$

The coordinates of D differ from those of C only in that k' replaces k.

If line AB is not vertical, the ordinary ratio r in which C divides $\overline{AB}$ can be found from the formula $r = (x - x_1)/(x_2 - x)$ by substituting the abscissas of A, B, C for x_1, x_2, x, respectively (see IV, §3, Ex. 6). This gives

$$r = \frac{\dfrac{a_1 + kb_1}{a_3 + kb_3} - \dfrac{a_1}{a_3}}{\dfrac{b_1}{b_3} - \dfrac{a_1 + kb_1}{a_3 + kb_3}}$$

or, after simplification,

$$r = \frac{kb_3}{a_3}.$$

If line AB is vertical this same result is obtained by use of the formula $r = (y - y_1)/(y_2 - y)$. Similarly, the ratio r' in which D divides $\overline{AB}$ is $r' = k'b_3/a_3$. Accordingly, we have

$$(AB, CD) = \frac{r}{r'} = \frac{k}{k'}.$$

Suppose, now, that just one of the points is ideal, say D. Then

* $\mathbf{c} = \mathbf{a} + k\mathbf{b}$ means that $c_1 = a_1 + kb_1$, $c_2 = a_2 + kb_2$, $c_3 = a_3 + kb_3$.

$a_3 + k'b_3 = 0$, or $a_3 = -k'b_3$. Also, by VIII 9.1 (AB, CD) equals the negative of the ordinary ratio in which C divides $\overline{AB}$, that is, $(AB, CD) = -kb_3/a_3$. Substituting $a_3 = -k'b_3$ in this, we obtain

$$(AB, CD) = \frac{k}{k'}.$$

It is left as an exercise to show that the same result is obtained if, instead of D, any one of the other points is ideal.

Finally, suppose A, B, C, D are ideal. Let a, b, c, d be four concurrent Euclidean lines associated with them, respectively, (that is, a belongs to the family of parallel lines associated with A, etc.) and let $\bar{a}, \bar{b}, \bar{c}, \bar{d}$ be the corresponding extended lines. Then $(AB, CD) = (\bar{a}\bar{b}, \bar{c}\bar{d})$ by VIII, 9.11 and $(\bar{a}\bar{b}, \bar{c}\bar{d}) = (ab, cd)$ by VIII, 9.5. Hence

$$(AB, CD) = (ab, cd). \tag{1}$$

If none of the lines a, b, c, d is vertical, let their slopes be m_1, m_2, m_3, m_4, respectively. Then, by V, §11, Ex. 11,

$$(ab, cd) = \frac{(m_3 - m_1)(m_4 - m_2)}{(m_3 - m_2)(m_4 - m_1)}. \tag{2}$$

Since A is the ideal point associated with a, we know that $m_1 = a_2/a_1$. Similarly, $m_2 = b_2/b_1$, and

$$m_3 = \frac{a_2 + kb_2}{a_1 + kb_1}, \qquad m_4 = \frac{a_2 + k'b_2}{a_1 + k'b_1}.$$

Substituting these values in (2) and simplifying, we obtain $(ab, cd) = k/k'$. If one of the lines a, b, c, d is vertical, say a, let the slopes of the others be as above. Then, by V, §11, Ex. 15,

$$(ab, cd) = \frac{m_4 - m_2}{m_3 - m_2}.$$

Substituting in this for m_2, m_3, m_4 as before, and noting that $a_1 = 0$, we again get $(ab, cd) = k/k'$. It is left as an exercise to show that this result is also obtained if b, c, or d is vertical. Thus, we see from (1) that $(AB, CD) = k/k'$.

The proof of the following theorem is now complete:

5.1 *If A, B, C, D are distinct collinear points, with coordinates* $\mathbf{a}, \mathbf{b}, \mathbf{a} + k\mathbf{b}, \mathbf{a} + k'\mathbf{b}$, *then* $(AB, CD) = k/k'$.

As an immediate consequence of this we have:

5.2 *Four distinct collinear points A, B, C, D, taken in that order, are a harmonic set if and only if they have coordinates of the form* $\mathbf{a}, \mathbf{b}, \mathbf{a} + k\mathbf{b}, \mathbf{a} - k\mathbf{b}$, *respectively.*

EXERCISES

1. Find the cross-ratios of the following sets of points in the given order:

(a) (1, 2, 3), (4, 5, 6), (5,7,9), (9, 12, 15);
(b) (1, 2, 3), (4, 5, 6), (9, 12, 15), (13, 17, 21);
(c) (1, 2, 3), (4, 5, 6), (5, 7, 9), (−3, −3, −3);
(d) (1, 2, 3), (4, 5, 6), (−7, −8, −9), (9, 12, 15);

2. Prove 5.1 if just one point is ideal and it is

(a) C, (b) B, (c) A.

3. Prove 5.1 if A, B, C, D are ideal and the following point is associated with a vertical line:

(a) B; (b) C; (c) D.

4. Deduce $(AB, CD) = 1/(AB, DC)$ from 5.1.

5. Deduce from 5.1 and §4, Ex. 6 that $(AB, CD) = (CD, AB)$, and hence that C, D, separate A, B harmonically if A, B separate C, D harmonically.

6. Equations of a Projectivity in the Projective Plane. It will be very helpful in determining these equations if we first recall some of the things done in Chapter VI, where we were concerned with geometry in the Euclidean plane. After defining a projective transformation of that plane into itself (VI, 2.1), we showed that its equations are (VI, §4)

(1) $$x' = \frac{a_1x + a_2y + a_3}{c_1x + c_2y + c_3}, \qquad y' = \frac{b_1x + b_2y + b_3}{c_1x + c_2y + c_3},$$

where

$$\Delta = \begin{vmatrix} a_1 & a_2 & a_3 \\ b_1 & b_2 & b_3 \\ c_1 & c_2 & c_3 \end{vmatrix} \neq 0,$$

and that the implicit form of (1) is

(1a) $$x = \frac{A_1x' + B_1y' + C_1}{A_3x' + B_3y' + C_3}, \qquad y = \frac{A_2x' + B_2y' + C_2}{A_3x' + B_3y' + C_3},$$

where A_1, B_1, C_1, . . ., are the cofactors of a_1, b_1, c_1, . . . in Δ, and

$$\delta = \begin{vmatrix} A_1 & B_1 & C_1 \\ A_2 & B_2 & C_2 \\ A_3 & B_3 & C_3 \end{vmatrix} \neq 0$$

since $\delta = \Delta^2$.

We also showed that, except perhaps for the points of two lines, equations (1) establish a 1-1 correspondence among the points of the Euclidean plane, and preserve collinearity and cross-ratio. It is useful to recall further that the exceptional points and lines occur only when c_1 and c_2 are not both zero, in which case A_3 and B_3 are not both zero. If c_1 and c_2 are not both zero, the straight line $c_1x + c_2y + c_2 = 0$ is not transformed into any line, no straight line is transformed into the straight line $A_3x_3' + B_3y_3' + C_3 = 0$, and each point of these lines is exceptional. Thus, an exceptional point is a point (x, y) making the denominator of (1) zero, or a point (x', y') making the denominator of (1a) zero.

One thing we did not learn previously about the transformation (1), and useful to us now, is how to express it in homogeneous coordinates. To do this we first make the substitutions

$$x = \frac{x_1}{x_3}, \quad y = \frac{x_2}{x_3} \qquad \text{and} \qquad x' = \frac{x_1'}{x_3'}, \quad y' = \frac{x_2'}{x_3'}$$

in (1) and, after simplification, obtain

$$\frac{x_1'}{x_3'} = \frac{a_1x_1 + a_2x_2 + a_3x_3}{c_1x_1 + c_2x_2 + c_3x_3}, \qquad \frac{x_2'}{x_3'} = \frac{b_1x_1 + b_2x_2 + b_3x_3}{c_1x_1 + c_2x_2 + c_3x_3}. \tag{2}$$

According to these equations, if $P(x_1, x_2, x_3)$ is a given Euclidean point, and it is sent into a Euclidean point P', the coordinates x_1', x_2', x_3' of P' are proportional to the values of the three polynomials $a_1x_1 + a_2x_2 + a_3x_3$, $b_1x_1 + b_2x_2 + b_3x_3$, and $c_1x_1 + c_2x_2 + c_3x_3$, respectively. To clarify this further, suppose that the given point and transformation are such that the first of these polynomials has the value 3, the second the value 0, and the third the value 4, in which case equations (2) reduce to $x_1'/x_3' = 3/4$, $x_2'/x_3' = 0/4$. One set of values of x_1', x_2', x_3' satisfying these equations is 3, 0, 4; any set proportional to these, but no other set, will likewise satisfy the equations. Hence the coordinates of P' are $3k$, $0 \cdot k$, $4k$, where k is any number not zero. In the general case, as we stated above, equations (2) imply that the coordinates of P' are as follows:

$$\begin{cases} x_1' = (a_1x_1 + a_2x_2 + a_3x_3)k \\ x_2' = (b_1x_1 + b_2x_2 + b_3x_3)k \\ x_3' = (c_1x_1 + c_2x_2 + c_3x_3)k, \end{cases} \tag{3}$$

where k is any number not zero. If one wishes, one can always take $k = 1$, obtaining

$$\begin{cases} x_1' = a_1x_1 + a_2x_2 + a_3x_3 \\ x_2' = b_1x_1 + b_2x_2 + b_3x_3 \\ x_3' = c_1x_1 + c_2x_2 + c_3x_3, \end{cases} \qquad \Delta = \begin{vmatrix} a_1 & a_2 & a_3 \\ b_1 & b_2 & b_3 \\ c_1 & c_2 & c_3 \end{vmatrix} \neq 0. \tag{4}$$

Thus we see that the transformation (1) can be expressed in homogeneous form by (2), or (3), or (4), the last of these being the simplest and, for our purposes, the most useful.

Being the expression of the transformation (1) in homogeneous coordinates, equations (4) are, of course, valid for those Euclidean points for which equations (1) are valid. Moreover, since equations (4) involve no denominators, they may also make sense for those Euclidean points for which equations (1) make no sense, namely, the vanishing points. In addition, equations (4), being homogeneous, lend themselves to substitution with ideal points, whereas equations (1) do not. What we are trying to suggest by these remarks is that equations (4) may represent a transformation of the entire projective plane into itself, and not merely of the Euclidean part of that plane. This is in fact the case, as we now show.

Let $P(x_1, x_2, x_3)$ be any projective point. Substitution of these coordinates in (4) reduces the right-hand sides to three numbers, which are the values of x_1', x_2', x_3'. Only if x_1', x_2', x_3' are all zero, do they fail to represent a point. If these values were all zero, the given numbers x_1, x_2, x_3 would satisfy the equations

$$\begin{aligned} 0 &= a_1x_1 + a_2x_2 + a_3x_3 \\ 0 &= b_1x_1 + b_2x_2 + b_3x_3 \qquad \Delta \neq 0. \\ 0 &= c_1x_1 + c_2x_2 + c_3x_3, \end{aligned}$$

This is impossible, for the only solution of these equations is $x_1 = 0$, $x_2 = 0$, $x_3 = 0$ (see Appendix, 5.2), whereas the given numbers, being the coordinates of a point, are not all zero. Thus, equations (4) always send a given projective point $P(x_1, x_2, x_3)$ into a projective point $P'(x_1', x_2', x_3')$. In particular, if P is a Euclidean point for which (1) is invalid, that is, if P is a vanishing point, it will be sent by (4) into an ideal point. To verify this we note that in this case P will have nonhomogeneous coordinates making the denominator of (1) zero, and hence homogeneous coordinates reducing $c_1x_1 + c_2x_2 + c_3x_3$ to zero. Thus $x_3' = 0$. The idea of a vanishing point going into an ideal point is familiar from Chapter VIII (see VIII, §§ 2, 3).

Conversely, if $P'(x_1', x_2', x_3')$ is any given projective point, there is a definite point $P(x_1, x_2, x_3)$ into which it is sent by (4). To prove this let us suppose that the coordinates of P' have been substituted in (4) and that we then try to solve for x_1, x_2, x_3. Since $\Delta \neq 0$, Cramer's Rule enables us to determine unique values for x_1, x_2, x_3. These values are not all zero inasmuch as x_1', x_2', x_3' are not all zero.

We have thus proved the following theorem:

6.1 If (x_1, x_2, x_3) *and* (x_1', x_2', x_3') *in equations* (4) *denote homogeneous coordinates of points in the projective plane, then these equations* *determine a* 1-1 *transformation,* $(x_1, x_2, x_3) \to (x_1', x_2', x_3')$, *of the plane into itself.*

We go on to show that this transformation also preserves collinearity and the cross-ratio of points, and hence is a projectivity (VIII, 19.1). For this purpose it is helpful to have the implicit form of the transformation. This is obtained by solving (4) for x_1, x_2, x_3 in terms of x_1', x_2', x_3' by Cramer's Rule.* The result is

$$(5)\qquad \begin{cases} x_1 = (A_1x_1' + B_1x_2' + C_1x_3')/\Delta \\ x_2 = (A_2x_1' + B_2x_2' + C_2x_3')/\Delta \\ x_3 = (A_3x_1' + B_3x_2' + C_3x_3')/\Delta, \end{cases}$$

where $A_1, B_1, C_1, \ldots$ have the same meanings as before. Since $(x_1\Delta, x_2\Delta, x_3\Delta)$ is the same point as (x_1, x_2, x_3), equations (5) can be written more simply as

$$(6)\qquad \begin{cases} x_1 = A_1x_1' + B_1x_2' + C_1x_3' \\ x_2 = A_2x_1' + B_2x_2' + C_2x_3' \\ x_3 = A_3x_1' + B_3x_2' + C_3x_3', \end{cases} \qquad \delta = \begin{vmatrix} A_1 & B_1 & C_1 \\ A_2 & B_2 & C_2 \\ A_3 & B_3 & C_3 \end{vmatrix} \neq 0.$$

Now let P, Q, R be any distinct collinear projective points, with images P', Q', R', and let the line containing P, Q, R have the equation

$$(7)\qquad ax_1 + bx_2 + cx_3 = 0.$$

The equation of the image of (7), found by substituting from (6) and simplifying the results, is

$$(8)\qquad (aA_1 + bA_2 + cA_3)x_1' + (aB_1 + bB_2 + cB_3)x_2' + (aC_1 + bC_2 + cC_3)x_3' = 0.$$

Since the coordinates of P, Q, R satisfy (7), those of P', Q', R' satisfy (8). Now, (8) will represent a line unless the coefficients of x_1', x_2', x_3' are all zero, that is, unless

$$\begin{aligned} aA_1 + bA_2 + cA_3 &= 0 \\ aB_1 + bB_2 + cB_3 &= 0 \\ aC_1 + bC_2 + cC_3 &= 0. \end{aligned}$$

This is impossible since $\delta \neq 0$ and a, b, c are not all zero (see Appendix, 5.2). Hence (8) always represents a line. Since P', Q', R' are on this line, it follows that the transformation (4) preserves collinearity.

---

* An alternative procedure is to express (1*a*) in homogeneous coordinates.

Using the fact that (4) is a 1-1 transformation of the projective plane into itself that preserves collinearity, we could go on to show that the image of a line is a line, that a 1-1 correspondence is established among the lines of the plane, and that the concurrence of lines is preserved. The details of the reasoning can be found at the end of VIII, §19. By way of illustration it may be noted that (4) transforms the projective line $c_1x_1 + c_2x_2 + c_3x_3 = 0$ into the ideal line, and the ideal line into the projective line $A_3x_1' + B_3x_2' + C_3x_3' = 0$.

Now let us consider cross-ratio in connection with the transformation (4). In view of the preceding discussion we know that (4) always sends four collinear points or concurrent lines into four collinear points or concurrent lines. In the preceding chapter cross-ratio was defined for all such sets of points and lines in the projective plane. That (4) preserves cross-ratio in case the four points and their images are Euclidean was proved in VI, §4 by use of (1), the nonhomogeneous form of (4). That (4) preserves cross-ratio in case the four lines and their images are extended lines follows from the fact that the cross-ratio of four extended lines equals the cross-ratio of the associated straight lines (VIII, 9.5), and the fact that (1) preserves the cross-ratio of straight lines (VI, 4.4). It then follows that (4) preserves cross-ratio in all other cases. For let A, B, C, D be any four collinear points, and A', B', C', D' their images under (4). Let E be a Euclidean point not collinear with A, B, C, D, and whose image is Euclidean. The concurrent lines AE, BE, CE, DE are then extended lines, as are the concurrent lines $A'E'$, $B'E'$, $C'E'$, $D'E'$. Since (4) transforms the former set of lines into the latter, we know that $(AE\ BE, CE\ DE) = (A'E'\ B'E', C'E'\ D'E')$ by what was stated above. Also, by VIII, 9.12, the cross-ratio on the left side equals (AB, CD), and that on the right equals $(A'B', C'D')$. Hence $(AB, CD) = (A'B', C'D')$. Thus, (4) always preserves the cross-ratio of points. It must then do likewise for lines. For let a, b, c, d be any four concurrent lines, and a', b', c', d' be their images under (4). Let a transversal meet a, b, c, d in A, B, C, D, and let (4) send these points into A', B', C', D'. Then $(AB, CD) = (A'B', C'D')$. Since, by VIII 9.12, $(ab, cd) = (AB, CD)$ and $(a'b', c'd') = (A'B', C'D')$,we conclude that $(ab, cd) = (a'b', c'd')$.

The proof of the following theorem is now complete:

6.2 *Every transformation* $(x_1, x_2, x_3) \to (x_1', x_2', x_3')$ *represented by equations* (4) *is a projectivity in the projective plane.*

We next prove the converse of this:

6.3 *Any given projectivity in the projective plane can be expressed by equations of the form* (4).

Proof. Let $\overline{T}$ be such a given transformation. Either it sends the ideal line into itself, and each extended line into an extended line, or else it sends the ideal line into an extended line v', some extended line $\overline{u}$ into the ideal line, and all other extended lines into extended lines. We shall consider only the second case, leaving the first as an exercise.

Let T denote that part of $\bar{T}$ which applies to Euclidean points. Then T establishes a 1-1 correspondence among the Euclidean points not on $\bar{u}$ and $\bar{v}'$, and preserves their collinearity and cross-ratio. Hence T is a projective transformation of the Euclidean plane as defined in VI, 2.1 and can therefore be represented by equations of the form (1), according to VI, 4.4. The denominators of (1) and (1a) equated to zero represent the straight lines u and v' which are associated with $\bar{u}$ and $\bar{v}'$, respectively. T can be expressed in homogeneous form by equations (4). According to 6.2, when both ideal and Euclidean points are considered, equations (4) represent a projectivity T' in the projective plane. Thus, T is a part of T', as well as of $\bar{T}$. Now, T' is identical with T for all straight lines except u and v'. Therefore T' is identical with $\bar{T}$ for all extended lines except, perhaps, $\bar{u}$ and $\bar{v}'$. The equations of $\bar{u}$ and $\bar{v}'$ are $c_1x_1 + c_2x_2 + c_3x_3 = 0$ and $A_3x_3' + B_3x_3' + C_3x_3' = 0$, respectively, and hence T', like $\bar{T}$, sends $\bar{u}$ into the ideal line and the ideal line into $\bar{v}'$. T' and $\bar{T}$ are therefore identical for all projective lines. They must then be identical for all projective points. For let $T'(P) = P'$, where P is any such point, and let g, h be two projective lines through P. Then T' sends g, h into two projective lines g', h' through P'. Since $\bar{T}$ also sends g, h into g', h', respectively, it follows that $\bar{T}(P) = P'$. Thus $\bar{T} = T'$. Hence $\bar{T}$ is expressed by equations (4).

It is useful to combine the main results of this section into a single statement:

6.4 *Every projectivity in the projective plane has equations of the type*

$$(9)\qquad \begin{cases} x_1' = a_1x_1 + a_2x_2 + a_3x_3 \\ x_2' = b_1x_1 + b_2x_2 + b_3x_3 \\ x_3' = c_1x_1 + c_2x_2 + c_3x_3, \end{cases} \qquad \Delta = \begin{vmatrix} a_1 & a_2 & a_3 \\ b_1 & b_2 & b_3 \\ c_1 & c_2 & c_3 \end{vmatrix} \neq 0,$$

and any given equations of this type, with $\Delta \neq 0$, *represent such a transformation. The implicit form of the projectivity* (9) *is*

$$(10)\qquad \begin{cases} x_1 = A_1x_1' + B_1x_2' + C_1x_3' \\ x_2 = A_2x_1' + B_2x_2' + C_2x_3' \\ x_3 = A_3x_1' + B_3x_2' + C_3x_3', \end{cases} \qquad \delta = \begin{vmatrix} A_1 & B_1 & C_1 \\ A_2 & B_2 & C_2 \\ A_3 & B_3 & C_3 \end{vmatrix} \neq 0,$$

where $A_1, B_1, C_1, \ldots$ *are the cofactors of* $a_1, b_1, c_1, \ldots$ *in* Δ.

It is clear from the discussions of this section that we have also proved the following:

6.5 *The projective transformations* (1) *of the Euclidean plane into* *itself are identical with the projectivities* (9) *of the associated projective plane insofar as ideal elements are not involved.*

Finally, combining 6.4 with the fact that the set of all projectivities in the projective plane is a group (VIII, 19.2), we obtain the theorem:

6.6 *The set of all transformations of the form* (9), *where* $\Delta \neq 0$, *is a group.*

We get the identity, for example, by taking all the coefficients in (9) to be zero except a_1, b_2, c_3, which we take equal to each other. The result is $x_1' = kx_1$, $x_2' = kx_2$, $x_3' = kx_3$, where $k \neq 0$. The inverse of (9) is the transformation $(x_1', x_2', x_3') \to (x_1, x_2, x_3)$ represented by (10). As we have seen, the inverse of (9) can also be expressed by equations (5), whose determinant has the value $1/\Delta$. For example, when the inverse of the projectivity

$$(11) \qquad \begin{cases} x_1' = 3x_1 + x_2 + x_3 \\ x_2' = x_1 - 4x_2 - 2x_3 \\ x_3' = x_1 - x_2 - x_3, \end{cases} \qquad \Delta = 8,$$

is expressed in the form (10) it is

$$(12) \qquad \begin{cases} x_1 = 2x_1' + 0 \cdot x_2' + 2x_3' \\ x_2 = -x_1' - 4x_2' + 7x_3' \\ x_3 = 3x_1' + 4x_2' - 13x_3', \end{cases}$$

the determinant of whose coefficients is 64, and when expressed in the form (5) its determinant is $\frac{1}{512} \cdot 64$, or $\frac{1}{8}$.

EXERCISES

1. Find the following under the projectivity (11):
(a) the image of the point (5, 1, 2); (b) the original of the point (1, 0, 0);
(c) the image of $3x_1 + x_2 + x_3 = 0$; (d) the image of the ideal line;
(e) the original of the ideal line; (f) the image of the line $2x_1 + x_2 - x_3 = 0$.

2. Verify that the determinant of the coefficients in (5) has the value $1/\Delta$.

3. Complete the proof of 6.3 by showing that a projectivity in the projective plane which sends the ideal line into itself and each extended line into an extended line can be expressed by equations of the form (4).

4. Another way of proving that (9) preserves collinearity and the cross-ratio of points is the following. Let A, B, C be distinct collinear points. Their coordinates are then of the form **r**, **s**, **r** + k**s**, respectively, where $k \neq 0$ (see §4). Let

$T(A, B, C) = A', B', C'$, where T denotes the transformation (9), and let A', B' have the coordinates $\mathbf{r}'$, $\mathbf{s}'$. Substitution in (9) then shows that C' has the coordinates $\mathbf{r}' + k\mathbf{s}'$. It follows from 4.2 that T preserves collinearity. On using a fourth point D, with coordinates $\mathbf{r} + j\mathbf{s}$, where $j \neq 0$, $j \neq k$, and applying 5.1, one sees that the cross-ratio of points is preserved. Carry out the details of this proof.

5. Verify that the point (6, 2, 2) is the image of the point (1, 0, 0) in the projectivity (11), but that these coordinates do not satisfy (11). (To avoid this we can express the projectivity by the more general equations

$$\begin{cases} kx_1' = 3x_1 + x_2 + x_3 \\ kx_2' = x_1 - 4x_2 - 2x_3 \\ kx_3' = x_1 - x_2 - x_3. \end{cases}$$

These are satisfied by the given coordinates if $k = \frac{1}{2}$. Similarly, they are satisfied by any coordinates of a point and its image if one chooses k suitably. These observations apply, more generally, to (9).)

6. By the *matrix* of the projectivity (9) is meant the 3×3 matrix whose elements are the coefficients of x_1, x_2, x_3 as they appear in (9). Two such matrices can be multiplied to yield another in the same way as two determinants are multiplied. Show that the resultant of two projectivities of the form (9) is a projectivity of that form whose matrix is the product of theirs, and hence whose determinant is not zero if neither of theirs is.

7. Using the fact stated in Ex. 6, verify that the resultant of (11) and (12) is the identity. (It may help to avoid confusion if double primes are added to x_1, x_2, x_3 in (12).)

8. Use the fact stated in Ex. 6 to prove that the transformation given explicitly by (10) is the inverse of (9). (It may be helpful to use (x_1'', x_2'', x_3'') in (10) instead of x_1, x_2, x_3.)

7. Geometries of the Projective Plane. In the early chapters where transformations of the Euclidean plane were considered, we saw that congruent, similarity, and affine transformations were special kinds of projective transformations of that plane and that metric and affine geometries were subgeometries of the projective geometry of that plane. We would now like to see to what extent this classification of transformations and geometries can be adapted to the projective plane.

According to 6.5, congruent, similarity, and affine transformations are identical with certain kinds of projectivities in the projective plane insofar as these projectivities involve Euclidean points. Considering the affine transformations first, let us recall that they send the Euclidean plane into itself without there being any exceptional points. Hence, they are related to those projectivities in the projective plane which send Euclidean points into Euclidean points and ideal points into ideal points, that is, to projectivities which leave the ideal line fixed. To the extent that they concern Euclidean elements, then, those transformations (9) of §6 which leave the ideal line fixed are identical with the affine transformations. We therefore call them **affine transformations of the projective plane,** and refer to those previously called affine as **affine transformations of the Euclidean plane.** Since the set of all projectivities (9) is a group, the projective group, and the resultant of two projectivities that leave the ideal line fixed is also a projectivity with this property, it follows that the affine transformations of the projective plane are a group. This group is called the **affine group in the projective plane.**

To determine the equations of this group requires that we impose on the transformation (9) the condition that if the point (x_1, x_2, x_3) goes into the point (x_1', x_2', x_3'), and $x_3 = 0$, then also $x_3' = 0$. Substitution of 0 for x_3 and x_3' then shows that $c_1x_1 + c_2x_2 = 0$ must hold for all possible values of x_1, x_2. This, in turn, implies that $c_1 = c_2 = 0$, as is easily verified. We are thus led to the equations

$$(1) \qquad \begin{cases} x_1' = a_1x_1 + a_2x_2 + a_3x_3 \\ x_2' = b_1x_1 + b_2x_2 + b_3x_3 \\ x_3' = \qquad\qquad\qquad\quad c_3x_3, \end{cases} \qquad \Delta \neq 0,$$

where $c_3 \neq 0$ since $\Delta \neq 0$. These equations clearly send a Euclidean point, that is, a point for which $x_3 \neq 0$, into a Euclidean point.

In summary, then, we have:

7.1 *The affine transformations of the projective plane are those special projectivities in this plane which are represented by equations* (1), *and they form a group. In their effect on Euclidean points they are identical with the affine transformations of the Euclidean plane.*

In practice no sharp distinction is made between the two kinds of affine transformations where Euclidean points are concerned, both being projective transformations sending the entire Euclidean plane into itself, and doing this in precisely the same way. It is this property of sending finite, that is, Euclidean, points into finite points that the word "affine" was intended to convey when first used by Euler, an early contributor to the development of projective geometry.*

In the Euclidean plane, the similarity transformations are affine transformations of the special kind that multiply distances by the same factor.

* In the past it was very common to refer to ideal points as *infinite points* or *infinitely distant points*, and to regard Euclidean points as *finite points*.

It follows from 7.1 that these similarity transformations are identical with certain affine transformations of the projective plane, and hence with certain special cases of equations (1), insofar as the latter involve Euclidean points. The special form which (1) takes in these cases can be found by expressing the equations of similarity transformations in homogeneous coordinates. Consider, for example, the direct similarities

$$(2)\qquad \begin{cases} x' = ax + by + c \\ y' = -bx + ay + d, \end{cases} \qquad a^2 + b^2 \neq 0.$$

Replacing x, y, x' y' by their equivalents in homogeneous coordinates, we obtain

$$\begin{cases} \dfrac{x_1'}{x_3'} = a\dfrac{x_1}{x_3} + b\dfrac{x_2}{x_3} + c \\ \dfrac{x_2'}{x_3'} = -b\dfrac{x_1}{x_3} + a\dfrac{x_2}{x_3} + d, \end{cases} \quad \text{or} \quad \begin{cases} \dfrac{x_1'}{x_3'} = \dfrac{ax_1 + bx_2 + cx_3}{x_3} \\ \dfrac{x_2'}{x_3'} = \dfrac{-bx_1 + a\,x_2 + dx_3}{x_3}. \end{cases}$$

From this, reasoning as in §6, we deduce

$$\begin{cases} x_1' = k(ax_1 + bx_2 + cx_3) \\ x_2' = k(-bx_1 + ax_2 + dx_3) \\ x_3' = kx_3, \end{cases}$$

where k is any number not zero. Finally, taking $k = 1$ for simplicity, we obtain

$$(3)\qquad \begin{cases} x_1' = ax_1 + bx_2 + cx_3 \\ x_2' = -bx_1 + ax_2 + dx_3 \\ x_3' = x_3 \end{cases}$$

as the special form of (1) that we have been seeking. We note that (1) reduces to (3) when $a_1 = b_2$, $a_2 = -b_1$, and $c_3 = 1$. Also we note that the value of Δ for (3) is $a^2 + b^2$, which is not zero.

Regarded as transformations of the projective plane, then, equations (3) are special kinds of affine transformations of that plane. Since they are identical with equations (2) insofar as Euclidean points are concerned, equations (3) are called *direct similarity transformations of the projective plane*. By the same algebraic procedure as that used above, we can put the opposite similarities of the Euclidean plane into homogeneous form and obtain what are called the *opposite similarity transformations of the projective plane*. Our combined results are then as follows:

7.2 Definition. *The similarity transformations of the projective plane are those special affine transformations of the projective plane that have the equations*

$$\begin{cases} x_1' = ax_1 + bx_2 + cx_3 \\ x_2' = \pm(-bx_1 + ax_2) + dx_3 \\ x_3' = x_3, \end{cases} \qquad a^2 + b^2 \neq 0, \tag{4}$$

and are called direct or opposite according as the + or − sign is used. (In their effect on Euclidean points these transformations are identical with the similarity transformations of the Euclidean plane.)

Finally, let us recall that motions in the Euclidean plane are those special similarity transformations that preserve distance. Hence, by 7.2, these motions are identical with certain similarity transformations of the projective plane, that is, with certain special cases of (4), to the extent that (4) involves Euclidean points. These special cases can be found by expressing in homogeneous coordinates the equations of motions,

$$\begin{cases} x' = ax + by + c \\ y' = \pm(-bx + ay) + d, \end{cases} \qquad a^2 + b^2 = 1. \tag{5}$$

Since these equations differ from the Cartesian equations of similarity transformations only in that we now have $a^2 + b^2 = 1$, their homogeneous form is

$$\begin{cases} x_1' = ax_1 + bx_2 + cx_3 \\ x_2' = \pm(-bx_1 + ax_2) + dx_3 \\ x_3' = x_3, \end{cases} \qquad a^2 + b^2 = 1. \tag{6}$$

These equations are, then, simply the special case of (4) when $a^2 + b^2 = 1$. This permits the following statement:

7.3 Definition. *The special similarity transformations of the projective plane represented by equations* (6) *are called motions* (*or congruent transformations*) *of the projective plane, and are called direct or opposite according as the + or — sign is used.* (*To the extent that they involve Euclidean points, these transformations are identical with the motions of the Euclidean plane.*)

It is clear now that projectivities in the projective plane can be subdivided successively just as they are in the Euclidean plane, into affine, similarity, and congruent transformations. Also, each of these classes of transformations is a group. This has already been verified for the set of all projectivities and the set of affine projectivities. We leave to the reader its verification for the similarities and motions. The latter two groups are known as **Euclidean metric groups of the projective plane.***

Accordingly, we have the same classification of geometries in the projective plane as in the Euclidean plane: **projective geometry**, associated with the projective group, **affine geometry**, associated with the affine group, and **metric geometry**, associated with the Euclidean metric groups. From what has been shown it is clear that each of these geometries is identical, as far as Euclidean points are concerned, with the same-named geometry in the Euclidean plane. Considered in their entirety, these geometries of the projective plane are, of course, distinct from their counterparts of the Euclidean plane since they deal with ideal as well as with Euclidean points. This distinction is of prime importance with regard to projective geometry, for it is only in the projective plane that this system of geometry really comes into its own. The same is not true of the affine and metric geometries, for while the projective plane is well-suited for their development, so is the Euclidean plane. It is only in the projective plane, then, that all three geometries can simultaneously attain their full development, and this is a reason that they are often called *geometries of the projective plane.*†

Affine geometry of the projective plane is the study of the properties of geometric figures which are preserved by the affine group in this plane, but not by the projective group. These properties are called affine and, insofar as Euclidean figures are concerned, are the same as the affine properties studied in Chapter IV. However, since we are now working in the projective plane, care must sometimes be taken in stating these properties to make it clear that we are talking about Euclidean figures. We cannot simply say, as we did previously, that the parallelism of lines is an affine property, for, as things stand, this statement makes no sense in the projective plane inasmuch as "line" is interpreted as "projective line" and any two projective lines meet. What we should say is that the parallelism of *Euclidean* lines is an affine property.‡ Similarly, we cannot say as we did earlier that the property of being a triangle is affine, but should say that the property of being a *Euclidean* triangle is affine. Actually, the property of being a triangle is projective. Often the statement of an affine property is clear without the addition of the word "Euclidean." This is the case with ratio of division, for example, since it is defined only for Euclidean points

* There is one more Euclidean metric group in the projective plane, the **equiareal group.** Its transformations are identical, as far as Euclidean points are concerned, with the transformations of the equiareal group in the Euclidean plane.

† Since there are other geometries of the projective plane besides these three, it may be well to mention that the metric geometry of this plane which we have been discussing is known more precisely as *Euclidean metric geometry*.

‡ Some books call two projective lines *parallel* if they meet in an ideal point. In this case it is correct to say that the parallelism of lines, meaning projective lines, is an affine property.

and lines. Likewise, since only Euclidean triangles have medians, there can be no doubt as to the meaning of the statement that the medians of a triangle are concurrent.

Euclidean metric geometry of the projective plane is the study of those properties that are preserved by at least one of the Euclidean metric groups in this plane, but not by the affine group in this plane. Insofar as Euclidean figures are concerned, these metric properties are the same as the metric properties studied in Chapter III. Stating them does not call for the kind of special care required in the case of affine properties, for we have never defined distance, angle-size, area-size, perpendicularity, congruence, and similarity for any figures other than Euclidean.

EXERCISES

1. Obtain the equations of the affine transformations of the projective plane by expressing in homogeneous coordinates the equations of the affine transformations of the Euclidean plane.

2. Show that the congruent, similarity, and equiareal transformations of the projective plane form groups, respectively.

3. Prove that if $c_1x_1 + c_2x_2 = 0$ for all pairs of values of x_1, x_2, then $c_1 = 0$ and $c_2 = 0$.

4. Define the *translations of the projective plane,* find their equations, and show that they leave each ideal point fixed. Illustrate the latter property, using the circular model of the plane.

5. (a) Express a rotation through angle α about the origin in homogeneous coordinates. (b) Show that the resulting *rotation of the projective plane* leaves no ideal point fixed if α is not 0 or π. (c) What is the effect on ideal points if $\alpha = \pi$?

6. (a) Express a reflection in the x-axis in homogeneous coordinates. (b) Show that the resulting *reflection of the projective plane* leaves two ideal points fixed, and find their coordinates. (c) Illustrate the preceding property in the circular model.

7. Prove that the parallelism of Euclidean lines is preserved by the affine group in the projective plane.

8. Equations of Projective Conics. To find the equations of projective lines we took the equations of straight lines in rectangular coordinates, changed to homogeneous coordinates, and studied the loci of the resulting equations in the projective plane. We follow the same procedure for projective conics, keeping in mind that the projective plane dealt with in this chapter is an extended plane.

Starting with an example, let us consider the parabola

(1) $$y = x^2.$$

Expressing this equation in homogeneous coordinates we obtain

(2) $$x_2 x_3 = x_1^2.$$

The locus of (2) differs from that of (1) only in that it may contain one or more ideal points. To determine them we substitute 0 for x_3 in (2), obtain $0 = x_1^2$, and hence find that x_1 is 0. Since the value of x_2 is immaterial, we thus obtain $(0, k, 0)$, where $k \neq 0$. In particular we may take $k = 1$. Hence the locus of (2) contains the single ideal point (0, 1, 0).* This is the ideal point associated with the y-axis, which is the axis of the given parabola. The locus of (2) in the projective plane is therefore an extended parabola tangent to the ideal line at (0, 1, 0) (see VIII, 18.1, 18.7). Fig. IX, 4 illustrates this in the circular model of the projective plane: the

* Since zero is a double root of $0 = x_1^2$, some writers prefer to say that the locus contains two coincident ideal points.

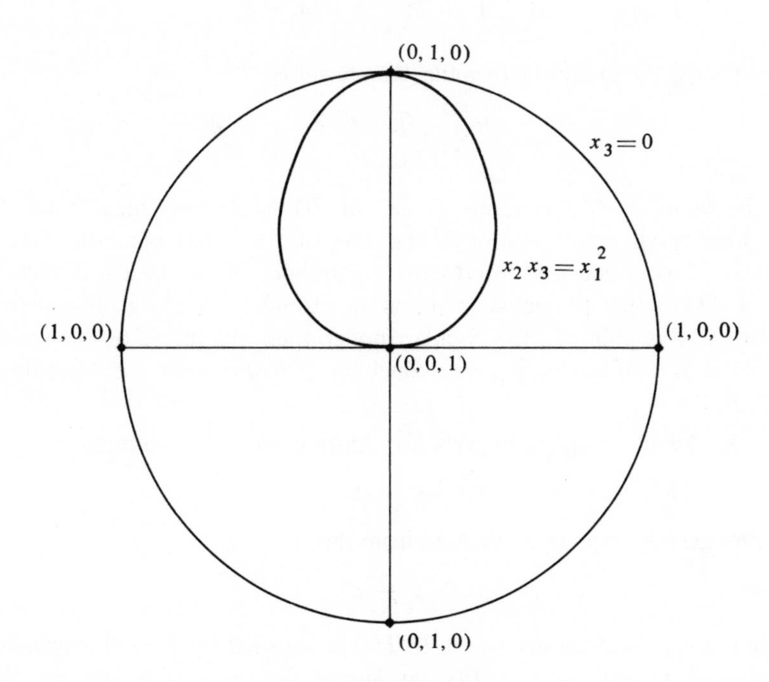

Fig. IX, 4

image of the extended parabola is a closed egg-shaped curve which is tangent to γ at the point corresponding to $(0, 1, 0)$. (This is the point A in Fig. VII, 19.) Had we used any other parabola whose axis is vertical we would again have been led to an extended parabola tangent to the ideal line at $(0, 1, 0)$.

The general parabola has the equation

$$Ax^2 + Bxy + Cy^2 + Dx + Ey + F = 0, \tag{3}$$

where $B^2 - 4AC = 0$, and it remains to consider the case in which its axis is not vertical, that is, B and C are not both zero. Then necessarily $C \neq 0$. Expressing (3) in homogeneous coordinates and clearing of fractions, we get

$$Ax_1^2 + Bx_1x_2 + Cx_2^2 + Dx_1x_3 + Ex_2x_3 + Fx_3^2 = 0. \tag{4}$$

The locus of (4) differs from that of (3) only in that it may contain ideal points. To find them we substitute 0 for x_3, obtaining

$$Ax_1^2 + Bx_1x_2 + Cx_2^2 = 0.$$

Since an ideal point is determined when we know the value of x_2/x_1, let us solve this equation for x_2/x_1, first writing it as

$$C\left(\frac{x_2}{x_1}\right)^2 + B\left(\frac{x_2}{x_1}\right) + A = 0,$$

and then using the quadratic formula. The result is

$$\frac{x_2}{x_1} = \frac{-B \pm \sqrt{B^2 - 4AC}}{2C} = \frac{-B}{2C}.$$

Thus the locus of (4) contains $(-2C, B, 0)$ as its only ideal point. This is the ideal point associated with the axis of the given parabola (see §2, Ex. 10). Thus, when (3) represents a parabola of the specified type, the locus of (4) in the projective plane is an extended parabola tangent to the ideal line at this point. In the circular model, then, the image of the extended parabola is a simple closed curve touching γ at the point corresponding to $(-2C, B, 0)$.

Next, let us consider hyperbolas, starting with the example

$$xy = 1. \tag{5}$$

The homogeneous equation obtained from this is

$$x_1x_2 = x_3^2. \tag{6}$$

Putting 0 for x_3 and we get $x_1x_2 = 0$. This is satisfied by $x_1 = 0$ regardless of the value of x_2, and by $x_2 = 0$ regardless of the value of x_1. We are therefore led to just two ideal points, (1, 0, 0) and (0, 1, 0), which are the ideal points associated with the x- and y-axes. Since these axes are the asymptotes of the given hyperbola, it follows that the locus of (6) in the projective plane is an extended hyperbola (VIII, 18.3). The ideal line, meeting the locus in two (distinct) points, is not tangent to it (VIII, 18.7). The image of this extended hyperbola in the circular model is shown in the earlier Fig. VIII, 18, where A and B correspond to (0, 1, 0) and (1, 0, 0), respectively, and γ corresponds to the line $x_3 = 0$. Had we started with any other hyperbola having one asymptote horizontal and one vertical, and hence with the equation $Bxy + Dx + Ey + F = 0$, where $B \neq 0$, we would again have been led to an extended hyperbola containing (1, 0, 0) and (0, 1, 0), and hence not tangent to the ideal line.

The remaining hyperbolas have equations of the form (3), where now $B^2 - 4AC > 0$ and A, C are not both zero. We suppose, then, that (3) represents such a hyperbola and, in particular, that $C \neq 0$. Under these conditions, the locus of equation (4), the homogeneous form of (3), in the projective plane is therefore such a hyperbola plus one or more ideal points. The algebraic details of finding these points are the same as we gave for the general parabola. Since $B^2 - 4AC \neq 0$ now, we obtain two (distinct) ideal points, $(2C, -B \pm \sqrt{B^2 - 4AC}, 0)$. They are the ideal points associated with the asymptotes of the hyperbola (see §2, Ex. 11). Under the specified conditions, the locus of (4) in the projective plane is therefore an extended hyperbola. The ideal line meets the locus in the two specified points and is therefore not tangent to it. The image of the locus in the circular model is a simple closed curve meeting γ in the points corresponding to the two specified ideal points. The reader can show that (4) is likewise an extended hyperbola if $C = 0$, $A \neq 0$, the ideal points now being $(0, 2A, 0)$ and $(-B, A, 0)$.

Finally, let us consider ellipses. Their equations are of the form (3), where now $B^2 - 4AC < 0$. In this case the attempt to find the ideal points on the locus of (4) in the projective plane leads to imaginary numbers, as the preceding algebraic work shows. Since only ideal points with real coordinates have been defined, this locus contains no ideal points and hence is identical with the locus of (3). According to VIII, §18, then, the locus of (4) is a projective conic.

We have now considered all possible nondegenerate conics and have found that they have equations of the form (4), where A, B, C are not all zero. All degenerate conics (VIII, §18) likewise, have equations of that form. Thus, the equation of the conic consisting of the single point (a, b, c) is $(bx_1 - ax_2)^2 + x_3^2 = 0$ if $c = 0$ and $(x_1 - ax_3)^2 + (x_2 - bx_3)^2 = 0$ if $c = 1$. Expansion of these equations reduces them to the form (4). The conic consisting of the two lines $a_1x_1 + a_2x_2 + a_3x_3 = 0$ and $b_1x_1 + b_2x_2 + b_3x_3 = 0$ has the equation $(a_1x_1 + a_2x_2 + a_3x_3)(b_1x_1 + b_2x_2 + b_3x_3) = 0$, which likewise can be reduced to the form (4). It should be mentioned that

A, B, C are all zero in the equations of some degenerate conics. An example is $x_1x_3 = 0$, representing the pair of lines $x_1 = 0$ and $x_3 = 0$.

We now show, conversely, that the loci of equations of the form (4), whenever such loci exist, are always projective conics. Let such an equation be given. If A, B, C are zero, the equation reduces to $Dx_1x_3 + Ex_2x_3 + Fx_3^2 = 0$, or $x_3(Dx_1 + Ex_2 + Fx_3) = 0$, whose locus is a degenerate conic consisting of two lines (which may coincide). If A, B, C are not all zero, the equation reduces to (3) on division by x_3^2, so that its locus consists of the locus of (3) and, perhaps, some ideal points. To analyze this situation further, suppose $B^2 - 4AC > 0$. Then the locus of (3) is either a hyperbola or two intersecting straight lines. In the first case the locus of (4) is an extended hyperbola, as shown earlier in this section. In the second case (4) is factorable since (3) is, and its locus consists of the two extended lines that are associated with the straight lines. Now suppose $B^2 - 4AC < 0$. The locus of (3) is then an ellipse or a point, but there may be no locus at all. The discussion of the ellipse earlier in this section actually covers these last two cases too, and hence shows that the locus of (4) is precisely the same as that of (3), and is either an ellipse, or a point, or is nonexistent. Finally, suppose $B^2 - 4AC = 0$. Then the locus of (3) is a parabola, or two parallel lines (which may coincide), or is nonexistent. In the first case the locus of (4) is an extended parabola, as shown earlier; in the second it is the extended lines associated with these parallel lines; in the third it is a single ideal point (Ex. 11).

The proof of the following is now complete:

8.1 *All projective conics have homogeneous equations of the second degree in x_1, x_2, x_3, that is, equations of the form (4). If the conic is nondegenerate, then A, B, C are not all zero. Conversely, the loci of such equations in the projective plane are never anything but projective conics. The latter are nondegenerate if and only if the loci of the corresponding nonhomogeneous equations are nondegenerate.*

For example, the locus of $x_1^2 + 3x_2^2 - x_3^2 = 0$ is a nondegenerate projective conic since the corresponding nonhomogeneous equation $x^2 + 3y^2 - 1 = 0$ represents an ellipse. The projective conic $x_1x_2 - x_1x_3 = 0$ is degenerate, consisting of the two projective lines $x_1 = 0$, $x_2 - x_3 = 0$, and the locus of the corresponding nonhomogeneous equation $xy - x = 0$ is likewise degenerate, consisting of the two straight lines $x = 0$, $y - 1 = 0$.

In analytic geometry, drawing the graph of (3) when $B \neq 0$ is usually a cumbersome problem. Our discussion suggests a way of facilitating it when the graph is a parabola (the case of the hyperbola is considered in the next section). Since $B \neq 0$ and $B^2 - 4AC = 0$, it follows that $C \neq 0$. The ideal point on the extended parabola is then $(-2C, B, 0)$, as we have seen. Accordingly, the slope of the axis of the parabola is $B/(-2C)$, which tells us something about the orientation of the parabola. The parabola $x^2 - 6xy + 9y^2 + 4x + y - 7 = 0$, for example, is seen immediately to be symmetrical to a line of slope ⅓.

In the preceding chapter we saw that every projectivity in the projective plane can be obtained as the resultant of projections of one plane on another (VIII, 19.4). Since the property of being a projective conic (of an extended plane) is preserved by projections (VIII, 18.5), it follows that every projectivity in the plane dealt with in this chapter always sends a projective conic into a projective conic. To prove this algebraically, and without using projections, let such a conic c be given by (4), and a projectivity T by equations (9) and (10) of §6. To obtain the equation (4′) into which (4) is transformed by T, we substitute the equations of T in (4). Since T is 1-1, (4′) must have a locus c', and $T(c) = c'$. Since (4) is a homogeneous second-degree equation and the equations of T are homogeneous of the first degree, (4′) is necessarily homogeneous and of not higher than second degree. If (4′) were of lower than second degree, (4) could not be obtained from it, as should be the case, by substitution of the equations of T^{-1}, which are also (9) and (10). (4′) is therefore of the second degree and c' is a projective conic. Moreover, c' is degenerate only if c is. For, the fact that T is 1-1 and sends lines into lines requires that c' consist of two lines or a single point only if c does.

8.2 *The property of being a degenerate or a nondegenerate projective conic, respectively, is preserved by all projectivities in the plane.*

It is convenient now also to prove the following:

8.3 *Any two nondegenerate projective conics in the plane are projectively equivalent.*

Proof. Let $\bar{c}$, $\bar{c}'$ be two nondegenerate projective conics in the plane, which we denote by $\bar{\alpha}$, and c, c' the nondegenerate Euclidean conics with which they are associated. Then $\bar{c}$ consists of c and perhaps one or two ideal points, and c', $\bar{c}'$ are similarly related. Since c, c' are projectively equivalent (VI, 8.1), there is a projectivity T in the Euclidean plane α associated with $\bar{\alpha}$ such that $T(c) = c'$. By 6.5 there is a projectivity $\overline{T}$ in $\bar{\alpha}$ that is identical with T insofar as ideal elements are not involved. Hence $\overline{T}(c) = c'$. Let A_i, $i = 1, 2, \ldots, 5$, be five points of c, and A_i', $i = 1, 2, \ldots, 5$, their images on c' under T and $\overline{T}$. Then no three points of either set are collinear. Also, the points A_i are on $\bar{c}$ and the points A_i' are on $\bar{c}'$. Let $\overline{T}(\bar{c}) = \bar{k}$. Then $\bar{k}$ is a nondegenerate projective conic by 8.2 and it contains the points A_i'. Since there is a unique nondegenerate projective conic through five points, no three collinear (VIII, 18.6), we conclude that $\bar{k} = \bar{c}'$. Hence $\overline{T}(\bar{c}) = \bar{c}'$ and $\bar{c}$, $\bar{c}'$ are projectively equivalent (VIII, 19.6).

EXERCISES

1. When does the projective conic (4) contain exactly the same points as the Euclidean conic (3)?

2. Express the following equations in homogeneous coordinates and describe the nature of the resulting loci in the projective plane:

(a) $x^2 + y^2 = 9$; (b) $3x^2 + 2y^2 = 6$; (c) $y^2 = 8x$;
(d) $xy = 2$; (e) $(x - y)(x + y) = 0$; (f) $2xy - 5x^2 = 0$;
(g) $x^2 + y^2 = 0$; (h) $x^2 + y^2 + 1 = 0$; (i) $(x - 2)(x - 3) = 0$.

3. Give the equation of a degenerate projective conic which (a) is a simple closed curve, (b) is not a simple closed curve.

4. Describe the loci (when they exist) of the following in the projective plane. In which cases are they nondegenerate conics?

(a) $x_1^2 = 0$; (b) $x_3^2 = 0$; (c) $x_1x_2 = 0$;
(d) $x_1x_3 = 0$; (e) $x_1^2 + x_2^2 = 0$; (f) $x_2^2 + x_3^2 = 0$;
(g) $x_1^2 + x_2x_3 = 0$; (h) $x_1x_2 + x_3^2 = 0$; (i) $x_1^2 + x_3^2 = x_2^2$;
(j) $x_1^2 + x_2^2 = x_3^2$; (k) $x_1^2 + x_2^2 + x_3^2 = 0$; (l) $x_1x_2 + x_2x_3 = 0$.

5. Show that each projective line through (0, 1, 0) other than the ideal line meets the extended parabola (2) in two distinct points, and hence is not tangent to the conic.

6. Give the equation of the family of parabolas whose axes are vertical and show that the equations of the corresponding extended parabolas are satisfied by (0, 1, 0) and no other ideal point.

7. Prove algebraically that the locus of (4) contains no ideal points when $B^2 - 4AC < 0$.

8. Prove algebraically that the extended asymptotes are tangent to the conic (6) at the points (1, 0, 0) and (0, 1, 0), and that they are the only lines tangent at these points.

9. Find the ideal points on the projective conic associated with the Euclidean conic $Bxy + Dx + Ey + F = 0$, where $B \neq 0$.

10. Find the ideal points on the projective conic $x_1^2 + 3x_1x_2 - 2x_2^2 + 4x_3^2 = 0$.

11. If (3) is a quadratic with no locus and $B^2 - 4AC = 0$, show that the locus of (4) consists of one point.

12. Show that the projective lines which are the extensions of the asymptotes of the following hyperbolas are tangent to the associated projective conics at their ideal points:

(a) $\dfrac{x^2}{a^2} - \dfrac{y^2}{b^2} = 1$; (b) $\dfrac{y^2}{a^2} - \dfrac{x^2}{b^2} = 1$.

13. If (3) represents a hyperbola, and $A \neq 0$, $C = 0$, prove that (4) represents an extended hyperbola without using 8.1.

14. Find the slope and inclination of the axis of the parabola

$$x^2 + 8xy + 16y^2 - 4x - 16y + 7 = 0.$$

15. The ideal point on an extended parabola is approached by a variable Euclidean point on the parabola whose x-value becomes infinite. Verify this when the equation of the associated Euclidean parabola is (a) $y = x^2$, (b) $y = ax^2 + bx + c$, (c) equation (3), where $B^2 - 4AC = 0$. (*Hint*. Homogeneous coordinates of the Euclidean point are $(x, f(x), 1)$, where $f(x)$ is a function obtained by solving the equation of the Euclidean parabola for y. Thus, in (a) the Euclidean point is $(x, x^2, 1)$.)

16. Each ideal point on an extended hyperbola is approached by a suitably chosen variable Euclidean point on the hyperbola whose x-value becomes infinite. Verify this when the equation of the associated Euclidean hyperbola is

(a) $xy = 1$,
(b) $Bxy + Dx + Ey + F = 0$,
(c) $b^2x^2 - a^2y^2 = a^2b^2$,
(d) equation (3), where $B^2 - 4AC > 0$.

(See *Hint* in Ex. 15.)

17. Prove that the extended parabola $x_2x_3 = x_1^2$ is a simple closed curve by mapping it topologically on the circle $x^2 + y^2 - 4y + 3 = 0$.

18. Prove that the extended hyperbola $x_1x_2 = x_3^2$ is a simple closed curve by mapping it topologically on the line $x_1 = 0$.

9. Equations of Tangents to Projective Conics. As was stated at the end of VI, §7, the straight line tangent to a nondegenerate Euclidean conic

$$Ax^2 + Bxy + Cy^2 + Dx + Ey + F = 0 \tag{1}$$

at the point (x_0, y_0) has the equation

$$2Ax_0x + B(y_0x + x_0y) + 2Cy_0y + D(x + x_0) + E(y + y_0) + 2F = 0. \tag{2}$$

We mentioned in VIII, §18 that this tangent meets the conic in no point except (x_0, y_0). As we have seen, when this conic is "extended" into the projective plane it becomes a nondegenerate projective conic

$$Ax_1^2 + Bx_1x_2 + Cx_2^2 + Dx_1x_3 + Ex_2x_3 + Fx_3^2 = 0, \tag{3}$$

acquiring a single ideal point if parabolic, two if hyperbolic, and none if elliptic. Since the single ideal point is associated with the axis of the parabola, and the two ideal points with the asymptotes of the hyperbola, none of these ideal points is associated with the line (2). It follows that the projective line associated with the line (2) meets the projective conic (3) only in the point (x_0, y_0) and is therefore tangent to this conic by definition (VIII, 18.7). To find the equation of this projective line we let (a_1, a_2, a_3) be homogeneous coordinates of the point (x_0, y_0), substitute in (2), and obtain

$$(4)\quad 2Aa_1x_1 + B(a_2x_1 + a_1x_2) + 2Ca_2x_2 + D(a_3x_1 + a_1x_3) + E(a_3x_2 + a_2x_3) + 2Fa_3x_3 = 0.$$

All the terms of this equation are actually of the same form, for $2Aa_1x_1$ can be written $A(a_1x_1 + a_1x_1)$, etc.

Whenever (3) represents a nondegenerate projective conic, and (a_1, a_2, a_3) is a Euclidean point on it, (4) is the equation of the tangent at this point. Although there is nothing about the method used to obtain (4) which would indicate that this statement is also correct when the point is ideal, the interesting fact is that it is correct. We now show this for the case in which the conic is an extended parabola. If the axis of the parabola is vertical, then $B = C = 0$, $E \neq 0$, and the ideal point on the extended parabola is $(0, 1, 0)$. Substituting for B, C, a_1, a_2, a_3 in (4), we get $Ex_3 = 0$, or $x_3 = 0$, which is the equation of the ideal line. If the axis of the parabola is not vertical, the ideal point on the extended parabola is $(-2C, B, 0)$, as was shown earlier. Substituting for a_1, a_2, a_3 in (4) we now get $(B^2 - 4AC)x_1 + (BE - 2CD)x_3 = 0$. Since $B^2 - 4AC = 0$ whenever the conic (1) is of the parabolic type (parabola or a degenerate case), and $BE - 2CD \neq 0$ in the nondegenerate case,* the preceding equation reduces to $x_3 = 0$.

9.1 *If (a_1, a_2, a_3) is any point on a nondegenerate projective conic given by equation* (3), *the tangent to this conic at this point has the equation* (4).

One application of this theorem is in analytic geometry, where it can be used to find the asymptotes of a hyperbola given by equation (1) when $B \neq 0$. Since the tangents to the extended conic (3) at its ideal points are the projective lines associated with the asymptotes of (1), Theorem 9.1 makes it possible for us to determine the equations of these asymptotes. We illustrate this for the hyperbola

$$4x^2 - 24xy + 11y^2 + 56x - 58y + 95 = 0.$$

The corresponding extended hyperbola is

$$4x_1^2 - 24x_1x_2 + 11x_2^2 + 56x_1x_3 - 58x_2x_3 + 95x_3^2 = 0.$$

To find its ideal points we put 0 for x_3 and obtain

$$4x_1^2 - 24x_1x_2 + 11x_2^2 = 0.$$

Dividing by x_2^2 gives

$$4\left(\frac{x_1}{x_2}\right)^2 - 24\left(\frac{x_1}{x_2}\right) + 11 = 0.$$

---

* See G. Salmon, *A Treatise on Conic Sections*, p. 198.

This quadratic in x_1/x_2 has the solutions $11/2$ and $1/2$. Hence the ideal points are $(11, 2, 0)$ and $(1, 2, 0)$. Substituting in equation (4), we obtain

$$2x_1 - x_2 + 3x_3 = 0 \qquad \text{and} \qquad 2x_1 - 11x_2 + 25x_3 = 0,$$

respectively, as the equations of the tangents at these points, and hence

$$2x - y + 3 = 0 \qquad \text{and} \qquad 2x - 11y + 25 = 0$$

as the equations of the asymptotes of the given hyperbola.

Our definition of a tangent to a nondegenerate projective conic $\bar{c}$, as a line meeting $\bar{c}$ in just one point, was made as a matter of convenience. Such a line also satisfies the definition of a tangent used in calculus, for it is the limiting position of a secant line, which is a line that meets the conic in two points. To prove this let A be any point on $\bar{c}$, and P be a variable Euclidean point on $\bar{c}$ which approaches A. Suppose, first, that A is Euclidean. Let c be the Euclidean conic with which $\bar{c}$ is associated. It is known from analytic geometry or calculus that the straight line AP approaches a definite straight line t as $P \to A$, and that t meets c only in A. Known also is that if c is a parabola, t is never its axis or parallel to its axis; and that if c is a hyperbola, t is never an asymptote or parallel to one. Hence the extended line associated with t meets $\bar{c}$ only in A and is the tangent at A according to our definition.

Now suppose that A is ideal. Let $P \to A$. If $\bar{c}$ is an extended parabola, we must show that the line AP approaches the ideal line; if $\bar{c}$ is an extended hyperbola we must show that line AP approaches the extended asymptote through A. Our proof is algebraic and rests on the fact that one line is, by definition, the limit of another line if the latter varies in such a way that the coefficients in its equation approach the coefficients in the equation of the first line, respectively. Consider, for example, the extended parabola $x_2x_3 = x_1^2$, which is associated with the parabola $y = x^2$. Then A is the point $(0, 1, 0)$, P has the coordinates $(b, b^2, 1)$, where b is variable, and the secant AP has the equation $x_1 - bx_3 = 0$ by §3, Ex. 8. The coordinates of P, taken as $(1/b, 1, 1/b^2)$, approach $(0, 1, 0)$ as $b \to \infty$. Thus $P \to A$ by 2.4. The equation of AP, written $(-1/b)x_1 + 0 \cdot x_2 + x_3 = 0$, has the coefficients $-1/b$, 0, 1. When line AP varies so that $b \to \infty$, these coefficients approach 0, 0, 1, which are the coefficients of $x_3 = 0$, the equation of the ideal line. Thus the secant AP approaches the ideal line, which is the tangent at A. As another example, consider the extended hyperbola $x_1x_2 = x_3^2$, which is associated with the hyperbola $xy = 1$. Take A and P to be $(0, 1, 0)$ and $(b, 1/b, 1)$, respectively. Then line AP has the equation $x_1 - bx_3 = 0$. The coordinates of P, taken as $(b^2, 1, b)$, approach $(0, 1, 0)$ as $b \to 0$. Thus $P \to A$. When line AP varies so that $b \to 0$, the coefficients $1, -b, 0$ in its equation respectively approach 1, 0, 0, which are the coefficients of $x_1 = 0$, the equation of the extended asymptote through A. Thus line AP approaches this asymptote, which is the tangent at A. The method illustrated in these two examples can be applied to all nondegenerate projective conics. This is left for the exercises and completes the proof of the theorem:

9.2 *The tangent to a nondegenerate projective conic at a point A is approached by the secant line AP as $P \to A$, where P is a variable point of the conic different from A.*

EXERCISES

1. Find the equation of the tangent to the ellipse
$$5x^2 + 4xy + 2y^2 - 24x - 12y + 25 = 0$$
at the point $(1, 3)$.

2. Use (4) to find the equations of the tangents at the specified points of the projective conics corresponding to the following Euclidean conics:

(a) $y = x^2$, at $(0, 1, 0)$; (b) $x = y^2$, at $(1, 0, 0)$;
(c) $y = ax^2 + bx + c$, at $(0, 1, 0)$; (d) $xy = k$, at $(1, 0, 0)$;
(e) $b^2x^2 - a^2y^2 = a^2b^2$, at $(a, b, 0)$; (f) $x^2 + y^2 = r^2$, at $(r, 0, 1)$.

3. Use (4) to find the equations of the tangents to the extended hyperbola
$$3x_1^2 - 4x_1x_2 - 4x_2^2 + 16x_1x_3 + 16x_2x_3 - 12x_3^2 = 0$$
at its ideal points.

4. Find the coordinates of the center of the hyperbola
$$3x^2 - 4xy - 4y^2 + 16x + 16y - 12 = 0.$$

5. Prove 9.1 when the conic is an extended hyperbola and the point is ideal. Consider the following cases: (a) A, C both zero; (b) A, C not both zero. (*Suggestion* in (b): Assuming $C \neq 0$, find the equations of the asymptotes of the corresponding hyperbola using the fact that they meet in the point $\left(\frac{2CD - BD}{B^2 - 4AC}, \frac{2AE - BD}{B^2 - 4AC}\right)$ and that the ideal points associated with them are $(2C, -B \pm \sqrt{B^2 - 4AC}, 0)$. Then show that the equations of the extended asymptotes are identical with (4) when the proper values are substituted for a_1, a_2, a_3.)

6. Show that (4) can be written so that its terms are the products, respectively, of A, B, C, D, E, F and second order determinants.

7. Prove 9.2 when A is ideal and the projective conics are associated with the following Euclidean conics:

(a) $y = ax^2 + bx + c$; (b) equation (3), §8, where $B^2 - 4AC = 0$;
(c) $Bxy + Dx + Ey + F = 0$; (d) $b^2x^2 - a^2y^2 = a^2b^2$;
(e) equation (3), §8, where $B^2 - 4AC > 0$.

10. Projective Curves of Higher Degree. In the opening section of the book we mentioned that the degree of a curve is one of its projective properties. Because of discussions already given, we can say that this has been verified for curves of the first and second degree, or, as they are more precisely named, *projective plane curves of the first and second degree.* By a **projective plane curve of degree** (or **order**) ***n***, briefly, a **projective *n*-ic,** is meant the locus in the projective plane of an equation in which a homogeneous polynomial of degree n in x_1, x_2, x_3 is equal to zero. That a projectivity preserves the property of being a first-degree projective curve follows from two facts: (1) a projectivity sends a line into a line (VIII, 19.5), and (2) the equation of a line is of the first degree. That a projectivity preserves the property of being a second-degree projective curve, or conic, was proved earlier in this chapter as Theorem 8.2.

The proof that a projective plane curve of third degree, or *projective cubic,* goes into another such curve under a projectivity is just like the proof that we gave for conics. The same can be said for *projective quartics, projective quintics,* etc. We therefore can state the theorem:

10.1 *The property of being a projective curve of degree n is projective, that is, it is preserved by all projectivities in the plane.*

As is the case with lines and conics, a projective n-ic when $n > 2$ often consists of Euclidean points and ideal points, and may then be regarded as an extended curve. The part composed of all the Euclidean points is called a **Euclidean curve of degree** (or **order**) ***n***, and its equation is found by expressing the projective n-ic in nonhomogeneous coordinates. When an n-ic is an extended curve, each ideal point on it is generally approached by a suitably chosen variable Euclidean point of the curve. When $n > 2$, **the** tangent to a projective n-ic at a point A is, by definition, the line which **is** approached (see §3, Ex. 15) by the line AP as $P \to A$, where P is a variable point of the curve different from A. (As was seen in §9, this definition can also be used when $n = 2$ instead of the one we worked with.) A tangent to an n-ic when $n > 2$ may meet it in more than one point.

An historical remark is in order here. The seventeenth century discovery that curves in Euclidean geometry could be generated or defined by algebraic equations was important for various reasons, among them that it (1) greatly extended the number and variety of curves, (2) offered a simple way of classifying such curves and hence of studying them systematically, and (3) made possible the introduction into Euclidean geometry of modern calculational methods. This algebraic approach naturally found its way into projective geometry, where, in time, one saw a further value for it, namely, as a means of eliminating metric and affine concepts from the subject. As long as two points, for example, are thought of intuitively as two positions in space, it is unavoidable that one should associate a distance with them, but no such association is necessary if points are thought of algebraically as number triples. Similarly, there is nothing metric about the individual letters and symbols in the equation of an extended line, whereas our definition of such a line, of course, does involve something that is metric.

EXERCISES

1. Put the following equations into homogeneous form, find the ideal points on the resulting cubics, and, when the latter are nondegenerate, show that these points are approached by variable Euclidean points on the cubics (see §8, Ex. 15). Describe the character of each of these cubics, viewed topologically.

(a) $y = x^3$; (b) $x = y^3$;
(c) $x^2 = y^3$; (d) $y^2 = x^3$;
(e) $y = ax^3 + bx^2 + cx + d$; (f) $x^3 - xy^2 = 0$;
(g) $xy(x + y - 1) = 0$; (h) $x^2(x - 1) = 0$.

2. The graph of $x^3 + y^3 - 3axy = 0$, known as a *folium of Descartes*, has an asymptote of slope -1, as shown in Fig. VII, 5(b). Find the ideal point of the corresponding projective cubic. Give an example of a Euclidean curve which has the same topological structure as this cubic.

3. (a) Write down the general equation of a projective cubic. (b) What is the greatest number of ideal points it can have? Why?

4. Prove that the property of being a cubic curve is projective.

5. Put the following equations into homogeneous form; find the ideal points on the resulting quartics; and, in (a)–(g), describe the nature of each quartic.

(a) $y = x^4$; (b) $x = y^4$;
(c) $x^4 + y^4 = 0$; (d) $x^4 + y^4 = 1$;
(e) $x^4 - y^4 = 0$; (f) $x^4 + y^4 + 1 = 0$;
(g) $x^4 + 1 = 0$; (h) $x^4 - x^3y = 2$;
(i) $x^4 - x^2y^2 = 3$; (j) $x^4 - 5x^2y^2 + 4y^4 = 1$.

6. What is the greatest number of ideal points possessed by a quartic? by an n-ic?

7. Show that the ideal line is tangent to the cubic $x_2x_3^2 = x_1^3$.

8. To which of the following curves is the ideal line tangent?

(a) $x_1^2x_3 = x_2^3$; (b) $x_2x_3^3 = x_1^4$; (c) $x_1^3 - x_1x_2^2 = 0$.

9. Determine whether the ideal line is tangent to the projective cubic associated with $y = ax^3 + bx^2 + cx + d$.

10. Given the cubic $x_1^2x_2 - x_3^3 = 0$, (a) find its ideal points, (b) show that each ideal point is approached by a variable Euclidean point on the cubic, (c) find the tangent at each ideal point.

11. Homogeneous Coordinates of Lines. According to the principle of duality, *point* and *line* have equal status in the projective plane. Considerable inequality, however, still remains in the way we have been dealing with them. We regard a line as a set of points, but do not think of a point as a set of lines, although we ought to in view of the principle of duality. Similarly, we define a curve as a locus of points, and not as a locus of lines. It is the points, not the lines, which have coordinates, and it is the lines, not the points, which have equations. All of these differences can be removed and in the rest of the chapter we show how this can be done.

Let us first find homogeneous coordinates for lines. Since the homogeneous coordinates of a point are three numbers, not all zero, which distinguish it from all other points, so the homogeneous coordinates of a line should be three numbers, not all zero, which distinguish it from all other lines. Also, each line should have infinitely many sets of such coordinates, each two sets being proportional. With these ideas in mind, consider the line

$$2x_1 + 4x_2 - 3x_3 = 0,$$

for example. Can we take the coefficients 2, 4, -3 as coordinates of this line? This seems reasonable since these numbers do distinguish this line from all other lines, and the proportional numbers $2k$, $4k$, $-3k$, where k is any number except 0, represent the same line, being the coefficients of the equation

$$2kx_1 + 4kx_2 - kx_3 = 0.$$

We are therefore led to the following definition:

11.1 Definition. *The homogeneous coordinates of a line*

$$a_1x_1 + a_2x_2 + a_3x_3 = 0$$

are the numbers a_1, a_2, a_3 or any proportional numbers ka_1, ka_2, ka_3, where $k \neq 0$. We enclose these numbers in square brackets to distinguish them from the numbers representing points.

According to this definition, the numbers 0, 0, 0 are not the coordinates of any line since the coefficients in the equation of a line are never all zero. Also, it follows from the definition that the point (b_1, b_2, b_3) is on the line $[a_1, a_2, a_3]$ if and only if $a_1x_1 + a_2x_2 + a_3x_3 = 0$ is satisfied by (b_1, b_2, b_3). This gives the basic theorem:

11.2 *The point (b_1, b_2, b_3) is on the line $[a_1, a_2, a_3]$, or the line is on the point, if and only if $a_1b_1 + a_2b_2 + a_3b_3 = 0$.*

In view of §3, Ex. 10, we can state the following:

11.3 *If $[A_1, A_2, A_3]$, $[B_1, B_2, B_3]$ are distinct lines, their point of intersection has the coordinates*

$$\begin{vmatrix} A_2 & A_3 \\ B_2 & B_3 \end{vmatrix}, \quad \begin{vmatrix} A_3 & A_1 \\ B_3 & B_1 \end{vmatrix}, \quad \begin{vmatrix} A_1 & A_2 \\ B_1 & B_2 \end{vmatrix}.$$

In previous sections we found the limit of a line by using the coefficients in its equation. The procedure can now be stated in terms of coordinates:

11.4 Definition. *A variable line g is said to approach a fixed line b, or to have b as limit line (in symbols, $g \to b$), if g varies so that its coordinates approach those of b, respectively.*

It is customary to refer to coordinates of lines as **line coordinates** and to coordinates of points as **point coordinates.**

EXERCISES

1. Give three sets of coordinates, including the general set, for each of the following lines:

(a) $x_1 - x_2 + 2x_3 = 0$; (b) $3x_1 + x_2 = 0$; (c) $x_3 = 0$;

(d) $\dfrac{x_1}{2} + \dfrac{x_2}{3} - x_3 = 0$; (e) $x_1 = 0$; (f) $3x_2 - 5x_3 = 0$.

2. Give equations of the lines having the following coordinates:

(a) $[-3, -2, -1]$; (b) $[\frac{1}{3}, \frac{1}{4}, \frac{1}{5}]$; (c) $[0, 3, 0]$.

3. Find coordinates for the following lines:

(a) the extended x-axis; (b) the ideal line;

(c) the extended line associated with the straight line through the origin with slope -2.

4. Determine whether

(a) the line $[-1, 4, 2]$ goes through the point $(3, 1, \frac{1}{2})$,

(b) the point $(3, 1, -\frac{1}{2})$ is on the line $[-1, 4, 2]$,

(c) the line $[0, 0, 1]$ contains the points $(1, 0, 0)$, $(0, 1, 0)$,

(d) the line $[0, 0, 1]$ is on all the points $(x_1, x_2, 0)$,

(e) the point $(0, 0, 1)$ is on all the lines $[a_1, a_2, 0]$.

5. Describe the following families of lines:

(a) $[a_1, a_2, 0]$; (b) $[a_1, 0, a_3]$; (c) $[0, a_2, a_3]$.

6. Describe the following lines:

(a) $[a_1, 0, 0]$; (b) $[0, a_2, 0]$; (c) $[0, 0, a_3]$.

7. Find the line which is approached by the line $[b, b^2, b^3]$ when $b \to 2$; when $b \to \infty$; when $b \to 0$.

8. If the distinct lines g and a meet line b in points P and A, respectively, and $g \to a$, prove that $P \to A$.

12. Equations of Points.

Having defined coordinates of lines, we now show how to associate an equation with each point. For this purpose it is convenient to represent the coordinates of the general line as $[u_1, u_2, u_3]$, which is analogous to using (x_1, x_2, x_3) as the coordinates of the general point. Thus, for the line $[2, 3, -1]$ we have $u_1 = 2$, $u_2 = 3$, $u_3 = -1$. Guided by the principle of duality we make the following definition:

12.1 Definition. *By an equation of a point is meant an equation in the variables u_1, u_2, u_3, representing line coordinates, which is satisfied by those, and only those, lines which are on the point.*

Since the equation of a line is a first-degree equation in x_1, x_2, x_3 whose coefficients are the coordinates of the line, we expect the equation of a point to be a first-degree equation in u_1, u_2, u_3 whose coefficients are the coordinates of the point. Under this view an equation of the point (b_1, b_2, b_3) is $b_1u_1 + b_2u_2 + b_3u_3 = 0$. This is correct according to the above definition, for a given line $[a_1, a_2, a_3]$ satisfies this equation if $b_1a_1 + b_2a_2 + b_3a_3 = 0$, and we know that this equality holds if and only if the line $[a_1, a_2, a_3]$ goes through the point (b_1, b_2, b_3).

12.2 *An equation of the point (b_1, b_2, b_3) is*

$$b_1u_1 + b_2u_2 + b_3u_3 = 0,$$

and so is any equation obtained from this by multiplication by a constant not zero. Conversely, every given homogeneous first-degree equation in u_1, u_2, u_3 is the equation of a point, namely, the point whose coordinates are the coefficients in the equation.

The proof of the converse part of this theorem is immediate and is left to the student. As an illustration of the theorem we note that one equation of the point $(3, -5, 8)$ is $3u_1 - 5u_2 + 8u_3 = 0$, while another is $6u_1 - 10u_2 + 16u_3 = 0$. The line $[2, 6, 3]$ satisfies these equations and hence is on the point represented by them.

Problems in analytic projective geometry which were previously solved in terms of coordinates of points and equations of lines can now be solved by using equations of points and coordinates of lines, no new algebraic techniques being needed. To illustrate this, let us find the coordinates of the line joining the points whose equations are $2u_1 + u_2 - u_3 = 0$ and $u_1 + 3u_2 + u_3 = 0$.

Inasmuch as the line goes through the given points, its coordinates satisfy their equations. These coordinates are therefore the common solution of these equations. Since the technique of solving equations does not depend on the geometrical meaning of the unknowns, we may obtain the common solution by using the formula stated in §3, Ex. 10. This gives for u_1, u_2, u_3 the values

$$\begin{vmatrix} 1 & -1 \\ 3 & 1 \end{vmatrix}, \quad \begin{vmatrix} -1 & 2 \\ 1 & 1 \end{vmatrix}, \quad \begin{vmatrix} 2 & 1 \\ 1 & 3 \end{vmatrix},$$

or any three numbers proportional to them. Hence the desired coordinates are $[4, -3, 5]$, or, more generally, $[4k, -3k, 5k]$, $k \neq 0$.

As a further illustration, let us find the equation of the point of intersection of the lines $[3, 1, 2]$, $[-1, 4, -3]$.

The desired equation is $b_1u_1 + b_2u_2 + b_3u_3 = 0$ and our problem is to find b_1, b_2, b_3. Since the lines go through the point represented by this equation, their coordinates must satisfy this equation. Accordingly, we have

$$3b_1 + b_2 + 2b_3 = 0$$

$$-b_1 + 4b_2 - 3b_3 = 0.$$

Since the values of b_1, b_2, b_3 that we are seeking are the common solution of these equations, we again use the formula of §3, Ex. 10. One common solution is therefore

$$b_1 = \begin{vmatrix} 1 & 2 \\ 4 & -3 \end{vmatrix} = -11, \quad b_2 = \begin{vmatrix} 2 & 3 \\ -3 & -1 \end{vmatrix} = 7, \quad b_3 = \begin{vmatrix} 3 & 1 \\ -1 & 4 \end{vmatrix} = 13,$$

and the desired equation corresponding to this solution is

$$-11u_1 + 7u_2 + 13u_3 = 0.$$

EXERCISES

1. Give equations of the following points:
(a) $(2, 1, -3)$; (b) the origin; (c) the ideal point on the extended x-axis; (d) the ideal point associated with lines of slope 2; (e) $(0, k, 0)$.

2. Using the circular model of the projective plane, plot the points represented by the following equations:
(a) $u_1 + 3u_2 = 0$; (b) $u_1 - 3u_2 = 0$; (c) $3u_2 = 0$.

3. Using a rectangular coordinate system, plot the points with the following equations:
(a) $6u_1 - 3u_2 - u_3 = 0$; (b) $2u_1 + \frac{1}{3}u_3 = 0$; (c) $3u_2 - u_3 = 0$.

4. Find the coordinates of the lines through the following pairs of points:
(a) $2u_1 - 3u_2 + u_3 = 0$ and $3u_1 + 4u_2 - 2u_3 = 0$;
(b) $3u_1 - 5u_2 = 0$ and $4u_2 - u_3 = 0$.

5. Find the equations of the points of intersection of the following pairs of lines:
(a) $[3, 4, -2]$, $[0, 4, -1]$; (b) $[3, -5, 0]$, $[2, -3, 1]$.

6. Find the equation of the point in which the line $[-1, 1, -2]$ meets the line through the points $3u_1 + 4u_2 - u_3 = 0$ and $5u_1 - 3u_2 + u_3 = 0$.

7. Find the coordinates of the line which goes through the point $2u_1 + 3u_2 + u_3 = 0$ and the point of intersection of the lines [1, 1, 1], [2, 1, 3].

8. State an algebraic test for each of the following: (a) the collinearity of three points when their equations are given; (b) the concurrence of three lines when their coordinates are given.

13. Linear Combination of Lines. In §4 we gave the definition of a linear combination of number-triples and applied it to triples which represent the coordinates of points. The definition applies equally well to triples representing coordinates of lines. In fact, if the words *point* and *line* as used in the proofs of 4.1 and 4.2 are interchanged, it will be found that what results is a proof of the following:

13.1 *If* $[a_1, a_2, a_3]$, $[b_1, b_2, b_3]$ *represent distinct lines, then* $[a_1 + kb_1, a_2 + kb_2, a_3 + kb_3]$, *where* $k \neq 0$, *represents a line different from them which passes through their point of intersection. Conversely, any line different from them which goes through their point of intersection has coordinates expressible as* $[a_1 + kb_1, a_2 + kb_2, a_3 + kb_3]$, *where* $k \neq 0$.

This theorem, like 4.1 and 4.2, can be stated more compactly if we denote the coordinates of the given lines by $\mathbf{a}$ and $\mathbf{b}$, and the specified linear combination of these coordinates by $\mathbf{a} + k\mathbf{b}$.

In §5 we proved two theorems on the cross-ratio of points whose coordinates are expressed in compact vector form. The analogous theorems hold for lines:

13.2 *If a, b, c, d are distinct concurrent lines with coordinates* $\mathbf{a}, \mathbf{b}, \mathbf{a} + k\mathbf{b}, \mathbf{a} + k'\mathbf{b}$, *then* $(ab, cd) = k/k'$.

13.3 *Four concurrent lines a, b, c, d, in that order, are a harmonic set if and only if they have coordinates of the form* $\mathbf{a}, \mathbf{b}, \mathbf{a} + k\mathbf{b}, \mathbf{a} - k\mathbf{b}$, *respectively.*

The above theorem is a corollary of 13.2. We omit the proof of 13.2.*

EXERCISES

1. If $\mathbf{a}$ represents the coefficients of the line $2x_1 + x_2 - x_3 = 0$, and $\mathbf{b}$ those of the line $x_1 - 2x_2 + x_3 = 0$, give the equation of the line whose coefficients are

(a) $\mathbf{a} + \mathbf{b}$, (b) $\mathbf{a} - \mathbf{b}$, (c) $\mathbf{a} + 2\mathbf{b}$, (d) $3\mathbf{a} + 2\mathbf{b}$.

---

* The proof cannot be obtained by interchanging *point* and *line* in the proof of the analogous theorem 5.1 since the latter proof is not purely projective.

2. If **a** is [3, -1, 2] and **b** is [4, 0, 5], verify that the following lines are distinct from them and go through their point of intersection: (a) $\mathbf{a} + \mathbf{b}$; (b) $\mathbf{a} - \mathbf{b}$; (c) $\mathbf{a} + 2\mathbf{b}$.

3. The following lines go through the specified point of intersection in Ex. 2 and hence have cordinates of the form $\mathbf{a} + k\mathbf{b}$. Find k in each case.

(a) $[1, -1/7, 1]$; (b) $[2, 2, 6]$; (c) $[10, -2, 9]$.

4. Verify that the lines $[1, 2, 3]$, $[-1, 2, -3]$, $[0, 4, 0]$, $[2, 0, 6]$, in that order, are a harmonic set. Are the lines $[1, 2, 3]$, $[-1, 2, -3]$, $[0, 1, 0]$, $[1, 0, 3]$ a harmonic set? Are the lines $[1, 2, 3]$, $[1, -2, 3]$, $[0, 4, 0]$, $[2, 0, 6]$ a harmonic set?

14. Projective Transformations in Line Coordinates. Our next task in giving lines equal status with points is to express a projective transformation in terms of line coordinates.

If g' is the image of a line g under such a transformation, and $[u_1, u_2, u_3]$ are coordinates of g, then the equations we seek should express the coordinates of g' in terms of u_1, u_2, u_3. To obtain these equations we first find the equation of g' by applying the transformation, expressed in point coordinates, to the equation of g, which is

$$u_1x_1 + u_2x_2 + u_3x_3 = 0.$$

That is, we substitute for x_1, x_2, x_3 in the above equation the expressions for them given by equations (10), §6, which in concise form are the following:

$$x_i = A_ix_1' + B_ix_2' + C_ix_3', \qquad i = 1, 2, 3.$$

After this substitution is made and terms are regrouped, the equation of g' is found to be

$$(u_1A_1 + u_2A_2 + u_3A_3)x_1' + (u_1B_1 + u_2B_2 + u_3B_3)x_2' + (u_1C_1 + u_2C_2 + u_3C_3)x_3' = 0.$$

The coefficients of x_1', x_2', x_3' in this equation being coordinates of g', let us denote them by u_1', u_2', u_3':

$$\text{(1)} \qquad \begin{cases} u_1' = A_1u_1 + A_2u_2 + A_3u_3 \\ u_2' = B_1u_1 + B_2u_2 + B_3u_3 \\ u_3' = C_1u_1 + C_2u_2 + C_3u_3, \end{cases} \qquad \delta = \begin{vmatrix} A_1 & A_2 & A_3 \\ B_1 & B_2 & B_3 \\ C_1 & C_2 & C_2 \end{vmatrix} \neq 0.$$

These are the desired equations of the transformation in terms of line coordinates. Like equations (9) and (10), §6, they are linear and homogeneous. Their determinant differs from the determinant in (10) only in that the rows and columns are interchanged, and hence has the same nonzero value δ.

To express the transformation (1) in implicit form, we solve for u_1, u_2, u_3. In the case of u_1 we get

$$u_1 = \frac{\begin{vmatrix} B_2 & B_3 \\ C_2 & C_3 \end{vmatrix} u_1' - \begin{vmatrix} A_2 & A_3 \\ C_2 & C_3 \end{vmatrix} u_2' + \begin{vmatrix} A_2 & A_3 \\ B_2 & B_3 \end{vmatrix} u_3'}{\delta}$$

On recalling the meanings of Δ, A_2, B_2, C_2, etc., from equations (9) and (10) of §6, one can show that the above equation reduces to

$$u_1 = \frac{a_1u_1'\Delta + b_1u_2'\Delta + c_1u_3'\Delta}{\delta} = \frac{a_1u_1' + b_1u_2' + c_1u_3'}{\Delta}$$

since $\delta = \Delta^2$. Similar results are obtained for u_2 and u_3. The implicit form of (1) is therefore

$$\begin{cases} u_1 = (a_1u_1' + b_1u_2' + c_1u_3')/\Delta \\ u_2 = (a_2u_1' + b_2u_2' + c_2u_3')/\Delta \\ u_3 = (a_3u_1' + b_3u_2' + c_3u_3')/\Delta, \end{cases}$$

or, since $[u_1, u_2, u_3]$ and $[u_1\Delta, u_2\Delta, u_3\Delta]$ represent the same line,

$$\text{(2)} \qquad \begin{cases} u_1 = a_1u_1' + b_1u_2' + c_1u_3' \\ u_2 = a_2u_1' + b_2u_2' + c_2u_3' \\ u_3 = a_3u_1' + b_3u_2' + c_3u_3', \end{cases} \qquad \Delta = \begin{vmatrix} a_1 & b_1 & c_1 \\ a_2 & b_2 & c_2 \\ a_3 & b_3 & c_3 \end{vmatrix} \neq 0.$$

14.1 *The projective transformation which sends points into points in accordance with equations* (9) *and* (10) *of* §6 *sends lines into lines in accordance with the above equations* (1) *and* (2).

Equations (1) and (2) permit the same ease in answering questions concerning lines in projective transformations as do equations (9) and (10) of §6 in answering questions concerning points. In fact, we have seen that (9) and (10) are awkward to use in problems involving lines.

Although equations (1) are the equations of a projective transformation in line coordinates, it might seem that they can be found only after we know the equations of the same transformation in point coordinates, for A_1, B_1, C_1, etc. are the cofactors of a_1, b_1, c_1, etc. in the determinant Δ. If this were really the case, it would mean that points are more basic than lines. But the Principle of Duality tells us otherwise. Actually, the whole procedure of finding equations for projective transformations can be reversed. Instead of starting out with equations in point coordinates and then deriving equations in line coordinates, we could, if we wished, start out with the above equations (1), where the coefficients are any numbers such that their determinant is not zero, prove that they represent a projective transformation, and then show that this transformation, expressed in point coordinates, is given by equations

(9), §6, where a_1, a_2, a_3, etc. would now denote the cofactors of A_1, A_2, A_3, etc.

EXERCISES

1. Express the transformation

$$\begin{cases} x_1' = x_1 - x_2 + x_3 \\ x_2' = 2x_1 + x_2 - x_3 \\ x_3' = x_1 - 3x_2 - 2x_3 \end{cases}$$

in implicit form, find its equations, both explicit and implicit, in line coordinates, and then find the images of the following:

(a) (4, 0, 1); (b) [2, −1, 0]; (c) $3x_1 - x_2 - x_3 = 0$; (d) $u_1 - 3u_2 + u_3 = 0$.

2. Find the images of the following lines under the transformation

$$\begin{cases} u_1' = u_1 + u_2 - u_3 \\ u_2' = u_1 - u_2 + u_3 \\ u_3' = 2u_1 + u_2 - 3u_3. \end{cases}$$

(a) [3, − 1, 2]; (b) [0, 0, 1]; (c) the ideal line; (d) $2x_1 + x_2 - 5x_3 = 0$.

3. (a) Find the implicit form of the transformation in Ex. 2. (b) Use the result in (a) to find the equation of the image of the point $u_1 + 2u_2 - u_3 = 0$.

4. Express the transformation of Ex. 2 in point coordinates. Then find the coordinates of the image of the point (1, 2, −1). Compare your answer with that in Ex. 3(b).

5. Prove that the transformation (1) establishes a 1-1 correspondence among the lines of the plane.

6. Assuming the fact stated in Ex. 5, prove by a purely geometric argument that the transformation establishes a 1-1 correspondence among the points of the plane.

7. Prove that the transformation (1) preserves the concurrence and cross-ratio of lines. (*Suggestion.* Follow the method of §6, Ex. 4.)

15. Line Curves. In all our previous discussions *curve* has meant a locus, or set, of points. The points in each set have satisfied some geometric condition and this condition was always expressed by an equation involving the coordinates of the points. Although the reader, in his study of mathematics, has also encountered sets of lines meeting some condition, such as that of being tangent to a circle or having a specified slope, he has perhaps never applied the name "curve" to a set of lines. Yet, in keeping with the Principle of Duality, we should do this when working in the projective plane. Henceforth, then, we shall refer to certain sets of projective lines as *line curves*. More precisely, a set of projective lines is called a **line curve** if the lines satisfy some geometric condition that can be expressed by an equation in the coordinates of the lines. Each line of the set will be said to be *on* or to *belong to* the line curve. Curves that are loci of points as encountered previously will be called **point curves.**

We have already had one example of a line curve, namely, the set of lines satisfying the condition of going through the same point, and have found that this line curve always has an equation of the first degree in u_1, u_2, u_3. Accordingly we shall call it a *line curve of the first degree* (or *class*). By a **line curve of degree** (or **class**) ***n*** we mean a line curve whose constituent lines satisfy an equation which results from setting equal to zero a homogeneous polynomial of degree n in u_1, u_2, u_3.

In order to learn something about the structure and location of a line curve when its equation is given, one naturally tries to draw some of its member lines. This presents no problem when the line curve is of the first degree: we simply determine the point represented by the equation of the curve and draw lines through it. Consider a line curve of the second degree, say $u_1^2 + u_2^2 - u_3^2 = 0$. To draw its graph by plotting lines we have to find sets of values of u_1, u_2, u_3 satisfying this equation and then plot the corresponding lines. Thus, one set of values being [3, 4, 5], we write down the equation $3x_1 + 4x_2 + 5x_3 = 0$ and draw the line it represents. Similarly, corresponding to the set [2, 1, −2] we draw the line $2x_1 + x_2 - 2x_3 = 0$, and so forth. This method of drawing the line curve is clearly very laborious. We therefore do not carry it through. Later, with a knowledge of *nonhomogeneous line coordinates* to aid us, we show how to cope more efficiently with this problem of determining a line curve from its equation. Nonhomogeneous line coordinates will also help us in solving the converse problem of deducing the equation of a line curve from its definition.

16. Nonhomogeneous Line Coordinates. Leaving lines aside for a moment, let us recall how we plot a point, say (1, 3, 2). We do not use these numbers directly since they are homogeneous coordinates, but obtain from them the nonhomogeneous coordinates $(\frac{1}{2}, \frac{3}{2})$, locate these numbers on the x- and y-axes, and thus plot the given point. In analogous fashion we plot a line by finding its nonhomogeneous line coordinates and then utilizing numbers on the x- and y-axes related to these coordinates.

16.1 Definition. *The nonhomogenous line coordinates of the line* $[u_1, u_2, u_3]$, *where* $u_3 \neq 0$, *are the two numbers* u, v *whose values are* $u = u_1/u_3$ *and* $v = u_2/u_3$. *These coordinates are written* $[u, v]$. *The line may be either Euclidean or projective.*

Thus, the line $8x_1 + 3x_2 + 2x_3 = 0$, whose homogeneous coordinates are $[8, 3, 2]$, has the nonhomogeneous coordinates $[4, \frac{3}{2}]$.

The lines for which u_3 is 0 do not have nonhomogeneous coordinates. They are the lines through the origin. All other lines have such coordinates. Thus, the x- and y-axes, extended or not, do not have nonhomogeneous coordinates. The ideal line has the nonhomogeneous coordinates $[0, 0]$.

Let g be a line not passing through the origin, and hence with the equation

$$u_1x_1 + u_2x_2 + u_3x_3 = 0, \tag{1}$$

where $u_3 \neq 0$. Dividing by u_3 we obtain

$$\frac{u_1}{u_3}x_1 + \frac{u_2}{u_3}x_2 + x_3 = 0,$$

or

$$ux_1 + vx_2 + x_3 = 0. \tag{2}$$

It is in this form of the equation of g, where the coefficient of x_3 is 1, that the nonhomogeneous coordinates of g are clearly exhibited as the coefficients of x_1 and x_2. Thus, the line $2x_1 + 3x_2 + x_3 = 0$ has the nonhomogeneous coordinates $[2, 3]$.

To find a geometric interpretation of u and v let us express equation (2) in nonhomogeneous point coordinates (which is possible only if g is not ideal). Dividing by x_3 we obtain

$$ux + vy + 1 = 0 \tag{3}$$

The locus of this equation is the straight line associated with g. If $u \neq 0$ the line (3) has the x-intercept $a = -1/u$. Hence $u = -1/a$. If $v \neq 0$ the line has the y-intercept $b = -1/v$. Hence $v = -1/b$. If $u = 0$, then $v \neq 0$, in which case the line has a y-intercept, no x-intercept, and is perpendicular to the y-axis. Similarly, if $v = 0$, then $u \neq 0$ and the line is perpendicular to the x-axis. We state these results as follows:

16.2 *The nonhomogeneous coordinates* $[u, v]$ *of a line (not ideal nor through the origin) are the negative reciprocals of its x- and y-intercepts, respectively, whenever these intercepts exist. There is no x-intercept when* $u = 0$, *the line being perpendicular to the y-axis, and there is no y-intercept when* $v = 0$, *the line then being perpendicular to the x-axis.*

The line $[2, -3]$, for example, has an x-intercept of $-\frac{1}{2}$ and a y-intercept of $\frac{1}{3}$. The line $[2, 0]$ has the x-intercept $-\frac{1}{2}$ and is perpendicular to the x-axis.

The idea of using the number scales on the axes in order to plot lines is not really new to the student, for he has long known how to draw a line by using its intercepts. What is new here is that this idea is being used in conjunction with the *coordinates* of lines.

EXERCISES

1. Which lines do not have nonhomogeneous coordinates?

2. Plot the lines having the following coordinates:
(a) $[\frac{1}{2}, -\frac{1}{3}]$; (b) $[\frac{1}{2}, 0]$; (c) $[0, -\frac{1}{3}]$; (d) $[-2, 1]$.

3. Are $[0, 0]$ the coordinates of any line? If so, which line?

4. Find the nonhomogeneous coordinates of the following lines whenever such coordinates exist:

(a) $2x_1 + 3x_2 - 4x_3 = 0$; (b) $5x_1 + 2x_3 = 0$; (c) $4x_1 - 3x_2 = 0$;
(d) $2x_3 = 0$; (e) $x_1 = 0$; (f) $x_2 = 0$;
(g) $3x_2 - x_3 = 0$; (h) $7x_1 - 2x_2 + x_3 = 0$; (i) $x_1 - x_2 - \frac{1}{4}x_3 = 0$.

17. Line Curves of the Second Degree. We now return to the problem of drawing the graph of

$$u_1^2 + u_2^2 - u_3^2 = 0, \tag{1}$$

which was considered in §16. Expressing this equation in nonhomogeneous line coordinates, we get

$$u^2 + v^2 = 1,$$

and hence the following partial table of values:

u	0	± 1	$\pm\frac{1}{2}\sqrt{2}$	$\pm\frac{3}{5}$	$\pm\frac{4}{5}$
v	± 1	0	$\pm\frac{1}{2}\sqrt{2}$	$\pm\frac{4}{5}$	$\pm\frac{3}{5}$

The first eight lines represented in this table have been drawn by the use of 16.2, and appear in Fig. IX, 5. In order to call attention to the fact that these lines, and all of the others represented by (1), are tangent to the circle with radius 1 and center at the origin, we have drawn the circle and shown only parts of the lines. We shall return to this after the following two examples.

A line curve whose equation is given can sometimes be drawn more conveniently from a table giving values of intercepts rather than values of u and v. Consider

$$4u_1u_2 - u_3^2 = 0,$$

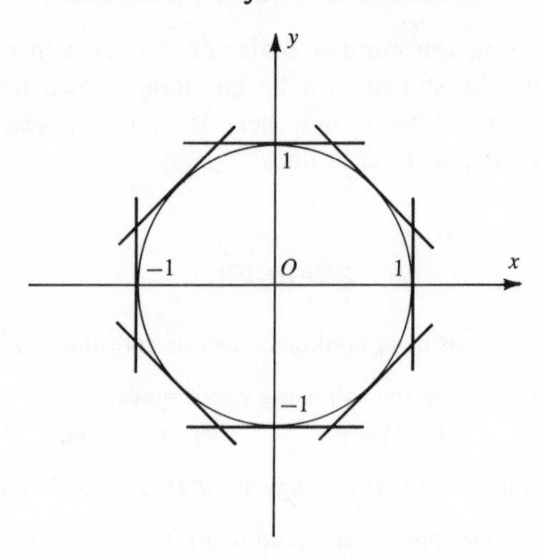

Fig. IX, 5

for example. In nonhomogeneous form this is

$$4uv = 1,$$

or

$$4 = (-1/u)(-1/v).$$

In other words, the product of the intercepts of each line of the curve is 4. The line curve is then easily drawn (Fig. IX, 6). Its member lines appear to be tangent to a hyperbola whose asymptotes are the coordinate axes. Similarly, to draw the graph of

$$5u_1u_2 + u_1u_3 + u_2u_3 = 0,$$

we first obtain

$$5uv + u + v = 0,$$

then

$$5 = (-1/u) + (-1/v),$$

which shows that the sum of the intercepts of each line of the curve is 5. The line curve is then readily drawn (Fig. IX, 7). Its member lines seem to outline a first-quadrant parabola symmetrical to the line bisecting that quadrant.

Let us now return to the line curve (1) and prove that its member lines are tangent to the circle with radius 1 and center at the origin. The equation of the circle is

$$x_1^2 + x_2^2 - x_3^2 = 0.$$

According to (4), §9, the equation of the tangent to the circle at the point (a_1, a_2, a_3) is

$$2a_1x_1 + 2a_2x_2 - 2a_3x_3 = 0$$

or

$$a_1x_1 + a_2x_2 - a_3x_3 = 0.$$

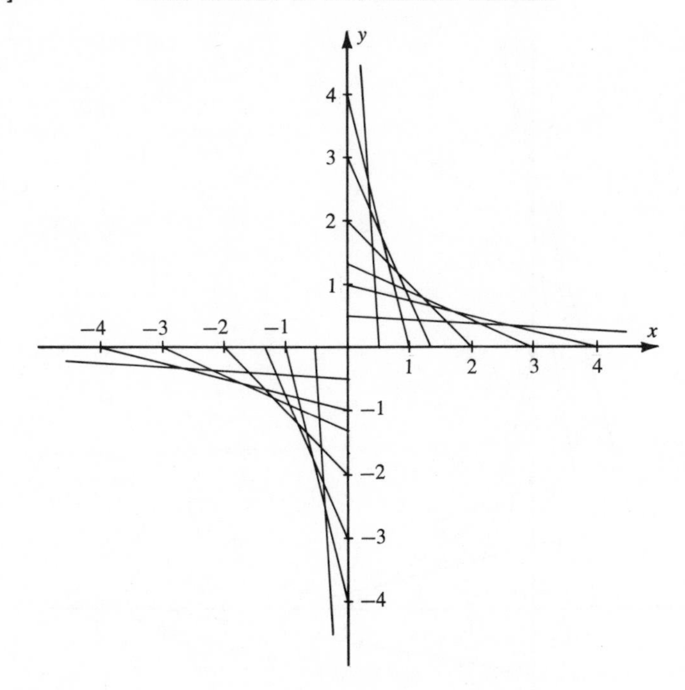

Fig. IX, 6

Hence, for this line we have $u_1 = a_1$, $u_2 = a_2$, $u_3 = -a_3$,

or
$$u_1^2 + u_2^2 - u_3^2 = a_1^2 + a_2^2 - a_3^2 = 0$$

since (a_1, a_2, a_3) is on the circle. Thus, each tangent is a member of the line curve (1). It is easily seen that the converse of this statement is also true. We conclude that the line curve (1) is the set of tangents to the circle. Similarly, it can be shown that the line curves in Figs. IX, 6 and 7 are the sets of tangents to a hyperbola and a parabola, respectively.

As these examples suggest, a simple way of obtaining line curves is to consider the sets of tangents to point curves. When such line curves lie in an extended plane and are drawn in a Cartesian diagram, they therefore present the same contours as do the related point curves. This is one reason why the name *line conic* is applied to the line curve consisting of the tangents to a conic, the latter then being referred to as a *point conic*. Another reason is that such line curves always have second-degree equations. We just proved this for the case in which the point conic is the circle $x^2 + y^2 = 1$ and now prove it in another special case. This time, however, we start out with the equation of the point conic and deduce from it the equation of the line curve consisting of its tangents.

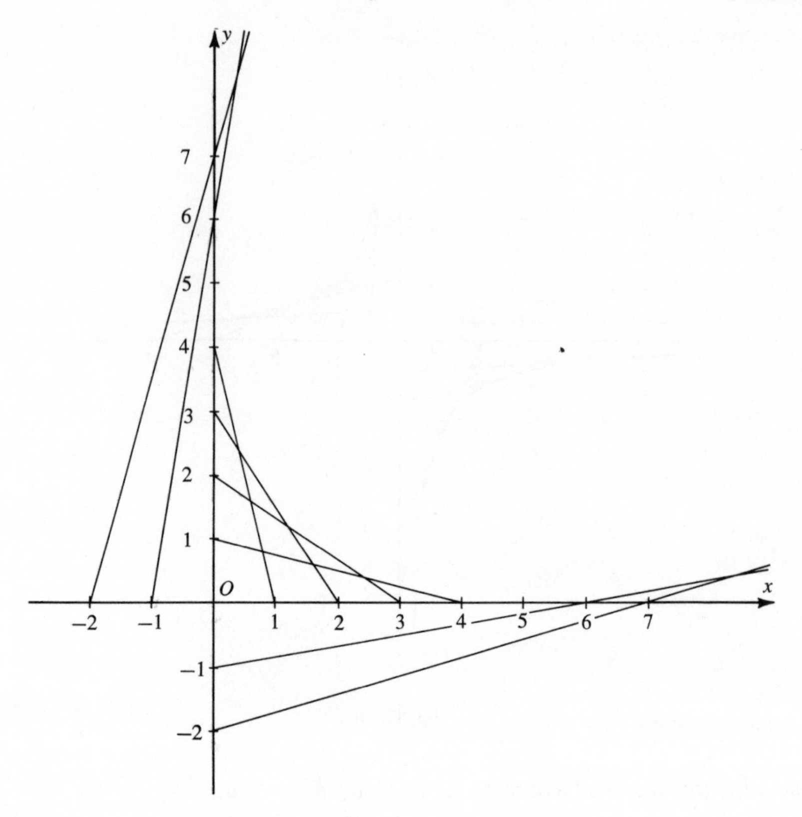

Fig. IX, 7

Consider the extended parabola

(2) $$x_2x_3 = x_1^2.$$

Since the algebraic work is simpler when done in nonhomogeneous coordinates, let us deal with the associated parabola

(3) $$y = x^2.$$

The line curve in which we are interested then consists of the tangents to this parabola and one additional tangent, the ideal line. According to (2), §9, the tangent to the parabola (3) at the point (x_0, y_0) has the equation

$$y + y_0 = 2x_0x.$$

The x- and y-intercepts of this line are $y_0/2x_0$ and $-y_0$, respectively. If the line does not go through the origin, by 16.2 we have

(4) $$y_0/2x_0 = -1/u \quad \text{and} \quad -y_0 = -1/v.$$

Also, since the point (x_0, y_0) is on the parabola we can write

$$y_0 = x_0^2. \tag{5}$$

On eliminating x_0 and y_0 from (4) and (5) we obtain

$$u^2 = 4v \tag{6}$$

as the relation which holds for all tangents to the parabola, excluding the one through the origin. Equation (6) also holds for the ideal line [0, 0]. Thus, (6) is satisfied by all the tangents to the extended parabola (2) except the one through the origin. Conversely, to each pair of values of u, v satisfying (6) there corresponds one such tangent. For, given such values of u and v, we can use (4) to compute a pair of values of x_0, y_0, and these values necessarily satisfy (5) and hence represent a point on the parabola. Now let us write (6)in homogeneous form:

$$u_1^2 = 4u_2u_3. \tag{7}$$

The locus of (7) differs from that of (6) only in that it contains a line through the origin, the line [0, 1, 0]. Since this is the tangent to the parabola which goes through the origin, we conclude that (7) is the equation of the line curve whose lines are tangent to the extended parabola (2). A diagram of this line curve in the circular model of the projective plane appears in Fig. IX, 8. The member lines of the line curve completely fill out the shaded area and outline the extended parabola.

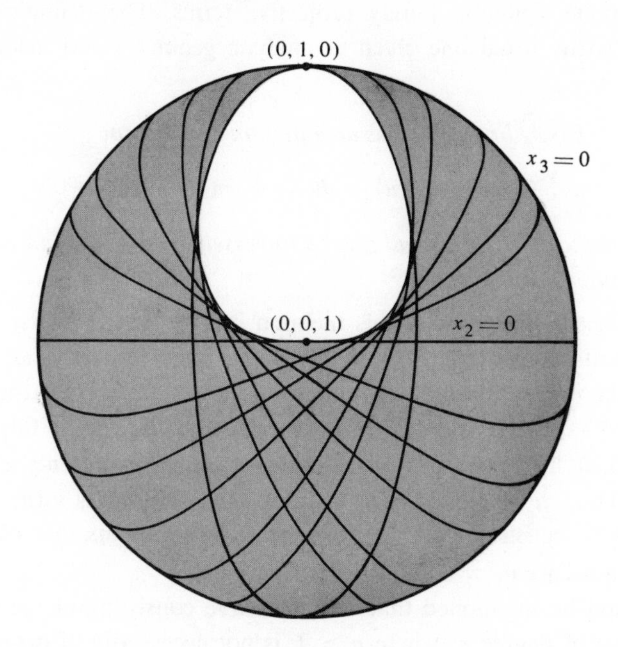

Fig. IX, 8

Let us now restate some of the ideas of this section in a more formal way than we have done thus far.

17.1 Definition. *Another name for conic, as this term was used in previous chapters, is point conic. The line curve which consists of the tangents to a nondegenerate point conic is called a nondegenerate line conic. The line conic is said to correspond to the point conic, and vice versa. A point at which a line of the line conic is tangent to the point conic is called a point of contact of the line conic.*

Clearly, a point of contact of a line conic has the property that through it there passes a unique line of the line conic. This is analogous to the fact that a tangent to a point conic contains a unique point of the point conic.

It will be recalled that a degenerate point conic consists of all the points on one or another of two lines (which may coincide), or else of a single point. On dualizing this we obtain:

17.2 Definition. *A degenerate line conic consists of all the lines on one or another of two points (which may coincide), or else of a single line.*

Thus, by this definition, a degenerate point conic and a degenerate line conic are dual figures. A nondegenerate line conic and a nondegenerate point conic, too, are dual figures, and this is proved in the next chapter after we redefine a point conic in purely projective terms. The definition we have been using is the usual one given in analytic geometry and involves metric notions.

17.3 *Every line conic has an equation of the form*

$$au_1^2 + bu_1u_2 + cu_2^2 + du_1u_3 + eu_2u_3 + fu_3^2 = 0, \tag{8}$$

where the coefficients are not all zero. Conversely, when such an equation has a locus the latter is a line conic.

We consider the proof of this theorem in §18. The converse part of the theorem merits some attention at this time. That (8) can fail to have a locus or that it can represent but a single line we see from the equations $u_1^2 + u_2^2 + u_3^2 = 0$ and $u_1^2 + u_2^2 = 0$, respectively, the single line being the ideal line [0, 0, 1]. When (8) is factorable it clearly represents a degenerate line conic. Thus, $u_1^2 + u_1u_2 + u_1u_3 = 0$ on being written as $u_1(u_1 + u_2 + u_3) = 0$ is seen to consist of all the lines through one or another of the points whose equations are $u_1 = 0$ and $u_1 + u_2 + u_3 = 0$.

It should be mentioned that the line curve consisting of the tangents to a point curve of degree n, where $n > 2$, is not necessarily of degree n.

EXERCISES

1. Draw the line conics represented by the following equations and identify the corresponding point conic in each case:

(a) $u_2^2 = u_1u_3$; (b) $12u_1u_2 - u_3^2 = 0$;
(c) $25u_1^2 - 9u_2^2 = u_3^2$; (d) $9u_1^2 + 9u_2^2 = u_3^2$;
(e) $25u_1^2 + 9u_2^2 = u_3^2$.

2. Find the homogeneous equations of the line conics whose lines have the following properties and draw the line conics:

(a) the sum of the intercepts is -5; (b) the product of the intercepts is -4; (c) the x-intercept is 1 more than the y-intercept; (d) the y-intercept is 2 more than the x-intercept.

3. Find the equations of the line conics corresponding to the following point conics:

(a) $x_1^2 - x_2^2 - x_3^2 = 0$; (b) $x_1^2 + x_2x_3 = 0$;
(c) $2x_1^2 + x_2^2 - x_3^2 = 0$; (d) $x_1x_2 + x_3^2 = 0$.

4. Describe the following degenerate line conics;

(a) $u_1^2 + u_3^2 = 0$; (b) $(u_1 - u_2)^2 + u_3^2 = 0$;
(c) $(u_1 + u_2)(u_1 + u_3) = 0$; (d) $u_1u_3 + u_2u_3 + u_3^2 = 0$.

5. Find the equation in line coordinates of the set of tangents to the parabola $y = x^2 - 2x$.

6. Find the lines of the conic $u_1^2 + u_2^2 - u_3^2 = 0$ which go through the following points:

(a) $2u_1 + u_2 + u_3 = 0$; (b) $u_1 + u_2 = 0$; (c) the origin; (d) $(1, 0, 0)$.

7. Give an algebraic argument to show that at most two tangents can be drawn to a parabola, ellipse, or hyperbola from a point not on it.

8. Find three lines of the conic

$$2u_1^2 - u_2^2 - 2u_3^2 + 5u_2u_3 - u_1u_3 - 3u_1u_2 = 0$$

which are concurrent and thus show that it is degenerate (See Ex. 7).

9. Taking for granted that a nondegenerate point conic and a nondegenerate line conic are dual figures, show that a point of contact of a line conic can be defined by dualizing the definition of a tangent to a point conic (VIII, 18.7).

18. Line Conics as Projective Figures. Thus far it has not been made clear why the study of line conics belongs to projective geometry. The following theorem settles this by showing that the property of being a line conic is projective.

18.1 *A line conic is always sent into a line conic by a projective transformation.*

An algebraic proof is left as an exercise. We give a nonalgebraic proof. Let L be any given line conic. If L is degenerate, it consists of a single line or else of all the lines through one or another of two points, and a projective transformation clearly sends it into a degenerate line conic. If L is nondegenerate, let K be the corresponding nondegenerate point conic. Each projective transformation T sends K into a nondegenerate point conic K' by 8.2. We need only show that T sends the tangents to K into the tangents to K' in a 1-1 manner. Let g be any tangent to K, say at the point P. Then T sends P into a point P' of K', and g into a line g' containing P'. Since T sends the points of K into those of K' in a 1-1 way, and P is the only point of K on g, P' must be the only point of K' on g'. Thus g' is tangent to K' at P'. Conversely, if h' is any tangent to K', say at the point Q', let Q be the point of K which goes into Q', and h the line which goes into h'. Then h must contain Q, but no other point of K, for if h contained another point of K, then h' must contain some other point of K' besides Q'. Thus h is tangent to K at Q.

Incidentally, we have also proved:

18.2 *The property of being a point of contact of a nondegenerate line conic is projective.*

This is analogous to the fact that the property of being a tangent to a nondegenerate point conic is projective (VIII, 18.9).

18.3 *All nondegenerate line conics in the plane are projectively equivalent.*

To prove this, let L_1, L_2 be any two nondegenerate line conics in the plane. The corresponding point conics C_1, C_2 are nondegenerate and therefore projectively equivalent by 8.3. A projectivity then exists which maps C_1 on C_2. Since the property of being a tangent to a nondegenerate point conic is projective, the tangents of C_1 go into the tangents of C_2 under this projectivity. In other words, L_1 goes into L_2.

Using this theorem we can now prove that every line conic L has a second-degree equation (this was stated earlier without proof, as theorem 17.3). Suppose that L is degenerate. If it consists of a single line $[a, b, c]$, its equation is $(bu_1 - au_2)^2 + u_3^2 = 0$ when $c = 0$, and $(u_1 - au_3)^2 + (u_2 - bu_3)^2 = 0$ when $c = 1$. (When $c \neq 0$ there is no loss of generality in supposing that $c = 1$.) If L consists of the lines through one or another of two points (which may coincide), its equation is

$$(a_1u_1 + a_2u_2 + a_3u_3)\ (b_1u_1 + b_2u_2 + b_3u_3) = 0,$$

where the points are represented, respectively, by the factors set equal to zero. Now suppose L is nondegenerate. By 18.3 it is therefore projectively equivalent to the nondegenerate line conic M consisting of the tangents to the circle $x^2 + y^2 = 1$. A projectivity exists, then, which maps M on L. The equations of this projectivity in line coordinates are of the form (1), §14, that is, they are homogeneous and of the first degree. When they are substituted in the equation of M, which we showed in §17 to be $u_1^2 + u_2^2 - u_3^2 = 0$, we obtain the equation of L. The latter equation is necessarily of the second degree. This proves the first part of 17.3. We omit the proof of the converse part.*

Finally, if we take for granted that a nondegenerate point conic and a nondegenerate line conic are dual figures (this is proved in the next chapter, as already mentioned) we obtain the following theorem by dualizing VIII, 18.6:

18.4 *There is a unique nondegenerate line conic containing five given lines of the plane, no three of which are concurrent.*

EXERCISES

1. Give an algebraic proof of 18.1.

2. If P is a point and L is a nondegenerate line conic, it can be shown algebraically, by solving simultaneously the equations of P and L, that the number of distinct lines of L passing through P is 0, 1, or 2. Illustrate these cases, taking as L some familiar line conic. In which of these cases is P a point of contact of L?

3. From the fact stated in Ex. 2 deduce that no three lines of a nondegenerate line conic are concurrent.

4. Prove that the property of being a line curve of the third degree is projective.

5. At most how many lines of a third-degree line curve go through a coplanar point? Justify your answer.

19. Correlations.

We mentioned in VIII, §19 that there are two types of projective transformations of the plane into itself, *collineations* and *correlations*. Collineations send points into points, lines into lines, and preserve cross-ratio; as we shall see, correlations send points into lines, lines into points, and preserve cross-ratio. Thus far, of course, we have considered only collineations. As the basis for now discussing correlations we shall use the following definitions:

19.1 Definition. *A correlation is a transformation of the plane into itself in which (a) points go into lines in a 1-1 manner, (b) collinear points*

* See G. Salmon, *A Treatise on Conic Sections*, §§151, 285.

go into concurrent lines, and (c) the cross-ratio of points is always equal to the cross-ratio of their image lines.

In view of the algebraic work we have done in connection with collineations it is now easy to show that correlations exist. For we need only consider the transformation represented by the equations

$$(1)\qquad \begin{cases} u_1' = a_1x_1 + a_2x_2 + a_3x_3 \\ u_2' = b_1x_1 + b_2x_2 + b_3x_3 \\ u_3' = c_1x_1 + c_2x_2 + c_3x_3, \end{cases} \qquad \Delta = \begin{vmatrix} a_1 & a_2 & a_3 \\ b_1 & b_2 & b_3 \\ c_1 & c_2 & c_3 \end{vmatrix} \neq 0,$$

where (x_1, x_2, x_3) are the coordinates of an arbitrary point, $[u_1', u_2', u_3']$ are the coordinates of the line which is its image, and the determinant Δ is not zero. Since these equations differ from equations (4), §6 only in that u_1', u_2', u_3' replace x_1', x_2', x_3', the reasoning by which we obtained 6.1 can now be applied to the above equations (1), yielding the conclusion that this transformation sends the points of the plane into the lines of the plane in a 1-1 manner, and that it can be written implicitly as

$$(2)\qquad \begin{cases} x_1 = A_1u_1' + B_1u_2' + C_1u_3' \\ x_2 = A_2u_1' + B_2u_2' + C_2u_3' \\ x_3 = A_3u_1' + B_3u_2' + C_3u_3', \end{cases} \qquad \delta = \begin{vmatrix} A_1 & B_1 & C_1 \\ A_2 & B_2 & C_2 \\ A_3 & B_3 & C_3 \end{vmatrix} \neq 0,$$

where $A_1, B_1, C_1, \ldots$ are the cofactors of $a_1, b_1, c_1, \ldots$ in Δ.

In §6 we proved that the transformation (4) of that section preserves collinearity by showing that the coordinates of three points P', Q', R' satisfy a first-degree equation in x_1', x_2', x_3' whenever the coordinates of their originals, $P, Q, R,$ satisfy a first-degree equation in x_1, x_2, x_3. The same reasoning, applied to the above transformation (1), shows that the coordinates of three lines p', q', r' satisfy a first-degree equation in u_1', u_2', u_3' whenever the coordinates of their originals, points $P, Q, R,$ satisfy a first-degree equation in x_1, x_2, x_3. Since a first-degree equation in u_1', u_2', u_3' represents a point, we see that p', q', r' meet in this point whenever P, Q, R are collinear. Thus, the transformation (1) meets condition (b) of Definition 19.1. To show that condition (c) is met, one need only apply the method of §6, Ex. 4, to the present situation. This completes the proof of the following:

19.2 *The transformations represented by equations* (1), *where* (x_1, x_2, x_3) *and* $[u_1', u_2', u_3']$ *denote a point and line of the plane, are correlations.*

The converse of this is also true. We shall state it after Theorems 19.3 and 19.4 since its proof depends on them.

19.3 *There is at least one transformation of the form* (1) *that sends four specified points, no three collinear, into four specified lines, no three concurrent.*

19.4 *There cannot be two different correlations that send four specified points, no three collinear, into four specified lines, no three concurrent.*

19.5 *All correlations have equations of the form* (1).

The proofs of the above three theorems are just like the proofs of the corresponding theorems for collineations and are therefore covered in exercises.

Besides the properties used to define it, a correlation has the following additional properties. We leave the proof of this as an exercise since it is much like the proof of 19.2.

19.6 *A correlation sends the lines of the plane into the points of the plane in a* 1-1 *manner, concurrent lines going into collinear points, with preservation of cross-ratio.*

We could have used these properties to define a correlation. Had we done so, then those mentioned in 19.1 could be deduced from them. It follows from this that *the inverse of a correlation is a correlation.* Thus the transformation $[u'_1, u', u'_3] \rightarrow (x_1, x_2, x_3)$ represented by (2) is a correlation since it is the inverse of (1). One concludes, furthermore, that *a correlation always sends noncollinear points into nonconcurrent lines.*

It is clear from their properties that correlations send any projective figure (one that is definable in terms of projective concepts) into the dual figure. Thus, the figure consisting of four points, no three collinear, and the six lines joining them in pairs goes into the figure consisting of four lines, no three concurrent, and their six points of intersection, that is, a complete quadrangle goes into a complete quadrilateral. Similarly, if we assume that a point conic is definable in projective terms (such a definition is given in the next chapter), then its image under a correlation is the dual figure, which is necessarily a line conic. For, (1) transforms the equation of a point conic into a second-degree equation in u'_1, u'_2, u'_3, and the locus of the latter equation is a line conic by 17.3.

According to the Principle of Duality, if a given projective statement (one involving only projective concepts) is true, so is the dual statement. We never proved this principle, although we have amply illustrated it. The existence of correlations constitutes a basis for proving the principle. To see why this is so, let us suppose that some theorem, say the Theorem of Pappus (VIII, 16.3), has been proved and that we wish to prove its dual by using correlations. We must therefore prove that if a, b, c and a', b', c' are two sets of distinct lines passing, respectively, through distinct points P and Q, then the three lines m, k, l determined by the pairs of points ab' and $a'b$, bc' and $b'c$, ac' and $a'c$ are concurrent. Let us subject the plane to a correlation. The figure formed by the aforementioned points and lines then goes into the figure formed by two sets of distinct points A, B, C and A', B', C' lying, respectively, on distinct lines p, q, and by the three points of intersection M, K, L of the pairs of lines AB' and $A'B$, BC' and $B'C$, AC' and $A'C$. By the Theorem of Pappus, the points M, K, L are collinear. Since they are the images of lines m, k, l in the correlation, these lines are concurrent and the proof is complete. The dual of any other projective theorem can be proved in like manner.

It is to be noted that the method just described requires the existence of but a single correlation, whereas we have seen that there are infinitely many correlations, each having equations of the form (1).

The set of all correlations is not a group. For, although the inverse of a correlation is a correlation as we saw earlier, the resultant of two correlations is a collineation. The set of all collineations and correlations, however, is a group. This group is called the **general projective group in the plane.** The collineations are a subgroup of the general projective group.

EXERCISES

1. Find the images of the point (1, 0, 2) and the line [2, 1, − 1] in the correlation

$$u_1' = 3x_1 + x_2 + x_3, \quad u_2' = x_1 - 4x_2 - 2x_3, \quad u_3' = x_1 - x_2 - x_3.$$

2. Express the correlation of Ex. 1 in a form from which the image (x_1', x_2', x_3') of a given line $[u_1, u_2, u_3]$ can be determined conveniently, that is, find the equations expressing x_1', x_2', x_3' in terms of u_1, u_2, u_3.

3. Deduce from 19.1 and 19.6 that a correlation sends noncollinear points into nonconcurrent lines, and vice versa.

4. Prove 19.6, using (1) as the correlation. (*Hint.* Use the equations in Ex. 5 to prove the cross-ratio property.)

5. Show that the point (x_1', x_2', x_3') into which the correlation (1) sends the line $[u_1, u_2, u_3]$ is given by the equations

$$\begin{cases} x_1' = A_1u_1 + A_2u_2 + A_3u_3 \\ x_2' = B_1u_1 + B_2u_2 + B_3u_3 \\ x_3' = C_1u_1 + C_2u_2 + C_3u_3, \end{cases} \qquad \delta = \begin{vmatrix} A_1 & A_2 & A_3 \\ B_1 & B_2 & B_3 \\ C_1 & C_2 & C_3 \end{vmatrix} \neq 0,$$

where $A_1, B_1, C_1, \ldots$ are the cofactors of $a_1, b_1, c_1, \ldots$ in Δ.

6. Prove 19.3. (*Suggestion.* Use the algebraic facts about (1) that follow from 6.4 and VIII, 19.3).

7. (a) Prove 19.4. (*Suggestion.* See proof of VIII, 19.3.) (b) Prove 19.5.

8. Prove that (a) the resultant of two correlations is a collineation, (b) the set of all collineations and correlations is a group.

9. Prove that the correlations given explicitly by (1) and (2) are mutually inverse by showing that their resultant is the identity.

10. A transformation of the form (1) in which $\Delta = 0$ is said to be *degenerate*. Show that (a) if the matrix of Δ is of rank 1, all points of the plane go into the same line, (b) if the matrix of Δ is of rank 2, all points of the plane go into a set of concurrent lines.

11. A correlation which is its own inverse called a **polarity.** (a) Verify that in a polarity the image and the original of a point are the same line. (This line is called the *polar* of the point in the polarity, and the point is called the *pole* of the line.) (b) If A, B are distinct points, and the polar of A in a given polarity passes through B, show that the polar of B in the polarity passes through A. (c) If T is a correlation that is not a polarity, and the image of a point A passes through a point B, verify that in general the image of B does not pass through A.

12. Show that the following correlation is a polarity (see Ex. 11):

$$u_1' = 2x_1 - x_2 + x_3,\ u_2' = -x_1 + x_2 + 3x_3,\ u_3' = x_1 + 3x_2 - x_3.$$

(*Suggestion.* Noting that the correlation can also be expressed as in Ex. 5, show that $T^2 = I$, where T denotes the correlation.)

20. The Summation Notation. We have confined ourselves in this chapter to as simple a notation as possible to avoid preoccupation with matters that are purely algebraic. Now that this danger is over it may be well to mention other notations. In this section we consider the *summation notation,* with which we presume the reader to be somewhat familiar, and in the next, vector and matrix notations.

If subscripts are used for the coefficients, as well as for the variables, the equation of a line can be written as

$$a_1x_1 + a_2x_2 + a_3x_3 = 0, \tag{1}$$

and hence, in the summation notation, as

$$\sum_{i=1}^{3} a_ix_i = 0. \tag{2}$$

If (r_1, r_2, r_3) is a specific point on the line, this fact is expressed by writing $\sum_{i=1}^{3} a_ir_i = 0$. If

$$\sum_{i=1}^{3} b_ix_i = 0 \quad \text{and} \quad \sum_{i=1}^{3} c_ix_i = 0$$

are two other lines, the three lines are concurrent if and only if the determinant

$$\begin{vmatrix} a_1 & a_2 & a_3 \\ b_1 & b_2 & b_3 \\ c_1 & c_2 & c_3 \end{vmatrix}$$

is zero. This determinant is often abbreviated $|abc|$. When it is clear that only three-termed expressions are being considered, the above lines can be expressed simply by writing

$$\sum a_i x_i = 0, \quad \sum b_i x_i = 0, \quad \sum c_i x_i = 0.$$

Remarks similar to the above apply to the equations of points.

In summation notation the collineation (9), §6 can be written as

$$x_1' = \sum_{i=1}^{3} a_i x_i, \qquad x_2' = \sum_{i=1}^{3} b_i x_i, \qquad x_3' = \sum_{i=1}^{3} c_i x_i,$$

or, more concisely,

$$x_1' = \sum a_i x_i, \qquad x_2' = \sum b_i x_i, \qquad x_3' = \sum c_i x_i,$$

and its determinant as

$$|abc|.$$

The collineation can be expressed even more concisely if we first write it as

$$(3) \qquad \begin{cases} x_1' = a_{11}x_1 + a_{12}x_2 + a_{13}x_3 \\ x_2' = a_{21}x_1 + a_{22}x_2 + a_{23}x_3 \\ x_3' = a_{31}x_1 + a_{32}x_2 + a_{33}x_3, \end{cases}$$

using double subscripts for the coefficients. For these equations can be written

$$x_1' = \sum_{j=1}^{3} a_{1j} x_j, \qquad x_2' = \sum_{j=1}^{3} a_{2j} x_j, \qquad x_3' = \sum_{j=1}^{3} a_{3j} x_j,$$

or, more compactly,

$$x_i' = \sum_{j=1}^{3} a_{ij} x_j, \qquad (i = 1, 2, 3).$$

The determinant of the collineation is commonly abbreviated $|a_{ij}|$.

If we write the inverse of (3) first as

$$(4) \qquad \begin{cases} x_1 = A_{11}x_1' + A_{21}x_2' + A_{31}x_3' \\ x_2 = A_{12}x_1' + A_{22}x_2' + A_{32}x_3' \\ x_3 = A_{13}x_1' + A_{23}x_2' + A_{33}x_3', \end{cases}$$

where $A_{11}, A_{12}, A_{13}, \ldots$ are the cofactors of $a_{11}, a_{12}, a_{13}, \ldots$ in $|a_{ij}|$, then these equations can be abbreviated

$$x_1 = \sum_{j=1}^{3} A_{j1} x_j', \qquad x_2 = \sum_{j=1}^{3} A_{j2} x_j', \qquad x_3 = \sum_{j=1}^{3} A_{j3} x_j',$$

or, more concisely,

$$x_i = \sum_{j=1}^{3} A_{ji}x'_j, \qquad (i = 1, 2, 3).$$

The determinant of (4) is denoted by $|A_{ji}|$.

Remarks like the above can also be made for correlations.

Although the summation notation, as used thus far, makes for great conciseness of algebraic expression, it has not helped to reveal any new geometric facts or relations. The situation is quite different, however, when it is applied to conics. To show this let us write the point conic

$$(5) \qquad Ax_1^2 + Bx_1x_2 + Cx_2^2 + Dx_1x_3 + Ex_2x_3 + Fx_3^2 = 0$$

as

$$Ax_1x_1 + \tfrac{1}{2}Bx_1x_2 + \tfrac{1}{2}Bx_2x_1 + Cx_2x_2 + \tfrac{1}{2}Dx_1x_3 + \tfrac{1}{2}Dx_3x_1 + \tfrac{1}{2}Ex_2x_3 + \tfrac{1}{2}Ex_3x_2 + Fx_3x_3 = 0,$$

and then, using coefficients with double subscripts, as

$$(6) \qquad a_{11}x_1x_1 + a_{12}x_1x_2 + a_{21}x_2x_1 + a_{22}x_2x_2 + a_{13}x_1x_3 + a_{31}x_3x_1 + a_{23}x_2x_3 + a_{32}x_3x_2 + a_{33}x_3x_3 = 0,$$

where $a_{12} = a_{21}$, $a_{13} = a_{31}$, $a_{23} = a_{32}$. Equation (6) can be written in the summation notation as

$$(7) \qquad \sum_{i,j=1}^{3} a_{ij}x_ix_j = 0, \qquad a_{ij} = a_{ji},$$

or, still more concisely, as

$$(8) \qquad \sum a_{ij}x_ix_j = 0, \qquad a_{ij} = a_{ji}.$$

On arranging the terms of (6) in the form

$$(9) \qquad \begin{aligned} &a_{11}x_1x_1 + a_{12}x_1x_2 + a_{13}x_1x_3 \\ &+ a_{21}x_2x_1 + a_{22}x_2x_2 + a_{23}x_2x_3 \\ &+ a_{31}x_3x_1 + a_{32}x_3x_2 + a_{33}x_3x_3 = 0, \end{aligned}$$

we find that the third order determinant

$$\begin{vmatrix} a_{11} & a_{12} & a_{13} \\ a_{21} & a_{22} & a_{23} \\ a_{31} & a_{32} & a_{33} \end{vmatrix}$$

comes into evidence. This determinant, which is abbreviated $|a_{ij}|$, is called the *discriminant* of the point conic.

An analogous discussion can be given for line conics.

We summarize these results, and then state several theorems as examples of the new insights into the algebra of conics provided by the summation notation.*

20.1 *Every point conic can be written in the form*

$$\sum a_{ij}x_ix_j = 0, \qquad a_{ij} = a_{ji},$$

and every line conic in the form

$$\sum b_{ij}u_iu_j = 0, \qquad b_{ij} = b_{ji},$$

Conversely, each of these equations represents a conic if not all of its coefficients are zero.

20.2 *A point or line conic is degenerate if and only if its discriminant is zero.*

20.3 *The tangent to the point conic*

$$\sum a_{ij}x_ix_j = 0, \qquad a_{ij} = a_{ji}, \quad |a_{ij}| \neq 0,$$

at the point (r_1, r_2, r_3) has the equation

$$\sum a_{ij}r_ix_j = 0.$$

This equation is obtainable from (9) by replacing the first x in each term by r.

20.4 *The point of contact on the line $[r_1, r_2, r_3]$ of the line conic*

$$\sum b_{ij}u_iu_j = 0, \qquad b_{ij} = b_{ji}, \quad |b_{ij}| \neq 0,$$

has the equation

$$\sum b_{ij}r_iu_j = 0.$$

20.5 *The line conic corresponding to the point conic*

$$\sum a_{ij}x_ix_j = 0, \qquad a_{ij} = a_{ji}, \quad |a_{ij}| \neq 0,$$

has the equation

$$\sum A_{ij}u_iu_j = 0, \qquad A_{ij} = A_{ji}, \quad |A_{ij}| \neq 0,$$

where A_{ij} is the cofactor of a_{ij} in $|a_{ij}|$.

20.6 *The point conic corresponding to the line conic*

$$\sum b_{ij}u_iu_j = 0, \qquad b_{ij} = b_{ji}, \quad |b_{ij}| \neq 0,$$

has the equation

$$\sum B_{ij}x_ix_j = 0, \qquad B_{ij} = B_{ji}, \quad |B_{ij}| \neq 0,$$

where B_{ij} is the cofactor of b_{ij} in $|b_{ij}|$.

---

* Proofs of the theorems are given in W. C. Graustein, *Introduction to Higher Geometry*, Chapter XII.

EXERCISES

1. Write the conic

$$3x_1^2 + x_2^2 - x_3^2 + 4x_1x_2 - 2x_1x_3 + 6x_2x_3 = 0$$

in the form (9) and determine whether it is degenerate.

2. Use 20.2 to show that the conic

$$x_1^2 - 2x_2^2 - x_3^2 - x_1x_2 - 3x_2x_3 = 0$$

is degenerate and find the lines on which its points lie.

3. Use 20.2 to show that the conic

$$2u_1^2 + u_2^2 + 3u_3^2 + 3u_1u_2 - 7u_1u_3 - 4u_2u_3 = 0$$

is degenerate and find the points through which its lines go.

4. Use 20.3 to find the equation of the tangent to the conic

$$5x_1^2 + 2x_2^2 + 25x_3^2 + 4x_1x_2 - 24x_1x_3 - 12x_2x_3 = 0$$

at the point (1, 3, 1).

5. Find the point of contact on the line [1, − 1, 4] of the conic

$$25u_1^2 - 9u_2^2 - u_3^2 = 0.$$

6. Use 20.5 to find the equations of the line conics corresponding to the following point conics:

(a) $x_1^2 - x_2x_3 = 0$; (b) $x_1^2 + x_2^2 - x_3^2 = 0$; (c) $x_1^2 - x_2^2 - x_3^2 = 0$.

7. Use 20.6 to find the equations of the point conics corresponding to the following line conics:

(a) $u_2^2 - u_1u_3 = 0$; (b) $u_3^2 - 12u_1u_2 = 0$; (c) $9u_1^2 + 9u_2^2 - u_3^2 = 0$.

8. Write the equation of a point so that the cofficients have subscripts; then write it in the summation notation.

9. Express in summation notation that the line $[u_1, u_2, u_3]$ goes through the point (x_1, x_2, x_3).

10. Write equations (1) and (2) of §19 in summation notation.

11. Rewrite equations (1) and (2) of §19 so that the coefficients have double subscripts; then apply the summation notation.

12. Using 20.2, show that $x^2 - 6xy + 9y^2 + 4x + y - 7$ is not factorable.

21. Vector and Matrix Notations.* Like some of the transformations considered in earlier chapters, collineations and correlations can be expressed conveniently in terms of vectors and matrices, the only difference

* In this section, familiarity with II, § § 6, 9 is presumed.

being that we now use 3×3 matrices and vectors with three components. We state the facts only for collineations, limiting ourselves to point coordinates. The facts in the other situations will then be apparent.

Consider the collineation

$$\text{(1)} \qquad \begin{cases} x_1' = a_1x_1 + a_2x_2 + a_3x_3 \\ x_2' = b_1x_1 + b_2x_2 + b_3x_3 \\ x_3' = c_1x_1 + c_2x_2 + c_3x_3, \end{cases} \qquad \Delta \neq 0.$$

The numbers on the left side are the components of a vector, which we write in the column form $\begin{pmatrix} x_1' \\ x_2' \\ x_3' \end{pmatrix}$ instead of the usual form (x_1', x_2', x_3'). Similarly, the right sides of (1), being three numbers in a certain order, are the components of a vector, which we shall write in column form. According to (1), the components of these two vectors are equal, respectively. Hence the vectors themselves are equal by definition and we may write

$$\text{(2)} \qquad \begin{pmatrix} x_1' \\ x_2' \\ x_3' \end{pmatrix} = \begin{pmatrix} a_1x_1 + a_2x_2 + a_3x_3 \\ b_1x_1 + b_2x_2 + b_3x_3 \\ c_1x_1 + c_2x_2 + c_3x_3 \end{pmatrix},$$

thus expressing the collineation by a single vector equation instead of the three given scalar equations.

According to the rule for multiplying a matrix and a vector, the vector on the right side of (2) equals the product

$$\begin{pmatrix} a_1 & a_2 & a_3 \\ b_1 & b_2 & b_3 \\ c_1 & c_2 & c_3 \end{pmatrix} \begin{pmatrix} x_1 \\ x_2 \\ x_3 \end{pmatrix}.$$

Hence (2) can be written

$$\text{(3)} \qquad \begin{pmatrix} x_1' \\ x_2' \\ x_3' \end{pmatrix} = \begin{pmatrix} a_1 & a_2 & a_3 \\ b_1 & b_2 & b_3 \\ c_1 & c_2 & c_3 \end{pmatrix} \begin{pmatrix} x_1 \\ x_2 \\ x_3 \end{pmatrix}.$$

Thus, *if we multiply the vector representing the original point by the matrix of the collineation, we obtain the vector representing the image point.*

Denoting the matrix of the collineation by A, the original vector by $\mathbf{X}$, and the image vector by $\mathbf{X}'$, we can write (3) more compactly as

$$\text{(4)} \qquad \mathbf{X}' = A\mathbf{X}.$$

Equations (3) and (4) are known as *matrix equations* of the collineation.

Using vectors and matrices, let us verify that the inverse of (1) has the equations

$$\begin{cases} x_1'' = A_1x_1' + B_1x_2' + C_1x_3' \\ x_2'' = A_2x_1' + B_2x_2' + C_2x_3' \\ x_3'' = A_3x_1' + B_3x_2' + C_3x_3', \end{cases} \qquad \delta = \Delta^2 \neq 0, \tag{5}$$

where primes are used for the original point, double primes for the image point, and the coefficients have their usual meaning. Our method of verification will be to find the resultant of (1) and (5) in matrix form and show that it is the identity. Writing (5) as

$$\begin{pmatrix} x_1'' \\ x_2'' \\ x_3'' \end{pmatrix} = \begin{pmatrix} A_1 & B_1 & C_1 \\ A_2 & B_2 & C_2 \\ A_3 & B_3 & C_3 \end{pmatrix} \begin{pmatrix} x_1' \\ x_2' \\ x_3' \end{pmatrix}, \tag{6}$$

and substituting from (3), we get for the resultant,

$$\begin{pmatrix} x_1'' \\ x_2'' \\ x_3'' \end{pmatrix} = \begin{pmatrix} A_1 & B_1 & C_1 \\ A_2 & B_2 & C_2 \\ A_3 & B_3 & C_3 \end{pmatrix} \left[\begin{pmatrix} a_1 & a_2 & a_3 \\ b_1 & b_2 & b_3 \\ c_1 & c_2 & c_3 \end{pmatrix} \begin{pmatrix} x_1 \\ x_2 \\ x_3 \end{pmatrix} \right]$$

$$= \left[\begin{pmatrix} A_1 & B_1 & C_1 \\ A_2 & B_2 & C_2 \\ A_3 & B_3 & C_3 \end{pmatrix} \begin{pmatrix} a_1 & a_2 & a_3 \\ b_1 & b_2 & b_3 \\ c_1 & c_2 & c_3 \end{pmatrix} \right] \begin{pmatrix} x_1 \\ x_2 \\ x_3 \end{pmatrix},$$

since the indicated multiplications are associative. On multiplying the above 3×3 matrices (see §6, Ex. 6), we obtain

$$\begin{pmatrix} x_1'' \\ x_2'' \\ x_3'' \end{pmatrix} = \begin{pmatrix} \Delta & 0 & 0 \\ 0 & \Delta & 0 \\ 0 & 0 & \Delta \end{pmatrix} \begin{pmatrix} x_1 \\ x_2 \\ x_3 \end{pmatrix}, \tag{7}$$

or, in scalar form,

$$x_1'' = \Delta \cdot x_1, \quad x_2'' = \Delta \cdot x_2, \quad x_3'' = \Delta \cdot x_3, \tag{8}$$

which is the identity.

The special case in which $\Delta = 1$ is of interest, for (6) then becomes

$$\begin{pmatrix} x_1'' \\ x_2'' \\ x_3'' \end{pmatrix} = \begin{pmatrix} 1 & 0 & 0 \\ 0 & 1 & 0 \\ 0 & 0 & 1 \end{pmatrix} \begin{pmatrix} x_1 \\ x_2 \\ x_3 \end{pmatrix} = \begin{pmatrix} x_1 \\ x_2 \\ x_3 \end{pmatrix}, \tag{9}$$

as multiplication of the matrix and vector shows, and we get the simple result

$$\mathbf{X}'' = \mathbf{X}.$$

The 3×3 matrix in (9) is called the *unit* 3×3 *matrix,* for if it is denoted by I, then $AI = IA = A$, where A is the matrix of (1). Also, $I\mathbf{Z} = \mathbf{Z}$, where $\mathbf{Z}$ is any 3-component column vector, as is seen from (9). (No confusion will result from the double use of I, as a matrix and as a transformation.)

Let us divide each coefficient on the right in (5) by Δ. This does not change the collineation (5), but it changes the matrix of the collineation to

$$\begin{pmatrix} A_1/\Delta & B_1/\Delta & C_1/\Delta \\ A_2/\Delta & B_2/\Delta & C_2/\Delta \\ A_3/\Delta & B_3/\Delta & C_3/\Delta \end{pmatrix} \tag{10}$$

When this is multiplied by the matrix A in either order, the result is the unit 3×3 matrix I. The matrix (10) is therefore called the *inverse* of A, and it is denoted by A^{-1}. Thus $AA^{-1} = A^{-1}A = I$.*

It is now possible to solve (4) for $\mathbf{X}$ in terms of $\mathbf{X}'$. First, we deduce from (4) that

$$A^{-1}\mathbf{X}' = A^{-1}(A\mathbf{X}),$$

for the equal vectors represented by the two sides of (4) are here multiplied by the same matrix. The remaining steps are:

$$\begin{aligned} A^{-1}\mathbf{X}' &= (A^{-1}A)\mathbf{X}, && \text{(by use of the associative law);} \\ &= I\mathbf{X}, && \text{(using the definition of } A^{-1}\text{);} \\ &= \mathbf{X}, && \text{(by a property of } I\text{).} \end{aligned}$$

Thus, from (4), a matrix equation of the collineation (1), we have obtained

$$\mathbf{X} = A^{-1}\mathbf{X}', \tag{11}$$

a matrix equation of the inverse collineation.†

Our algebraic descriptions of collineations thus far have all employed equations. This is not necessary. As long as it is clear what kind of coordinates are being used, one can designate a collineation simply by giving its matrix, for different collineations obviously cannot have the same matrix, and, as can be shown, two different matrices can represent the same collineation

*Although the collineation (5) is the inverse of (1), its matrix is not the inverse of the matrix of (1) unless $\Delta = 1$. Its matrix is called the *adjoint* of the matrix of (1).

† Equation (6), which involves the adjoint of A, is another matrix equation of the inverse collineation.

only when the elements of one matrix are proportional to those of the other. Thus the collineation (1) may be referred to as *the collineation*

$$(12) \qquad \begin{pmatrix} a_1 & a_2 & a_3 \\ b_1 & b_2 & b_3 \\ c_1 & c_2 & c_3 \end{pmatrix}, \qquad \Delta \neq 0,$$

and its inverse as *the collineation*

$$(13) \qquad \begin{pmatrix} A_1/\Delta & B_1/\Delta & C_1/\Delta \\ A_2/\Delta & B_2/\Delta & C_2/\Delta \\ A_3/\Delta & B_3/\Delta & C_3/\Delta \end{pmatrix}.$$

This use of matrices lends itself nicely to the classification of collineations which was given in §7. For if the elements of the above matrix (12) have all possible values subject to the specified condition, then the matrix represents the group of all collineations in the plane, and in like manner the matrix

$$\begin{pmatrix} a_1 & a_2 & a_3 \\ b_1 & b_2 & b_3 \\ 0 & 0 & c_3 \end{pmatrix}, \qquad \Delta \neq 0,$$

represents the affinities, the matrices

$$\begin{pmatrix} a & b & c \\ -b & a & d \\ 0 & 0 & 1 \end{pmatrix} \text{ and } \begin{pmatrix} a & b & c \\ b & -a & d \\ 0 & 0 & 1 \end{pmatrix}, \qquad a, b \text{ not both } 0,$$

the direct and opposite similarities, respectively, and the matrices

$$\begin{pmatrix} a & b & c \\ -b & a & d \\ 0 & 0 & 1 \end{pmatrix} \text{ and } \begin{pmatrix} a & b & c \\ b & -a & d \\ 0 & 0 & 1 \end{pmatrix}, \qquad a^2 + b^2 = 1,$$

the direct and opposite motions, respectively.

EXERCISES

1. Using line coordinates, express the collineation (1) by (a) an equation involving only vectors, (b) an equation involving a 3×3 matrix.

2. Express the correlation (1), § 19, by an equation involving a 3×3 matrix.

3. Find the product of $\begin{pmatrix} 2 & 0 & 2 \\ -1 & -4 & 5 \\ 3 & 4 & 7 \end{pmatrix}$ and $\begin{pmatrix} 1 \\ 5 \\ 0 \end{pmatrix}$.

4. Find the product of $\begin{pmatrix} 1 & 2 & 3 \\ 4 & 5 & 6 \\ 0 & 1 & 2 \end{pmatrix}$ and $\begin{pmatrix} 0 & 2 & 1 \\ 3 & 1 & 2 \\ 4 & 6 & 5 \end{pmatrix}$ first in one order, then in the other.

5. Find the inverse and the adjoint of the matrix $\begin{pmatrix} 2 & 1 & 0 \\ 3 & 4 & 0 \\ 0 & 0 & 1 \end{pmatrix}$.

6. Verify, without using the associative law, that

$$\left[\begin{pmatrix} 2 & 1 & 0 \\ 3 & 4 & 0 \\ 0 & 0 & 1 \end{pmatrix}\begin{pmatrix} 0 & 2 & 1 \\ 3 & 1 & 2 \\ 4 & 6 & 5 \end{pmatrix}\right]\begin{pmatrix} 1 \\ 5 \\ 0 \end{pmatrix} = \begin{pmatrix} 2 & 1 & 0 \\ 3 & 4 & 0 \\ 0 & 0 & 1 \end{pmatrix}\left[\begin{pmatrix} 0 & 2 & 1 \\ 3 & 1 & 2 \\ 4 & 6 & 5 \end{pmatrix}\begin{pmatrix} 1 \\ 5 \\ 0 \end{pmatrix}\right].$$

7. Prove, without using the associative law, that

$$\left[\begin{pmatrix} a & b & c \\ d & e & f \\ g & h & k \end{pmatrix}\begin{pmatrix} A & B & C \\ D & E & F \\ G & H & K \end{pmatrix}\right]\begin{pmatrix} x \\ y \\ z \end{pmatrix} = \begin{pmatrix} a & b & c \\ d & e & f \\ g & h & k \end{pmatrix}\left[\begin{pmatrix} A & B & C \\ D & E & F \\ G & H & K \end{pmatrix}\begin{pmatrix} x \\ y \\ z \end{pmatrix}\right].$$

8. If $\begin{pmatrix} 6 & 0 & 0 \\ 0 & 2 & 1 \\ 0 & 3 & 4 \end{pmatrix}$ denotes a collineation in point coordinates, give another matrix representing (a) the same collineation, (b) the inverse collineation.

9. Specify the nature (affinity, direct or opposite similarity, etc.) of each of the following collineations if point coordinates are being used:

(a) $\begin{pmatrix} 2 & 1 & 0 \\ 6 & 4 & 0 \\ 0 & 0 & 2 \end{pmatrix}$; (b) $\begin{pmatrix} 1 & 1 & 0 \\ -1 & 1 & 0 \\ 0 & 0 & 2 \end{pmatrix}$; (c) $\begin{pmatrix} 1 & 1 & 0 \\ 1 & -1 & 0 \\ 0 & 0 & -1 \end{pmatrix}$;

(d) $\begin{pmatrix} 0 & 1 & 0 \\ 1 & 0 & 0 \\ 0 & 0 & 3 \end{pmatrix}$; (e) $\begin{pmatrix} 0 & 2 & 0 \\ -2 & 0 & 0 \\ 0 & 0 & 2 \end{pmatrix}$; (f) $\begin{pmatrix} 3 & 4 & 0 \\ 4 & -3 & 0 \\ 0 & 0 & 5 \end{pmatrix}$.

10. Give two different matrices representing the identity collineation in point coordinates.

11. Prove that $AI = IA$, where I is the unit 3×3 matrix and A is any 3×3 matrix.

12. Prove that the product of the matrices (12) and (13) in either order is the unit 3×3 matrix I.

22. Collineations As Topological Transformations. Since we know what it means to say that a point P in the projective plane approaches another point H in this plane (see 2.4), the definitions of a continuous transformation and a topological transformation for the projective plane are precisely the same as for the Euclidean plane (VII, 2.1). Thus, a transformation T of the projective plane into itself is *continuous* if $P' \to H'$ whenever $P \to H$, where $T(H, P) = H', P'$, and T is *topological* if it is 1-1, continuous, and has a continuous inverse.

A collineation is a 1-1 mapping of the projective plane onto itself, as we know, and its equations in point coordinates show that it is continuous. For consider any *specific* collineation as represented by equations (1), §21. Since all the coefficients are then constants, x_1', x_2', x_3' are polynomials in x_1, x_2, x_3, and hence are continuous functions of these variables over the entire projective plane. Let $P(x_1, x_2, x_3)$ be a variable point which approaches the point $H(h_1, h_2, h_3)$. Then $x_1 \to h_1$, $x_2 \to h_2$, $x_3 \to h_3$ by 2.4. Let the images of P and H be $P'(x_1', x_2', x_3')$ and $H'(h_1', h_2', h_3')$. When $x_1 \to h_1$, $x_2 \to h_2$, $x_3 \to h_3$ in equations (1), it follows from the specified continuity that $x_1' \to h_1'$, $x_2' \to h_2'$, $x_3' \to h_3'$. Thus, $P' \to H'$ by 2.4. The collineation is therefore a continuous transformation of the plane into itself. The inverse of the collineation, being a collineation, is likewise continuous. Thus we see that collineations are topological transformations of the projective plane into itself.

The topological properties of curves in the projective plane are the properties that are preserved by all topological transformations of this plane into itself, and they are precisely the same as the topological properties of curves in the Euclidean plane (VII, §4). We have already shown, for example, that projective lines and projective conics are simple closed curves. This was done on the basis of an agreement that involved using the circular model of the projective plane (VIII, §10). Let us now do it by using the fact that collineations are topological transformations. Consider projective lines. In §2, Ex. 15, we saw that the extended x-axis can be mapped topologically on a circle. This line is therefore a simple closed curve. Since all projective lines are projectively equivalent (VIII, 19.7), there is a collineation which sends the extended x-axis into an arbitrary projective line g. Since the collineation is a topological transformation, g must be a simple closed curve. Similarly, since (1) a circle is a nondegenerate projective conic that is a simple closed curve, (2) all nondegenerate projective conics are projectively equivalent (8.3), and (3) collineations are topological transformations, we conclude that all nondegenerate projective conics are simple closed curves.

In the discussion of cubics in §10 nothing was said about their being projectively equivalent. Using the above type of argument we can show that nondegenerate cubics are not all projectively equivalent. Consider the cubics $x_2x_3^2 = x_1^3$ and $x_1^3 + x_2^3 - 3x_1x_2x_3 = 0$. The first of these, being the Euclidean cubic $y = x^3$ plus the ideal point (0, 1, 0), is a simple closed curve (§10,

Ex. 1a). The second, which consists of the folium of Descartes $x^3 + y^3 - 3xy = 0$ and the point $(1, -1, 0)$, is not a simple closed curve (§10, Ex. 2). These curves are not projectively equivalent. For if they were, there would be a collineation carrying one into the other, and this would imply that they have the same topological structure, which is not true.

From an algebraic standpoint, collineations are the simplest homeomorphisms of the projective plane inasmuch as their equations are of the first degree. Thus, two-dimensional projective geometry is seen to be a subject which, viewed algebraically, is concerned with homeomorphisms of the simplest sort. By contrast, topological geometry of the projective plane is concerned with the properties of figures that are preserved by *all* homeomorphisms of this plane. These properties, as noted earlier, are called topological, and no property we called projective is among them. The distinction between a line and a nondegenerate conic, which exists in projective geometry and is important there, disappears in topological geometry since these curves have exactly the same topological properties, both being simple closed curves. This relation between topological geometry and projective geometry, in which the latter is a subgeometry of the former, is analogous to the relation between affine and metric geometry, or between projective and affine geometry. We recall, for example, that parabolas, ellipses, and hyperbolas are essentially different types of curves in affine geometry since they cannot be transformed into one another by affine transformations, but that this distinction disappears in projective geometry, where they are all projectively equivalent.

Now that we see that the collineations of a projective plane are a subgroup of the set of topological transformations of that plane into itself, it is clear that we could amend our definition of a projective property to read as follows: *a projective property is a property that is preserved by all collineations but not by all topological transformations.* This definition is more precise than the original one since it enables us to distinguish between projective properties and topological properties, and more in keeping with our definitions of metric and affine properties. Its use would require no change in the list of projective properties given previously.

EXERCISES

1. A transformation, T, of lines is called *continuous* if $p' \to h'$ when $p \to h$, where $T(h, p) = h', p'$ and the small letters denote lines, and is called *topological* if it is also 1-1 and has a continuous inverse. Prove that a collineation, regarded as a transformation of lines, is topological.

2. Formulate definitions that state when a transformation of points into lines is (a) continuous, (b) topological, and use them to prove that correlations are topological transformations.

3. Prove that nondegenerate projective quartics are not all projectively equivalent.

4. Prove that $x_1' = 2x_1$, $x_2' = x_2^3$, $x_3' = x_3$ is a homeomorphism of the projective plane into itself.

5. Show that the homeomorphism of Ex. 4 may send a line into a line, but that it does not always do this.

6. Why is $x_1' = 2x_1^3$, $x_2' = 5x_2$, $x_3' = x_3^2$ not a topological transformation of the projective plane into itself?

7. Describe the images of the following figures under a collineation and also under a homeomorphism of the projective plane: (a) a point; (b) a set of three points; (c) three collinear points; (d) four points, no three of which are collinear; (e) a simple closed curve; (f) a figure shaped like the number 8; (g) two nondegenerate conics meeting in four points; (h) a line and a nondegenerate conic to which it is tangent; (i) a line and a nondegenerate conic of which it is a secant.

CHAPTER X

Projective Descriptions of Conics

1. Introduction. Although projective conics go into one another under projective transformations, and so are a suitable subject of study in projective geometry, it should not be overlooked that our definition of them relies heavily on metric concepts. Projective point conics, we have seen, are obtained by adding ideal points to Euclidean conics, and projective line conics consist of the tangents to projective point conics. A basic concept in our entire study of projective conics is therefore that of a Euclidean conic, whose definition involves distance, even if one regards it as a plane section of a cone. What we now seek are geometric descriptions of point conics and line conics that are more projective in character and apply equally well to ideal and non-ideal elements. In the preceding chapter we achieved algebraic descriptions of this kind. Using the new geometric descriptions, we shall be able to prove the theorems of Pascal and Brianchon.

In this chapter, except when the contrary is indicated, "conic" means "projective conic," "plane" means "extended plane," and all the Euclidean and projective conics mentioned are nondegenerate. The entire discussion is confined to a single extended plane.

2. A Projective View of Point Conics. The Euclidean points of a conic satisfy a certain condition involving distance, but the ideal points of that conic, of course, do not satisfy this condition. On the other hand, *all* the points of the conic obey certain formulas involving cross-ratio. It is therefore reasonable to try to base the new description of a point conic on cross-ratio. Indeed, we already know that Euclidean conics can be described in these terms, for in Chapter V it was shown that if straight lines are drawn joining any four points A, B, C, D of such a conic to a fifth point P of the conic, the cross-ratio $(AP\ BP, CP\ DP)$ of the four lines has the same value for all positions of P (see V, 14.1). This property is easily extended to projective point conics, for let K' be such a conic, and A', B', C', D', P' any distinct points on it. If K is any other point conic, it can be mapped on K' by a projectivity T since all (nondegenerate) point conics are projectively equivalent (IX, 8.3). Let A, B, C, D, P be the points of K which T sends into A', B', C', D', P'. Then $(AP\ BP, CP\ DP) = (A'P'\ B'P', C'P'\ D'P')$. In particular, K may be chosen to be an ellipse, in which case V, 14.1 applies, telling us that the cross-ratio on the left side is constant as P ranges over K and A, B, C, D remain fixed. Correspondingly, P' will range over K' and the cross-ratio on the right side will maintain this same constant value. This proves the theorem:

2.1 *If four points of a point conic are given, a chosen cross-ratio of the lines joining them to a fifth point of the conic has the same value for all positions of this point.*

The following corollary of this theorem can be proved just as we proved V, 14.2:

2.2 *A chosen cross-ratio formed by the tangent to a point conic and the lines joining the point of tangency to three other points of the conic equals the corresponding cross-ratio of the lines joining these four points to a fifth point on the conic.*

These important theorems are due to Steiner (1796–1863), who was much concerned with the generation of curves by projective methods. The property stated in 2.1 does, in fact, provide us with a method of generating a point conic, for no other curve has this property, as the following converse of 2.1 shows:

2.3 *If a point moves so that a chosen cross-ratio of the lines joining it to four given points, no three collinear, is constant, its locus and the four points form a point conic.*

We omit the proof of this theorem. (See E. A. Maxwell, *Generalized Homogeneous Coordinates*, p. 114.) The theorem provides us with the kind of description of a point conic which we have been seeking. As a corollary of the theorem we have the following:

2.4 *There is a unique point conic that passes through four given points, no three collinear, and is tangent at one of them to a given line (not containing any of the other given points).*

Proof. Let the given points be A, B, C, D, let the given line be t, and let t contain D. Denote by r the value of $(AD\ BD, CD\ t)$. By 2.3 there is a unique point conic through A, B, C, D such that $(AP\ BP, CP\ DP) = r$, where P is any point of the conic except A, B, C, D. If t' is the tangent at D, then

$$(AD\ BD, CD\ t') = (AP\ BP, CP\ DP)$$

by 2.2. It follows that $(AD\ BD, CD\ t) = (AD\ BD, CD\ t')$ and hence, since t and t' both go through D, that $t = t'$. The conic is thus tangent to t at D.

EXERCISES

1. Prove 2.2, justifying each step by specific reference to a previous theorem or definition.

2. An alternative method of proving 2.1 makes use of the given conic and limits. Prove 2.1 in this way for the case in which the conic is (a) an extended parabola, (b) an extended hyperbola.

3. Projective Correspondences. In the projective view of a point conic provided by 2.1 and 2.2 the four given points come in for special attention. Closely related to this view is another, in which no points of the conic stand out in any particular way. The discussion in this section will prepare the ground for this second view, which is presented in §4.

When a line g is projected on another line g' from a point O, a 1-1 correspondence is established between the points of the lines in which, as we know, the cross-ratio of any four points on g equals the cross-ratio of the corresponding four points on g'. This is a very simple example of a *projective correspondence*. More generally, by a **projective correspondence** we mean 1-1 correspondence between two sets of collinear points or two sets of concurrent lines in which the cross-ratio of any four elements in one set equals the cross-ratio of the four corresponding elements in the other set. A projective correspondence, like any 1-1 correspondence, determines two transformations which are mutually inverse (I, §3). These transformations are called (one-dimensional) *projectivities;* we studied them in VIII, §16, including the exercises of that section.

Fig. X,1 offers further examples to help clarify the concept of a projective correspondence. If we project g on g' from O, and then g' *on* g'' from O', the correspondence between the points of g and g'' in which $A, B, C, \ldots$ correspond to $A'', B'', C'', \ldots$, respectively, is clearly a projective correspondence. It is established by the projectivity which is the resultant of the perspectivities with centers O and O', in that order. The *same* projective correspondence between the points of g and g'' results from first projecting g'' on g' from O', and then projecting g' on g from O. In this case the correspondence is established by the projectivity which is the resultant of the perspectivities with centers O' and O, in that order. These two projectivities are mutually inverse. In Fig. X, 1 we also note that the correspondence between the lines through O and those through O' in which lines OA', OB', OC', . . . correspond to lines $O'A'$, $O'B'$, $O'C'$, . . ., respectively, is a projective correspondence. It is established either by the *perspectivity with axis* g' which maps the pencil through O on the pencil through O', or the inverse of this perspectivity (see VIII, §16, Ex. 4).

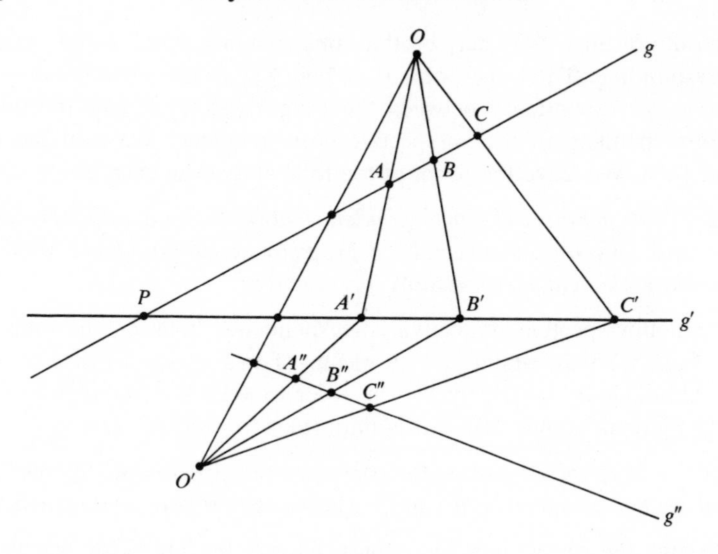

Fig. X, 1

In view of VIII, 9.15, the following theorem is immediate:

3.1 *A projective correspondence between two sets of collinear points or concurrent lines is completely determined when three pairs of corresponding elements are specified.*

A projective correspondence between the points of two lines is particularly simple if the lines joining pairs of corresponding points are concurrent, as is the case for the points of g and g' in Fig. X,1. Likewise, the simplest projective correspondence between two sets of concurrent lines occurs when the points of intersection of pairs of corresponding lines are collinear, as in the correspondence between the lines through O and those through O' in Fig. X, 1. These special projective correspondences are said to be *perspective,* and are called **perspective correspondences.** The projective correspondence between the points of g and g'' in Fig. X,1 is nonperspective. The student can verify that if the points of g are joined to a point O_1 not on g and different from O and O', a projective correspondence between the two sets of lines O_1A, O_1B, O_1C, . . . and $O'A'$, $O'B'$, $O'C'$, . . . can be obtained which is nonperspective.

The set of all points on a line g is called, briefly, a **range with axis** g, and denoted by $\{g\}$. Similarly, the set of all coplanar lines through a point O is called a **pencil with center** $\boldsymbol{O}$, and denoted by $\{O\}$. **Projective ranges** and **projective pencils** are ranges and pencils which are in projective correspondence.

Although two corresponding elements in a projective correspondence are generally distinct, they may be the same element, which is then said to be **self-corresponding.** Thus, the point P in Fig. X,1 is self-corresponding in the perspective correspondence between the ranges $\{g\}$, $\{g'\}$, and the line OO' is self-corresponding in the perspective correspondence between the pencils $\{O\}$ and $\{O'\}$. We leave the proof of the following as an exercise.

3.2 *A perspective correspondence always has a self-corresponding element, and only one. Conversely, a projective correspondence with a self-corresponding element is necessarily perspective.*

The nonperspective projective correspondence between the ranges $\{g\}$, $\{g''\}$ in Fig. X,1 was established by means of two perspectivities, as already shown. Although we used only two in order to keep the example simple, the following theorem shows that two is important for another reason:

3.3 *If a given projective correspondence between different ranges or pencils is not perspective, it can be established by two perspectivities.*

We give the proof only for ranges, leaving the proof for pencils as an exercise. Suppose that the given correspondence is between the ranges $\{g\}$, $\{g'\}$, and that it is determined by the distinct points A, B, C of g and their correspondents A', B', C' on g' (Fig. X, 2). Let g'' be a line through A', distinct from g'. If g is projected on g'' from a point O_1 which is on line AA', then A, B, C go into A', B'', C''. If g'' is projected on g' from O_2, the intersection of lines $B'B''$ and $C'C''$, then A', B'', C'' go into A', B', C'. We thus obtain the two perspectivities

$$ABC \ldots \overset{O_1}{\underset{\wedge}{=}} A'B''C'' \ldots \overset{O_2}{\underset{\wedge}{=}} A'B'C' \ldots .$$

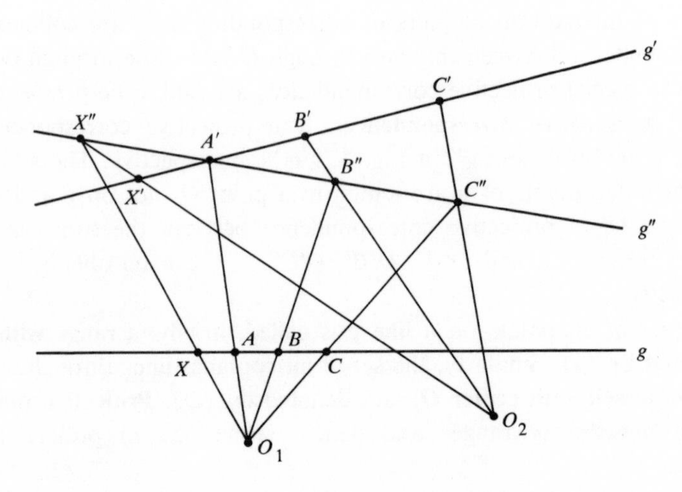

Fig. X, 2

Their resultant is seen to be

$$ABC \ldots \overline{\wedge}\, A'B'C' \ldots,$$

a projectivity between $\{g\}$ and $\{g'\}$ in which A, B, C go into A', B', C'. By 3.1, the projective correspondence established by this projectivity is identical with the given one.

Incidentally, the above proof provides a method of constructing pairs of corresponding points in a given projective correspondence. For if X is an arbitrary point of g, we need only apply the two specified projections in order to obtain the corresponding point X' on g' (Fig. X, 2).

It may be helpful to point out that 3.3 does not say we cannot establish a projective correspondence by using more than two perspectivities. But, according to 3.3, when the ranges or pencils are distinct the same correspondence can be obtained by using just two perspectivities.

EXERCISES

1. Prove 3.1.

2. Prove 3.2.

3. Prove 3.3 for pencils by dualizing the proof given for ranges. Verify that the proof provides a method of constructing pairs of corresponding lines in a projective correspondence between distinct pencils.

4. Show that no three points of intersection of corresponding lines of two projective nonperspective pencils are collinear.

5. State and prove the dual of the proposition in Ex. 4.

6. How many projective correspondences are there between two given pencils or ranges? Justify your answer.

7. Show how to establish a perspective correspondence, given (a) two ranges and a center of perspectivity, (b) two pencils and an axis of perspectivity.

8. Given three points A, B, C on a line g, three points A', B', C' on a different line g', and an arbitrary fourth point D on g, construct D' on g' so that $(AB, CD) = (A'B', C'D')$.

9. A range or pencil can be put into projective correspondence with itself. (a) How many pairs of corresponding elements determine such a correspondence? (b) How many self-corresponding elements, at most, can such a correspondence have without every element being self-corresponding? (c) How many perspectivities, at most, are needed to establish a projective correspondence of a range or pencil with itself?

10. By using three perspectivities, establish a nonperspective projective correspondence between different ranges. Then verify by actual construction that the same correspondence is obtained by using two suitably chosen perspectivities.

11. Using two perspectivities, establish a projective correspondence between two pencils.

4. A Second Projective View of Point Conics. Using the ideas of the preceding section, let us now take a point conic K, choose any two of its points O and O', and establish a projective correspondence between the pencils $\{O\}$ and $\{O'\}$. If X is any point of K other than O and O', we pair line OX with line $O'X$. Thus, in Fig. X, 3, lines a, b, c, d, etc. correspond to lines a', b', c', d', etc. This disposes of all the lines of the pencils which meet K in two points. The remaining lines are tangents. We pair t, the tangent at O, with OO', regarded as a line through O', and pair t', the tangent at O',

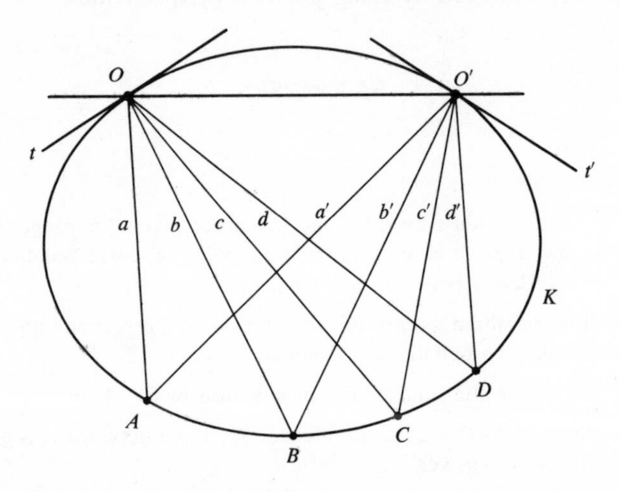

Fig. X, 3

with OO' regarded as a line through O. The 1-1 correspondence thus established is projective, for according to 2.1 and 2.2 any four lines of one pencil and the corresponding four of the other have equal cross-ratios. Thus, $(ab, cd) = (a'b', c'd')$ and $(ab, ct) = (a'b', c'OO')$. Also, the correspondence is not perspective since the points of intersection of corresponding lines are not collinear. We have thus proved:

4.1 *The points of any point conic are the intersections of corresponding lines of two projective nonperspective pencils with centers on the conic. Exactly one such projective correspondence is determined for each choice of centers O and O', namely, the correspondence in which line OP corresponds to line $O'P$, where P is any point of the conic. In this corre-*

spondence, OO' regarded as a line through O corresponds to the tangent at O', and regarded as a line through O' it corresponds to the tangent at O.

Having used K to generate two projective nonperspective pencils, we may say, conversely, that these pencils generate K in the sense that the points of K are the points of intersection of corresponding lines of the pencils. Thus A is the intersection of a and a', B of b and b', O of t and OO', O' of t' and OO', etc. Let us now generalize this by starting out with any two projective nonperspective pencils and showing that they generate a conic. Let the pencils be $\{O\}$, $\{O'\}$ and denate by L the locus of the points of intersection of corresponding lines. Then O is a point of L since OO', regarded as a line through O', can meet its corresponding line only in O. Similarly, O' is a point of L. If A, B, C are any other distinct points of L, no three of the points O, O', A, B, C are collinear (§3, Ex. 4). Hence there is a unique point conic K through these points (VIII, 18.6). Since the projective correspondence between $\{O\}$ and $\{O'\}$ can be regarded as determined by the three pairs of corresponding lines OA and OA', OB and OB', OC and OC', and since this same correspondence generates K according to 4.1, we conclude that K and L are identical. This proves:

4.2 *The points of intersection of corresponding lines of any two projective nonperspective pencils form a point conic containing the centers of the pencils. The relation of this projective correspondence to the conic is as described in* 4.1.

Taken together, 4.1 and 4.2 represent our second projective description of a point conic. In addition, 4.2 provides a new method of constructing one. Perhaps nothing said or done thus far to emphasize the difference between metric and projective geometry is quite as effective as the actual drawing of a point conic by this new method. One need only set up a projective correspondence having no self-corresponding element, say by using two perspectivities, and then apply 4.2. The process involves no measurement at all, only the drawing of lines and the marking of their points of intersection. Fig. X, 4 shows a point conic drawn in this way. The generating correspondence is between the pencils $\{O\}$, $\{O''\}$ and is established by the perspectivity with axis g_1 between $\{O\}$, $\{O'\}$, and the perspectivity with axis g_2 between $\{O'\}$, $\{O''\}$.

From this second projective description of a point conic we deduce the following theorem:

4.3 *There is a unique point conic which goes through three given noncollinear points and is tangent at two of them to given lines not containing the third point.*

Proof. Let O, O', A be the given points, and t, t' the given lines, with O on t and O' on t'. (We assume that t, t' are distinct and do not contain A.)

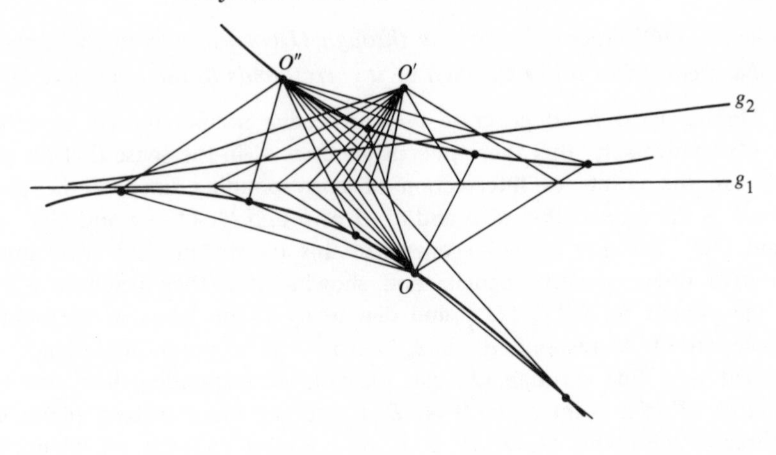

Fig. X, 4

Consider the pencils $\{O\}$, $\{O'\}$. By 3.1 there is a projective correspondence between them in which the lines t, OO', OA through O correspond, respectively, to the lines $O'O$, t', $O'A$ through O'. This correspondence is nonperspective since O, O', A are noncollinear. Hence it determines a point conic K by 4.2. This conic clearly contains O, O', A. Also, previous discussions show that t, t' are the tangents at O, O'. (If t, for example, were a secant, it would contain a point B of the conic distinct from O, and hence correspond to line BO', contradicting that it corresponds to line OO'.) Suppose K' were another point conic meeting these conditions. Its points, by 4.1, are the intersections determined by a certain projective correspondence between $\{O\}$ and $\{O'\}$, and this correspondence, in view of the aforementioned conditions, is necessarily one in which t, OO', OA correspond to $O'O$, t', $O'A$. By 3.1 it is therefore identical with the correspondence used to generate K. Hence K' cannot be different from K. From this contradiction we see that K is unique.

EXERCISES

1. By the method of the text draw a point conic in which O, O', O'', g_1, g_2, instead of being situated as in Fig. X, 4, are as shown in Fig. X, 5.

2. By combining metric notions with the method of the text one can draw point conics of *special* types. Verify that two nonperspective pencils through Euclidean points are projective when the angle between each two lines of one pencil agrees in size and sense with the angle between the corresponding lines of the other pencil, and draw the conic determined by two such pencils. (It should be a circle.) Note that the correspondence is completely determined as soon as we specify one pair of corresponding lines.

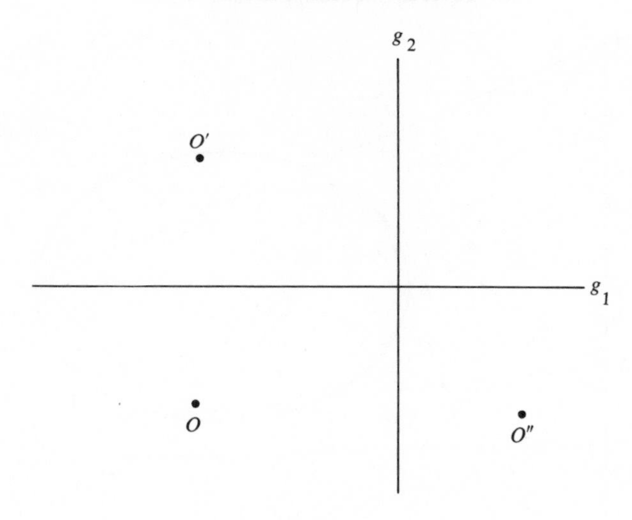

Fig. X, 5

3. Given that the point conic in Ex. 2 is a circle, show, by using projective pencils, how to construct any number of points on a circle when three of its points are given.

4. Prove that the point conic in Ex. 2 is a circle.

5. In Ex. 2 the pencils are *directly* congruent. Now let them be oppositely congruent, that is, corresponding angles have opposite senses. (The conic is a hyperbola.)

6. Show that the locus of the points of intersection of corresponding lines of two perspective pencils can be regarded as a degenerate point conic.

5. The Theorem of Pascal. This theorem was stated without proof in Chapter VIII. We now restate it in a broadened form to include the converse and give the proof.

5.1 Theorem of Pascal. *If the vertices of a simple hexagon are on a point conic, the pairs of opposite sides meet in collinear points. Conversely, if the pairs of opposite sides of a simple hexagon, no three of whose vertices are on a line, meet in collinear points, the six vertices are on a point conic.*

Proof. Let a point conic K be given, and let a simple hexagon $ABCDEF$, with consecutive sides 1, 2, 3, 4, 5, 6, be inscribed in it, the pairs of sides 1 and 4, 2 and 5, 3 and 6 meeting in $L, M, N,$ respectively (Fig. X, 6). By 2.1

$$(1) \qquad (DA\ DB, DC\ DE) = (FA\ FB, FC\ FE).$$

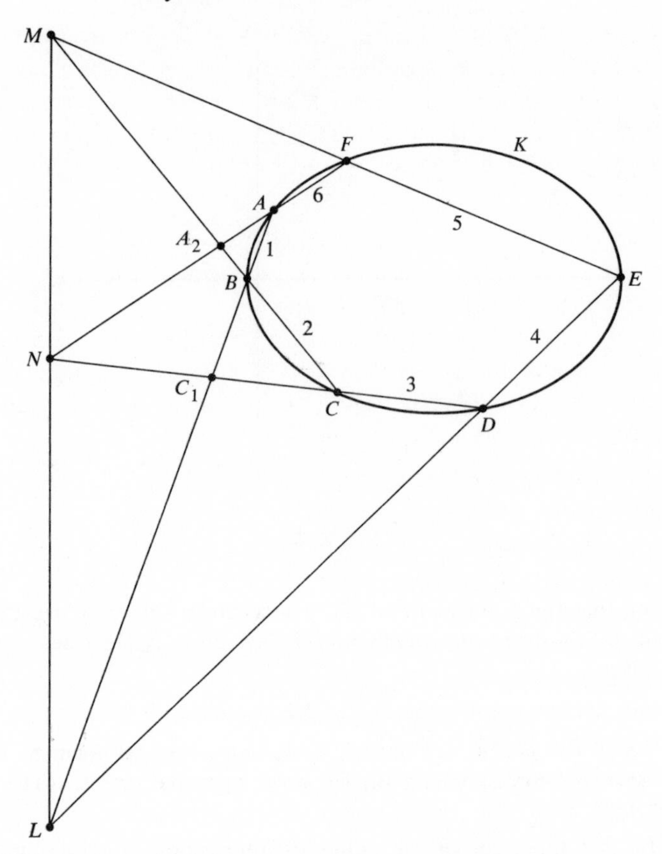

Fig. X, 6

The four lines on the left in this equation meet side 1 in points A, B, C_1, L; the four lines on the right meet side 2 in A_2, B, C, M. Hence

$$(AB, C_1L) = (A_2B, CM). \tag{2}$$

There is a unique projective correspondence between the ranges on lines 1 and 2 in which A, B, C_1 correspond to A_2, B, C, respectively, and equation (2) shows that L corresponds to M in this correspondence. Also, the correspondence is perspective since B is self-corresponding. Hence lines AA_2, C_1C, LM are concurrent. But AA_2 and C_1C, being sides 6 and 3 of the hexagon, meet in N. Thus N is on line LM, which proves the first part of the theorem.

To prove the converse part, let a simple hexagon labeled as above be given, with no three vertices on a line, and with L, M, N collinear. There is a unique projective correspondence between the ranges on lines 1 and 2 in which A, C_1, L correspond to A_2, C, M, respectively, and it is perspective since lines AA_2, C_1C, LM are concurrent in N. Therefore B, the intersection of lines 1 and 2, is self-corresponding. Thus equation (2) is true, and hence also equation (1). According to 2.3, if r is the value of the cross-ratios in (1), the locus of the point P such that $(PA\ PB, PC\ PE) = r$ is a point conic through A, B, C, E. This conic contains D, F in view of (1). The proof is then complete.

If we are given five points, no three collinear, we know that they determine a unique point conic. Pascal's theorem is important because it gives a condition which *six* points, no three collinear, must satisfy if they are to be on a point conic. In order to state this condition as simply as possible, let us first note that a simple hexagon whose pairs of opposite sides meet in collinear points is called a *Pascal hexagon,* and that the line of collinearity is called a *Pascal line.* We can then state the condition as follows:

5.2 Alternative Statement of Pascal's Theorem. *Six given points are on a point conic if and only if they are the vertices of a Pascal hexagon.*

By using this condition we can construct a sixth point of a point conic when five points of the conic are given, and hence can construct any number of points of the conic (see Ex. 1).

If six points, no three collinear, are given, it is possible by joining them in all possible orders to form 60 simple hexagons. If one of these is a Pascal hexagon, the given points lie on a point conic by 5.2, and hence all are Pascal hexagons according to this same theorem. The configuration formed by these 60 Pascal hexagons has many remarkable properties, as is suggested by the name *hexagrammum mysticum,* or mystic hexagram, which has been given to it. For example, its 60 Pascal lines pass by threes through 20 points (called *Steiner points*) and lie by fours on 15 lines (called *Plücker lines*).

Besides enabling us to construct points on a conic, Pascal's theorem, or, more exactly, the special case of it we shall now deduce, shows how to construct tangents to the conic. In Fig. X, 6 let F approach A along the curve. Side 6 then approaches the tangent at A and side 5 approaches line AE (see IX, §3, Ex. 15), so that the hexagon approaches the inscribed pentagon $ABCDE$ (Fig. X, 7). At the same time, L remains fixed, M approaches the point M' in which line AE meets line BC, and N approaches the point N' in which the tangent at A meets line CD (see IX, §11, Ex. 8). The points L, M', N' are collinear since the same is always true of L, M, N as $F \to A$. In the pentagon, side 3 is opposite A. If this side is disregarded, it is clear that sides 1 and 4 may be called opposite, and likewise 2 and 5. Since any inscribed pentagon $ABCDE$ can be regarded as the limiting position approached by an inscribed hexagon $ABCDEF$ when $F \to A$, we have proved the first part of the following theorem:

5.3 *If a simple pentagon is inscribed in a point conic, the point in which the tangent at one vertex meets the opposite side and the two points of*

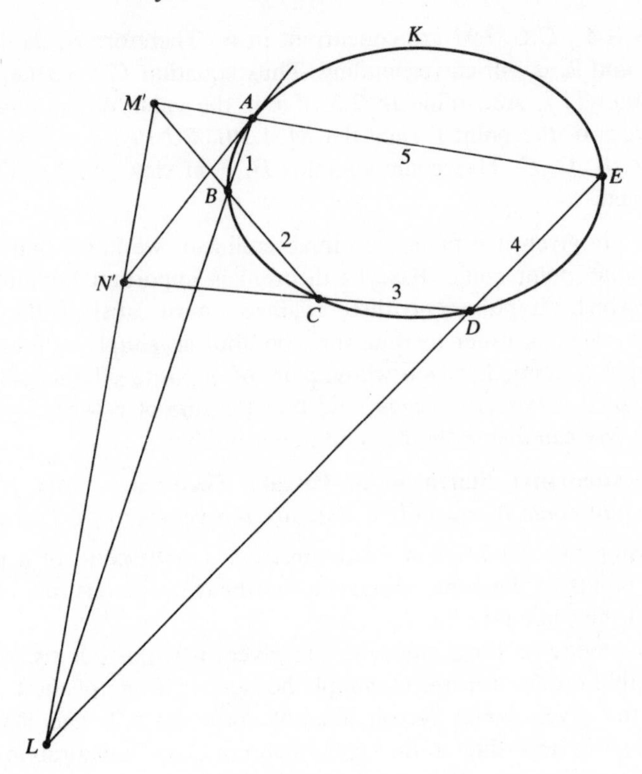

Fig. X, 7

intersection of the remaining pairs of opposite sides are collinear. Conversely, if a line through a vertex of a simple pentagon inscribed in a point conic meets the opposite side in a point collinear with the intersections of the remaining pairs of opposite sides, the line is tangent to the conic.

The construction of tangents by means of this theorem and the proof of the converse part are left as exercises. The theorem is easily checked intuitively by using a diagram in which the conic is a circle.

Pascal's theorem can also be specialized so as to yield a proposition about quadrangles. For convenience, consider an inscribed simple hexagon $ABCDEF$ whose pairs of opposite sides AB and DE, BC and EF, CD and FA meet in collinear points L, M, N (Fig. X, 8). Letting B approach A, and E approach D, and reasoning as above, we find that L, M, N approach three collinear points L', M', N', where L' is the intersection of the tangents at A and D (Fig. X, 9). If we regard A, C, D, F as the vertices of a complete quadrangle, and speak of this quadrangle as being inscribed in the conic, we can state the preceding result as follows:

5.4 *If a complete quadrangle is inscribed in a point conic, the tan-*

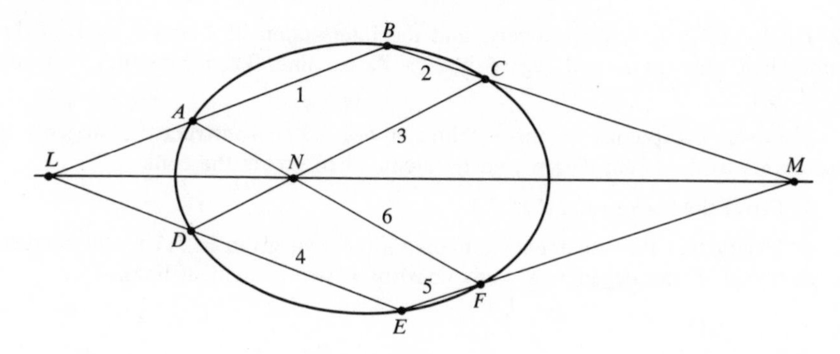

Fig. X, 8

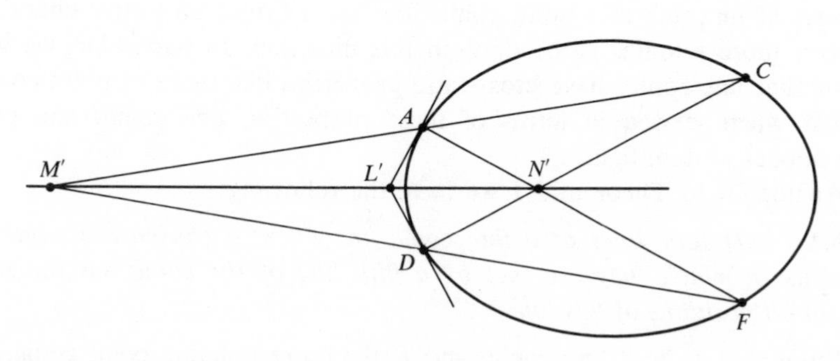

Fig. X, 9

gents at two vertices meet on the line joining the diagonal points which are not on the side through these vertices.

Analogous to Pascal's theorem, in which, of course, we have assumed the point conic to be nondegenerate, is the following proposition for a degenerate point conic:

5.5 *If a simple hexagon is inscribed in a degenerate point conic, consisting of two distinct lines, so that its vertices lie alternately on these lines, the pairs of opposite sides meet in collinear points.*

We proved this proposition in Chapter VIII, where we saw that it is simply another way of stating the Theorem of Pappus (see VIII, 16.4).

EXERCISES

1. Given five points A, B, C, D, E, no three collinear, use 5.2 to construct another point F on the conic determined by them. (*Hint.* Denote lines AB, BC,

CD, DE by 1, 2, 3, 4, respectively, and the intersection of 1 and 4 by L. Take another line through L and regard it as a Pascal line. Then construct F, using Fig. X, 6.)

2. Given five points, no three collinear, use 5.3 to construct the tangents at these points to the conic determined by them. Then sketch the conic.

3. Prove the converse part of 5.3.

4. Verify that the statement of Pappus's theorem given in 5.5 is still correct, though trivial, if the degenerate conic consists of two coincident lines.

6. Line Conics and Cross-Ratio. In view of what has been done with point conics in the preceding sections, our definition of a line conic as the set of tangents to a point conic now has a firmer projective character. However, more remains to be done in this direction. In particular, we wish to show that line conics have cross-ratio properties like those of point conics, and that when viewed in terms of these properties, line conics and point conics appear as dual figures.

Analogous to Theorem 2.1 we have the following:

6.1 *If four lines of a line conic are given, a chosen cross-ratio of the points in which they are met by a fifth line of the conic has the same value for all positions of this line.*

Proof. Let L be a line conic, and K the corresponding point conic. We first suppose that K is a circle with center O. Let a, b, c, d be four given lines of L, and t an arbitrary fifth line of L. These lines are then tangent to K at certain points P, Q, R, S, T, respectively (Fig. X, 10). If a, b, c, d meet t in A, B, C, D we must show that the value of (AB, CD) is the same for all positions of t. Since (AB, CD) can be expressed in terms of the angles formed at O by the lines AO, BO, CO, DO (see V, §6, Ex. 6), we need only show that these angles are independent of t. From elementary geometry we know that

$$\angle TOA = \tfrac{1}{2}\angle TOP = \text{any inscribed angle subtending arc } TP,$$
$$\angle TOB = \tfrac{1}{2}\angle TOQ = \text{any inscribed angle subtending arc } TQ.$$

Hence

$$\angle TOA - \angle TOB = \text{any inscribed angle subtending arc } TP - \text{arc } TQ,$$

or

$$\angle AOB = \text{any inscribed angle subtending arc } PQ.$$

In other words, although changing the position of t changes the positions of A and B, the value of $\angle OAB$ is fixed since P and Q are fixed.

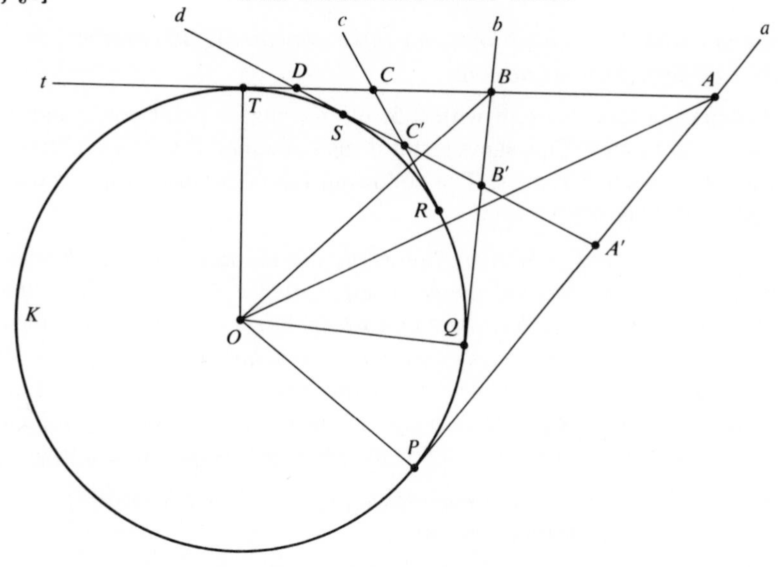

Fig. X, 10

Similarly, the values of the other angles at O can be shown to be independent of t. Thus the theorem is true if K is a circle. Suppose K is not a circle. Since all (nondegenerate) point conics are projectively equivalent, there is a circle in the same plane as K which goes into K under some projectivity of that plane, and the present theorem is true for that circle, as shown above. Since the property of being a tangent to a point conic is projective, the tangents to the circle go into the tangents to K under this projectivity. Also, the projectivity preserves cross-ratio. Hence the theorem is true for K. This completes the proof.

If line t in Fig. X, 10 approaches one of the four given lines, say d, then D approaches S, and we obtain the following special case of 6.1, which is analogous to 2.2:

6.2 *A chosen cross-ratio of the points in which four given lines a, b, c, d of a line conic meet a fifth line of the conic equals the corresponding cross-ratio of the three points in which a, b, c meet d and the point of contact on d.*

Thus, in Fig. X, 10, $(AB, CD) = (A'B', C'S)$.

According to 6.1, a given line conic is a locus of lines satisfying a certain condition involving cross-ratio. Although we shall not prove it, one can show, conversely, that any line curve meeting this condition is necessarily a line conic. The precise statement is as follows:

6.3 *If a line moves so that a chosen cross-ratio of its points of* *intersection with four given lines, no three concurrent, is constant, its locus and the four lines form a line conic.*

Comparing this theorem with 2.3, we see that a point conic and a line conic are dual figures. This same result is also obtained if we view a line conic in terms of projective ranges. The following theorems, which are converses, enable us to do this:

6.4 *The lines of a given line conic are the lines joining corresponding points of two projective, nonperspective ranges whose axes are lines of the conic. Exactly one such projective correspondence is determined for each choice of axes o and o', namely, the correspondence in which the point og corresponds to the point $o'g$, where g is any line of the conic. In this correspondence the point of contact on o corresponds to $o'o$, regarded as a point on o', and the point of contact on o' corresponds to oo', regarded as a point on o.*

6.5 *The lines joining corresponding points of two given projective, nonperspective ranges form a line conic.*

These theorems are the duals of 4.1 and 4.2. Their proofs are left as exercises.

When we drew line conics in Chapter IX, measurement played a part inasmuch as we used rectangular coordinates. Now, by means of 6.5, we can construct them projectively, without measurement of any sort. This is illustrated in Fig. X, 11. By projecting g on g'' from O, and g'' on g' from O_1, we obtain a projective correspondence between the points of g and g' which is not perspective since it has no self-corresponding point. The lines AA', BB', CC', . . . that join the pairs of corresponding points are the lines of a line conic according to 6.5. The diagram offers visual support of the fact that g and g' are lines of the conic.

EXERCISES

1. Draw a figure to illustrate 6.2 and give a detailed proof.

2. Restate 6.1 and 6.2 so as to obtain theorems about tangents to a point conic.

3. Verify that, according to their definitions, degenerate point conics and degenerate line conics are dual figures.

4. Prove 6.4 by dualizing the proof of 4.1.

5. Prove 6.5 by dualizing the proof of 4.2.

6. Restate 6.4 and 6.5 so as to obtain theorems about tangents to a point conic.

7. Draw more of the line conic in Fig. X,11 by using points of g below g''.

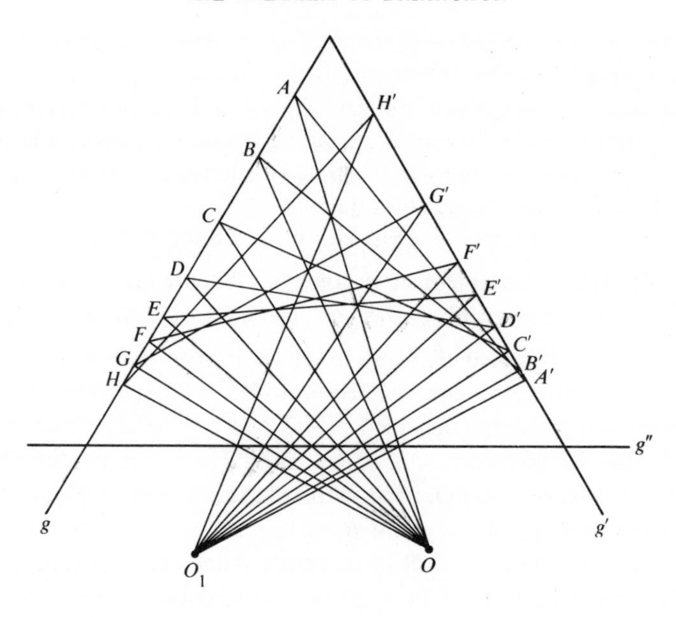

Fig. X, 11

8. Show that the lines joining corresponding points of two perspective ranges can be regarded as a degenerate line conic.

7. The Theorem of Brianchon. This theorem was stated without proof in an earlier chapter (see VIII, 18.11). At that time the concept of a line conic had not been introduced, and so the theorem was presented in terms of tangents to a point conic. Now we restate the theorem, broadened by inclusion of the converse, and show how it can be proved.

7.1 Brianchon's Theorem. *If the sides of a simple hexagon are on a line conic, the lines joining the pairs of opposite vertices are concurrent. Conversely, if the lines joining the pairs of opposite vertices of a simple hexagon, no three sides of which go through a point, are concurrent, the six sides are on a line conic.*

This theorem is seen to be the dual of Pascal's theorem if we recall that a simple hexagon is a self-dual figure, in which "vertex" and "side" are dual elements. The proof of Brianchon's theorem can therefore be obtained by dualizing the proof of Pascal's theorem. This is left as an exercise.

It is of interest to note that Pascal's theorem was discovered about 1640, before the Principle of Duality was known, and Brianchon's theorem not until 1806, in a period when the principle was first being recognized.

A hexagon with the property that the lines joining the pairs of opposite vertices go through a point is called a *Brianchon hexagon,* and the point is called a *Brianchon point*. From six given lines, no three concurrent, one can form sixty different simple hexagons by taking the lines as sides in all possible orders. If one of these hexagons is a Brianchon hexagon, the six lines are on a line conic by 7.1, and hence the other fifty-nine are also Brianchon hexagons by this same theorem. We therefore have the following:

7.2 Alternative Statement of Brianchon's Theorem. *Six given lines are on a line conic if and only if they are the sides of a Brianchon hexagon.*

We have already noted that any five lines a, b, c, d, e of the plane, no three concurrent, are on a unique line conic L (IX, 18.4). Additional lines on L must satisfy the condition specified in 7.2 and can be constructed accordingly. To do this, let us regard a, b, c, d, e, in that order, as the first five sides of a Brianchon hexagon, number the vertices 1, 2, 3, 4 as in Fig. X, 12, and choose as Brianchon point any point B on line 1,4, which is the line through the points 1 and 4. Since we wish to construct a line different from the given lines, B should not be point 1 or 4, or the intersection of lines 1,4 and 2,3. The points in which lines 2,B and 3,B meet sides e and a, respectively, are then vertices 5 and 6 of the hexagon. Line 5,6 is the sixth side of the hexagon and hence is on the conic L. By changing the position of B we can obtain other lines of L. If the lines thus obtained do not clearly outline L, we may arrange the five given lines in other orders and repeat the construction.

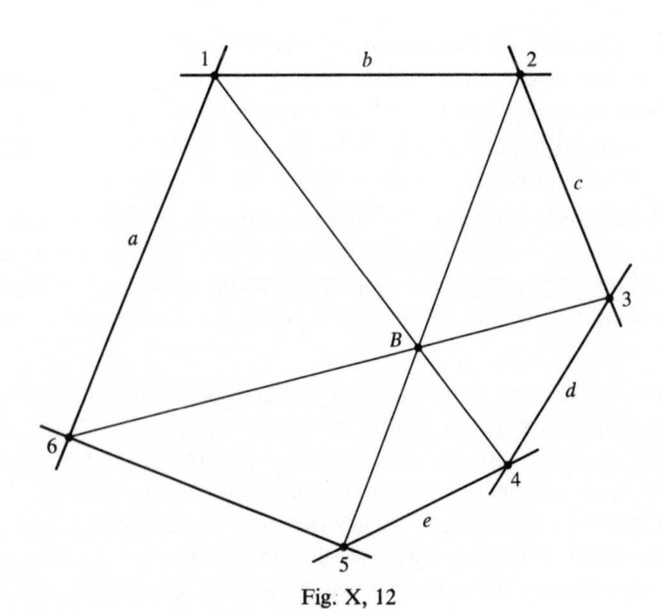

Fig. X, 12

Analogous to the special case of Pascal's theorem which shows how to construct the tangent at a point of a point conic (see 5.3), we have the following proposition, which enables one to construct the point of contact on a line of a line conic:

7.3 *In a simple pentagon whose sides belong to a given line conic, the line joining the point of contact on one side to the opposite vertex and the two lines joining the remaining pairs of opposite vertices are concurrent.* Conversely, if the line joining a vertex to a point A on the opposite side of the pentagon goes through the intersection of the lines joining the remaining pairs of opposite vertices, then A is the point of contact on that side.*

This theorem can be checked intuitively by taking as the line conic one which corresponds to a circle and drawing a figure. The proof of the theorem and its use in constructing points of contact are covered in the exercises.

EXERCISES

1. Restate 7.1, 7.2, 7.3 as propositions concerning a point conic.

2. Prove Brianchon's theorem as stated in 7.1.

3. A Pascal hexagon is not necessarily a Brianchon hexagon. To verify this, consider a hexagon $ABCDEF$ inscribed in a circle so that segments $\overline{AD}$ and $\overline{BE}$, but not segment $\overline{CF}$, are diameters.

4. Prove 7.3.

5. Given five lines of a line conic, show how to construct the points of contact on them by using 7.3, and draw the conic.

6. Show that 5.3 and 7.3 are duals.

---

* In a simple pentagon $ABCDE$, for example, A is opposite side CD, and the remaining pairs of opposite vertices are B, D and C, E.

Appendix: Determinants

1. Definitions of Determinants. A *determinant* is a polynomial in any number of letters of such special form that it can be symbolized concisely by a square array of these letters, or *elements*. The symbol, too, is called a determinant.

For a **second-order determinant,** whose symbol has two rows and two columns of elements, we have the definition:

$$(1) \qquad \begin{vmatrix} a_1 & a_2 \\ b_1 & b_2 \end{vmatrix} = a_1b_2 - a_2b_1.$$

We call a_1, a_2 the *elements of the first row*, and a_1, b_1 the *elements of the first column*. Also, a_1, b_2 form the *main diagonal*, and a_2, b_1 form the *secondary diagonal*. The polynomial represented by the symbol is therefore the product of the elements in the main diagonal minus the product of the elements in the secondary diagonal.

For a **third-order determinant,** which has three rows and three columns, we have the definition:

$$(2) \qquad \begin{vmatrix} a_1 & a_2 & a_3 \\ b_1 & b_2 & b_3 \\ c_1 & c_2 & c_3 \end{vmatrix} = a_1b_2c_3 + a_2b_3c_1 + a_3b_1c_2 - a_1b_3c_2 - a_2b_1c_3 - a_3b_2c_1.$$

Here, a_1, b_2, c_3 form the main diagonal, and a_3, b_2, c_1 the secondary diagonal, but the polynomial cannot, of course, be obtained from its symbol by the rule stated above for second order determinants. Instead, we repeat the first two columns as shown below, obtaining the six indicated diagonals:

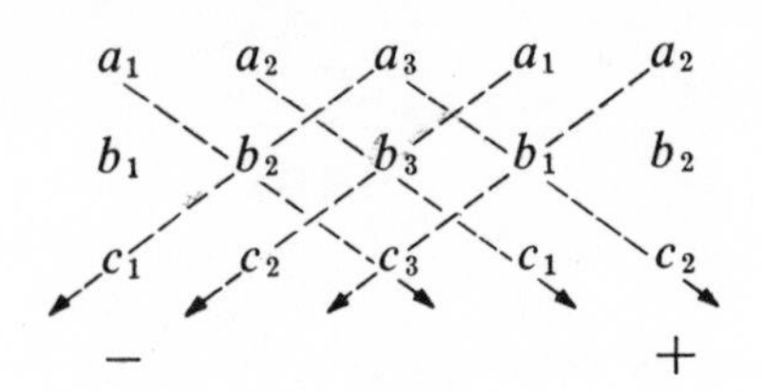

We form the product of the elements in each diagonal. The products obtained from the diagonals descending from left to right are the first three terms of the polynomial; the negatives of those obtained from the diagonals descending from right to left are the last three terms of the polynomial.

The **fourth-order determinant**

$$(3) \qquad \begin{vmatrix} a_1 & a_2 & a_3 & a_4 \\ b_1 & b_2 & b_3 & b_4 \\ c_1 & c_2 & c_3 & c_4 \\ d_1 & d_2 & d_3 & d_4 \end{vmatrix}$$

contains 24 terms, too many to write down, so we shall define it verbally. Each term has the form $\pm\, a_h b_i c_j d_k$, where the subscripts are the numbers 1, 2, 3, 4 in some order. There are as many terms as there are permutations of these numbers, which is 4!, or 24. The sign before a term is $+$ or $-$ according as the *departure* of the order of its subscripts from the natural order 1, 2, 3, 4 is even or odd. The order 1, 2, 4, 3, for example, has the departure 1 since it differs from the natural order in only 1 way: 4 precedes 3. The order 1, 4, 2, 3 has the departure 2 since it differs from the natural order in 2 ways: 4 precedes 2, and 4 precedes 3. The polynomial represented by the symbol (3) therefore contains the following among its terms:

$$a_1b_2c_3d_4 + a_1b_4c_2d_3 + a_4b_3c_2d_1 + \cdots - a_1b_2c_4d_3 - a_1b_3c_2d_4 - a_4b_3c_1d_2 - \cdots.$$

Determinants of higher order are defined in an analogous way. Thus, a fifth-order determinant has 5!, or 120, terms. Second and third order determinants can also be defined in this way, with results that agree with the alternative definitions given above. A *first order determinant* is defined to be a single element: $|a| = a$.

The determinants (1), (2), (3) are often abbreviated $|a\ b|$, $|a\ b\ c|$, $|a\ b\ c\ d|$, respectively.

The following examples illustrate determinants whose elements are specific numbers:

$$\text{(a)} \qquad \begin{vmatrix} 3 & 2 \\ -4 & 7 \end{vmatrix} = 21 - (-8) = 29;$$

$$\text{(b)} \qquad \begin{vmatrix} 3 & 1 & 4 \\ -1 & 2 & 5 \\ 0 & 6 & -1 \end{vmatrix} = (-6) + 0 + (-24) - 0 - 90 - 1 = 121.$$

Although determinants and matrices are different things, they have many connections with each other. Corresponding to any determinant, for example, there is a square matrix which has the same elements. It is called the **matrix of the determinant.** Thus, the matrix of the determinant (a) in the preceding paragraph is

$$\begin{pmatrix} 3 & 2 \\ -4 & 7 \end{pmatrix}$$

Conversely, from a given matrix one can generally form determinants of various orders by suppressing rows and columns. From

$$\begin{pmatrix} 3 & 6 & 2 \\ 4 & 3 & 1 \end{pmatrix},$$

for example, we can form three determinants of order 2, and six of order 1. This procedure is involved in the important concept of the *rank of a matrix.* A matrix is said to be of **rank r** if it contains at least one r-rowed determinant that is not zero and all determinants of higher order than r contained in it are zero; and it is said to be of **rank 0** if all of its elements are zero. Thus, the matrices

$$\begin{pmatrix} 3 & 6 & 2 \\ 4 & 3 & 1 \end{pmatrix}, \quad \begin{pmatrix} 3 & 1 & 4 \\ 6 & 2 & 8 \end{pmatrix}, \quad \begin{pmatrix} 0 & 0 & 0 \\ 0 & 0 & 0 \end{pmatrix}$$

have the ranks 2, 1, 0, respectively. The rank of the matrix of determinant (b) in the preceding paragraph is 3.

2. Basic Theorems. The evaluation of a determinant of order 4 or more by use of its definition is usually very laborious. The same may be true of a determinant of order 3 when its elements are algebraic expressions, or numbers of large absolute value. There are are theorems which facilitate these evaluations and at the same time provide a basis for the general handling of determinants. We state some of these theorems, which hold for determinants of all orders, and illustrate them for order 2 or 3. Their proofs, like the proofs of all theorems stated in this appendix, are omitted.

2.1 *If the rows and columns of a determinant are interchanged, the value of the determinant is unaltered.*

$$\begin{vmatrix} a & b \\ c & d \end{vmatrix} = \begin{vmatrix} a & c \\ b & d \end{vmatrix}.$$

Because of this theorem, any statement that can be proved for the rows of a determinant can also be proved for the columns. We therefore usually restrict the statements of subsequent theorems to rows alone.

2.2 *If we multiply the elements of a row of a determinant by k, the new determinant is k times the original. Hence, a common factor k can be removed from all the elements in a row of a determinant provided that we multiply the resulting determinant by k.*

$$\begin{vmatrix} ka & kb \\ c & d \end{vmatrix} = k \begin{vmatrix} a & b \\ c & d \end{vmatrix}.$$

2.3 *The sign of a determinant is reversed by the interchange of two rows.*

$$\begin{vmatrix} a & b \\ c & d \end{vmatrix} = -\begin{vmatrix} c & d \\ a & b \end{vmatrix}.$$

2.4 *The value of a determinant is zero if every element in a row is zero, or if the elements in one row are proportional to the elements in another row.*

$$\begin{vmatrix} a & b \\ ka & kb \end{vmatrix} = a\,b\,k - a\,b\,k = 0.$$

2.5 *If two determinants of the same order are identical, except perhaps for the elements of a certain row m, their sum is a determinant of that order in which the elements of row m are the sums of the corresponding elements of the rows m in the given determinants, and which is the same as these determinants with regard to all other elements.*

$$\begin{vmatrix} a & b \\ c & d \end{vmatrix} + \begin{vmatrix} e & f \\ c & d \end{vmatrix} = \begin{vmatrix} a+e & b+f \\ c & d \end{vmatrix}.$$

2.6 *The product D_1D_2 of two determinants D_1 and D_2 of the same order is a determinant of that order in which the element in row i and column j is the sum of the products of the corresponding elements of row i in D_1 and column j in D_2.*

$$\begin{vmatrix} a & b \\ c & d \end{vmatrix} \cdot \begin{vmatrix} e & f \\ g & h \end{vmatrix} = \begin{vmatrix} ae+bg & af+bh \\ ce+dg & cf+dh \end{vmatrix}.$$

2.7 *If to each element of a row of a determinant we add the same multiple of the corresponding element of another row, the value of the determinant is unchanged.*

$$\begin{vmatrix} a & b \\ c & d \end{vmatrix} = \begin{vmatrix} a+kc & b+kd \\ c & d \end{vmatrix}.$$

The following example illustrates the use of some of the above **theorems** in reducing the size of the numbers in a determinant.

$$\begin{vmatrix} 18 & 12 & 41 \\ 6 & -9 & 19 \\ 15 & 30 & 23 \end{vmatrix} = 9\begin{vmatrix} 6 & 4 & 41 \\ 2 & -3 & 19 \\ 5 & 10 & 23 \end{vmatrix} = 9\begin{vmatrix} 6 & 4 & 5 \\ 2 & -3 & 7 \\ 5 & 10 & -7 \end{vmatrix}.$$

The second of these determinants is obtained from the first by 2.2, applied to columns. The third is obtained from the second by multiplying the elements of the latter's first column by -6 and adding the results to the elements of its last column.

3. Minors and Cofactors. As we shall now see, the reduction of a determinant to others of lower order can further facilitate its evaluation.

By the **minor of an element** in a determinant of order n is meant the determinant of order $n-1$ obtained by striking out the row and column in which the element appears. For example, let Δ be a third-order determinant:

$$\Delta = \begin{vmatrix} a_1 & a_2 & a_3 \\ b_1 & b_2 & b_3 \\ c_1 & c_2 & c_3 \end{vmatrix}.$$

Then the minors of a_1 and b_2 are the second-order determinants

$$\begin{vmatrix} b_2 & b_3 \\ c_2 & c_3 \end{vmatrix} \quad \text{and} \quad \begin{vmatrix} a_1 & a_3 \\ c_1 & c_3 \end{vmatrix},$$

respectively.

Now let us multiply each element in the first row of Δ by its minor and form the algebraic sum

$$(1) \qquad a_1\begin{vmatrix} b_2 & b_3 \\ c_2 & c_3 \end{vmatrix} - a_2\begin{vmatrix} b_1 & b_3 \\ c_1 & c_3 \end{vmatrix} + a_3\begin{vmatrix} b_1 & b_2 \\ c_1 & c_2 \end{vmatrix}.$$

It will be found, on working out the second-order determinants which appear here, that this sum is Δ. Similarly, if the second row is used, one can verify that

$$(2) \qquad -b_1\begin{vmatrix} a_2 & a_3 \\ c_2 & c_3 \end{vmatrix} + b_2\begin{vmatrix} a_1 & a_3 \\ c_1 & c_3 \end{vmatrix} - b_3\begin{vmatrix} a_1 & a_2 \\ c_1 & c_2 \end{vmatrix} = \Delta.$$

A like result is obtained when the third row, or any column, is used, the rule for the signs being the following: if $i+j$ is even, where i and j denote the row and column in which an element appears, use the + sign before every product of that element and its minor; if $i+j$ is odd, use the — sign. Thus for a_1, a_2, a_3 we obtain $i + j = 2, 3, 4$, respectively; the corresponding signs are therefore +, −, +, which agrees with (1). For b_1, b_2, b_3 we obtain $i + j = 3, 4, 5$, so that the signs are −, +, −, in agreement with (2). The following table gives all the signs:

$$\begin{matrix} + & - & + \\ - & + & - \\ + & - & + . \end{matrix}$$

Like results are obtained for determinants of all orders. Since $(-1)^{i+j}$ is positive or negative according as $i+j$ is even or odd, we have the following theorem:

3.1 *If each element in a row (or column) of a determinant is multiplied by its minor and by $(-1)^{i+j}$, where i and j denote the row and column in which the element appears, the sum of all the products thus formed is the value of the determinant.*

This method of evaluating a determinant is called **expansion** (or **development**) **by minors.** By its use, for example, the calculation of a fourth-order determinant is reduced to that of third-order determinants. Thus,

$$\begin{vmatrix} 4 & 3 & 5 & -1 \\ 6 & 2 & 1 & 0 \\ 3 & 0 & -4 & 8 \\ 5 & 1 & 2 & 3 \end{vmatrix} = 4\begin{vmatrix} 2 & 1 & 0 \\ 0 & -4 & 8 \\ 1 & 2 & 3 \end{vmatrix} - 3\begin{vmatrix} 6 & 1 & 0 \\ 3 & -4 & 8 \\ 5 & 2 & 3 \end{vmatrix}$$

$$+ 5\begin{vmatrix} 6 & 2 & 0 \\ 3 & 0 & 8 \\ 5 & 1 & 3 \end{vmatrix} + 1\begin{vmatrix} 6 & 2 & 1 \\ 3 & 0 & -4 \\ 5 & 1 & 2 \end{vmatrix}.$$

The **cofactor of an element** in a determinant is obtained by multiplying the minor of the element by $(-1)^{i+j}$, where i and j denote the row and column in which the element appears. Theorem 3.1 can therefore be restated as follows:

3.2 *If each element in a row (or column) of a determinant is multiplied by its cofactor, the sum of all the products thus formed is the value of the determinant.*

This method of evaluating a determinant is called **expansion** (or **devel-**

opment) by cofactors. Thus, if A_1, A_2, A_3 are the cofactors of a_1, a_2, a_3 in the determinant Δ considered above, we have

$$\Delta = a_1A_1 + a_2A_2 + a_3A_3.$$

Expansion by cofactors is useful when it is not necessary to keep track of signs, which is the case in much theoretical work. Otherwise, one usually expands by minors.

Corresponding to any given determinant D, let us form another determinant D' of the same order in the following way: the element in row i and column j of D' is the cofactor of the element in row i and column j of D. We call D' the **adjoint** of D.

3.3 *If D is any determinant of order n, and D' is its adjoint, then $D' = D^{n-1}$.*

Thus, if D is the determinant Δ considered above, and A_1, B_1, C_1, . . . are the cofactors of a_1, b_1, c_1, . . ., then

$$\begin{vmatrix} A_1 & A_2 & A_3 \\ B_1 & B_2 & B_3 \\ C_1 & C_2 & C_3 \end{vmatrix} = \begin{vmatrix} a_1 & a_2 & a_3 \\ b_1 & b_2 & b_3 \\ c_1 & c_2 & c_3 \end{vmatrix}^2.$$

4. Laplace's Development. This generalizes the development of a determinant by minors. By a **minor** we now mean any subdeterminant of the original determinant, that is, any determinant of lower order than the original which can be formed from its elements without changing their positions. Thus,

$$(1) \qquad \begin{vmatrix} a_1 & a_3 \\ b_1 & b_3 \end{vmatrix}, \quad \begin{vmatrix} c_2 & c_4 \\ e_2 & e_4 \end{vmatrix}, \quad \begin{vmatrix} c_2 & c_4 & c_5 \\ d_2 & d_4 & d_5 \\ e_2 & e_4 & e_5 \end{vmatrix}$$

are minors in the determinant

$$\begin{vmatrix} a_1 & a_2 & a_3 & a_4 & a_5 \\ b_1 & b_2 & b_3 & b_4 & b_5 \\ c_1 & c_2 & c_3 & c_4 & c_5 \\ d_1 & d_2 & d_3 & d_4 & d_5 \\ e_1 & e_2 & e_3 & e_4 & e_5 \end{vmatrix}.$$

If we strike out the rows and columns in which a given minor appears, we always obtain another minor. It is called the **complement** of the given minor. Thus, the third determinant in (1) is the complement of the first. If we multiply the complement of a minor by $(-1)^h$, where h is the sum of the numbers denoting the rows and columns in which the minor appears, we obtain the **algebraic complement** of the minor. For the minor

$$\begin{vmatrix} a_1 & a_3 \\ b_1 & b_3 \end{vmatrix}$$

in our illustration, $h = 7$, $(-1)^7 = -1$, and the algebraic complement of this minor is

$$-\begin{vmatrix} c_2 & c_4 & c_5 \\ d_2 & d_4 & d_5 \\ e_2 & e_4 & e_5 \end{vmatrix}.$$

Laplace's development of a determinant is now described in the following theorem:

4.1 *If m rows (or columns) of a determinant are chosen, and each minor of order m situated in these rows is multiplied by its algebraic complement, the sum of all the products thus formed is the value of the determinant.*

To give a simple but important type of example, let us develop the determinant

$$\begin{vmatrix} a & b & 0 & 0 \\ c & d & 0 & 0 \\ 0 & 0 & e & f \\ 0 & 0 & g & h \end{vmatrix}$$

by using its first two rows. There are six minors of order 2 in these rows. Five of them have a column of zeros, and hence are equal to zero. The sixth and its algebraic complement are

$$\begin{vmatrix} a & b \\ c & d \end{vmatrix} \quad \text{and} \quad \begin{vmatrix} e & f \\ g & h \end{vmatrix},$$

respectively. Hence their product is the value of the given determinant.

5. Application to Systems of Equations. If the linear equations

$$\begin{cases} a_1x + b_1y = k_1 \\ a_2x + b_2y = k_2 \end{cases} \tag{1}$$

are solved for x and y by elementary algebra, and $a_1b_2 - a_2b_1 \neq 0$, the result is

$$x = \frac{k_1b_2 - k_2b_1}{a_1b_2 - a_2b_1}, \qquad y = \frac{a_1k_2 - a_2k_1}{a_1b_2 - a_2b_1}.$$

In determinant notation, this is

$$(2) \qquad x = \frac{\begin{vmatrix} k_1 & b_1 \\ k_2 & b_2 \end{vmatrix}}{\begin{vmatrix} a_1 & b_1 \\ a_2 & b_2 \end{vmatrix}}, \qquad y = \frac{\begin{vmatrix} a_1 & k_1 \\ a_2 & k_2 \end{vmatrix}}{\begin{vmatrix} a_1 & b_1 \\ a_2 & b_2 \end{vmatrix}},$$

or, further abbreviated,

$$x = \frac{|k \quad b|}{|a \quad b|}, \qquad y = \frac{|a \quad k|}{|a \quad b|}.$$

The determinant

$$\begin{vmatrix} a_1 & b_1 \\ a_2 & b_2 \end{vmatrix},$$

whose elements are the coefficients of x and y in the system of equations (1), is called the **determinant of the system.** If this determinant is not zero, the equations are satisfied by only one pair of values of x and y. Equations (2) are formulas for finding these values.

Using elementary algebra, one can likewise show that the system of three first-degree equations in three unknowns,

$$a_1x + b_1y + c_1z = k_1$$
$$a_2x + b_2y + c_2z = k_2$$
$$a_3x + b_3y + c_3z = k_3,$$

has

$$x = \frac{|k\ b\ c|}{|a\ b\ c|}, \qquad y = \frac{|a\ k\ c|}{|a\ b\ c|}, \qquad z = \frac{|a\ b\ k|}{|a\ b\ c|}$$

as its only solution when $|a\ b\ c|$, the **determinant of the system**, is not zero. Here, $|a\ b\ c|$ symbolizes

$$\begin{vmatrix} a_1 & b_1 & c_1 \\ a_2 & b_2 & c_2 \\ a_3 & b_3 & c_3 \end{vmatrix},$$

and $|k\ b\ c|$, for example, is the determinant obtained from this by substituting the constant terms k_1, k_2, k_3 for a_1, a_2, a_3, respectively.

These results, together with the entire terminology, can be generalized. We therefore have the theorem:

5.1 *If the determinant* Δ *of a system of n linear equations in n unknowns* $x_1, x_2, \ldots, x_n$ *is not zero, there is only one set of values of the unknowns that satisfy the equations. These values are given by the formulas*

$$x_1 = \frac{\Delta_1}{\Delta}, \quad x_2 = \frac{\Delta_2}{\Delta}, \quad \ldots, \quad x_n = \frac{\Delta_n}{\Delta},$$

where Δ_i *is the determinant obtained from* Δ *by substituting the constant terms for the coefficients of* x_i.

The formulas stated in this theorem are known as **Cramer's Rule.**

If the constant term in a linear equation is zero, the equation is said to be **homogeneous.** We therefore have the following important corollary of 5.1:

5.2 *If the determinant of a system of n homogeneous linear equations in n unknowns is not zero, the values* 0, 0, 0, . . ., 0 *are the only ones that satisfy the equations.*

Answers to Selected Exercises

CHAPTER I

§4.

3. I, R, R.

§5.

1. (a), (b), (e), (f).
3. 0°, 180°; 0°, 120°, 240°.

§7.

1. (a) $(-1, 1), (\frac{4}{3}, -\frac{1}{4})$; (b) line $y' = -7$, line $x' = 2$; (c) $(-1, \frac{7}{3})$.

§8.

1. $x = \frac{23}{22}, y = -\frac{4}{11}$.
3. $(22, 0), (\frac{23}{22}, -\frac{4}{11})$.
7. (6, 1) has many originals; (7, 9) has none. No. Yes. No.

CHAPTER II

§4.

1. (a) $(-2, 3)$; (b) $(-3, -4)$.
3. $y = 2x + 2$; $y = 4x/3$.
7. $x' = x + 5, y' = y - 1$.
9. $-53°$.
17. $x' = x + 2\pi, y' = y + 2\pi$.

§5.

1. (a) both are $x' = x - 2, y' = y + 4$; (b) length $2\sqrt{5}, \theta = 117°$.
11. Those in Exs. 8, 10.

§6.

1. (a) $\begin{pmatrix} x' \\ y' \end{pmatrix} = \begin{pmatrix} x \\ y \end{pmatrix} + \begin{pmatrix} 3 \\ 1 \end{pmatrix}$;

   (d) $\begin{pmatrix} x' \\ y' \end{pmatrix} = \begin{pmatrix} x \\ y \end{pmatrix} + \begin{pmatrix} 0 \\ 0 \end{pmatrix}$ or $\begin{pmatrix} x' \\ y' \end{pmatrix} = \begin{pmatrix} x \\ y \end{pmatrix}$.

3. $\begin{pmatrix} a \\ b \end{pmatrix} + \begin{pmatrix} c \\ d \end{pmatrix} = \begin{pmatrix} c \\ d \end{pmatrix} + \begin{pmatrix} a \\ b \end{pmatrix}$.

§7.

7. No.
13. It is left fixed by a rotation through $2\pi/n$ about its center.

§8.

3. $\alpha = \pi/4$. (a) $x' = y'$; (b) $x' = 0$; (c) $x' = -3/\sqrt{2}$;
(d) $x' + y' = 3\sqrt{2}$; (e) $x'^2 + y'^2 = 1$; (f) $y'^2 - x'^2 = 2$;
(g) $x'^2 + y'^2 - \sqrt{2}\,(x' + y') = 1$; (h) $(x' + y')^2 = \sqrt{2}\,(y' - x')$.
9. (a) $\alpha = \pi$. $x' = -x$, $y' = -y$;
(b) $\alpha = \arctan(-\frac{24}{7})$ and is between 90° and 180°.
$x' = (-7x - 24y)/25$, $y' = (24x - 7y)/25$;
(c) $\alpha = \arctan(-\frac{24}{7})$ and is between 0° and −90°.
$x' = (7x + 24y)/25$, $y' = (-24x + 7y)/25$.

§9.

1. $$\begin{pmatrix} x' \\ y' \end{pmatrix} = \begin{pmatrix} \frac{1}{2} & -\frac{\sqrt{3}}{2} \\ \frac{\sqrt{3}}{2} & \frac{1}{2} \end{pmatrix} \begin{pmatrix} x \\ y \end{pmatrix};$$
$$\begin{pmatrix} x' \\ y' \end{pmatrix} = \begin{pmatrix} \frac{1}{2} & \frac{\sqrt{3}}{2} \\ -\frac{\sqrt{3}}{2} & \frac{1}{2} \end{pmatrix} \begin{pmatrix} x \\ y \end{pmatrix}.$$
5. $$\begin{pmatrix} -\frac{5}{2} & \frac{3}{2} \\ 2 & -1 \end{pmatrix}; \begin{pmatrix} 2 & 3 \\ 4 & 5 \end{pmatrix}.$$
7. (a), (c), (d).

§10.

1. $x = (4 - 5\sqrt{3})/2$, $y = (5 + 4\sqrt{3})/2$.
9. Value 1. Inverse is $x' - h = (x - h)\cos\alpha + (y - k)\sin\alpha$,
$y' - k = -(x - h)\sin\alpha + (y - k)\cos\alpha$, where (x', y') is image point.

§11.

3. (a) $x'' = -y + 4$, $y'' = x - 2$, rotation;
(b) $x'' = -x + 2$, $y'' = -y + 2$, rotation;
(c) $x'' = x - 2$, $y'' = y$, translation.
7. One answer is: the rotations through π about $(0, 0)$ and $(1, -\frac{3}{2})$.

§12.

1. (a) Rotation, $\alpha = -53°$; (b) Rotation, $\alpha = 127°$; (c) Rotation, $\alpha = 3\pi/2$;
(d) Not a displacement; (e) Rotation, $\alpha = \pi/2$.
7. $x = Ax' - By' + BD - AC$, $y = Bx' + Ay' - BC - AD$.

§13.

5. (a) a displacement; (b) an opposite motion.

§14.

1. (a) $x' = -x, y' = y$; (b) $x' = y, y' = x$; (c) $x' = -y, y' = -x$.

§15.

3. The triangle with vertices (2, 1), $(\frac{7}{5}, \frac{9}{5})$, $(\frac{14}{5}, \frac{8}{5})$.
7. $x' = x, y' = -y + 6$, a reflection in line $y = 3$, and $x'' = x' + 1$, $y'' = y'$, a translation.

§16.

1. Converse not true.
5. (a) none; (b) rotations, reflections; (c) translations, glide-reflections.

§17.

11. (a) $x' = y - 1, y' = 1 - x$; (b) $x' = 1 - y, y' = x - 2$; (c) $x' = 3 + y, y' = -x$; (d) $x' = 5 - y, y' = 4 - x$.

CHAPTER III

§3.

5. Yes.
7. Each is a direct similarity, the resultant of a radial transformation with center O, ratio $-k$, and a rotation through π about O.

§4.

3. The areas of all figures are multiplied by an amount equal to the absolute value of each determinant.
5. $x' = (2x + 3y + 1)/\sqrt{13}$, $y' = (3x - 2y + 4)/\sqrt{13}$ and $x'' = \sqrt{13}\, x'$, $y'' = \sqrt{13}\, y'$.
7. $x' = 13x + 15, y' = 13y - 1$, a similarity with ratio 13.

§5.

5. Yes.

§6.

1. distance.
3. It is not possible to mention such properties on the basis of our discussion. They exist, however, and are nonmetric, as will be seen in Chap. IV.

CHAPTER IV

§2.

1. $x = (3x' + y' - 1)/7, y = (-x' + 2y' + 5)/7$.
3. No. Yes.

§3.

1. (a) -2; (b) $-\frac{1}{2}$; (c) $\frac{1}{2}$.
3. $r \to -1$ in both cases.

§4.

9. $(16x'^2 - 8x'y' + y'^2 - 21x' + 5y' + 6)x' = 0$.

§5.

1. $x' = 2x - 3y$, $y' = y$ and $x'' = x'$, $y'' = (5x' + 23y')/2$.
5. (a) parallelogram with one vertex at O;
   (b) parallelogram with center O;
   (c) ellipse with center O;
   (d) a triangle and its medians, the latter meeting in O.

§6.

1. affine in (a), (c), (e), (g); metric in (b), (d), (f), (h).

§7.

1. affine in (a), (d), (g), (h), (i), (j), (k); metric in (b), (c), (e), (f).

§8.

1. (a) noncollinear; (b) collinear.

§12.

7. $bx + ay = ab$.
9. (c), (d), (f), (g).

CHAPTER V

§2.

1. (a) $(0, -1), (0, -2), (0, -3)$; (b) $(0, 1), (0, 2), (0, 3)$;
   (c) $(0, -2), (0, -4), (0, -6)$.
3. If g, g' are parallel, all parallel projections preserve distance.

§5.

3. $(2, 0), (6, 0)$.

§6.

5. Each cross-ratio is $\frac{4}{3}$.

§7.

3. An open 5-sided figure separated into 2 branches.

§9.

1. (a) $\frac{6}{5}$; (b) $-\frac{1}{8}$; (c) $-\frac{5}{3}$; (d) $\frac{8}{3}$.
3. (a) $(BA, CD), (CD, BA), (DC, AB)$;
   (b) $(CA, DB), (DB, CA), (BD, AC)$;
   (c) $(CA, BD), (BD, CA), (DB, AC)$.

§11.

1. $-\sqrt{2}$.
5. (a) 6; (b) $1/R$, $1 - R$, $1/(1 - R)$, $(R - 1)/R$, $R/(R - 1)$.
9. $\frac{4}{3}$.

§12.

7. They do, in the given order. If this is *abcd*, the others are *abdc*, *bacd*, *badc*, *cdab*, *cdba*, *dcab*, *dcba*.

§13.

5. No.

§14.

1. (a) $2(\sin 20°)^2/\sin 10°$; (b) $\sin 70° \sin 20°/\sin 10° \sin 60°$.

CHAPTER VI

§3.

3. (a) $y = x, x' + y' = 1$; (b) $y = x, y' = x' - 1$; (c) $y = x, y' = x'$.

§4.

9. $(1, 0), (\frac{1}{5}, 0), (\frac{1}{2}, 0), (\frac{1}{3}, 0)$. Each cross-ratio is $\frac{1}{3}$.
11. (a) $x - y + 1 = 0$; (b) $3x' + 4y' - 13 = 0$; (c) $(0, 1)$;
    (d) Any line parallel to line $x - y + 1 = 0$.

§5.

1. I is affine.
3. $x'' = (x - y + 1)/(3x + y - 1), y'' = (x - 4y + 2)/(3x + y - 1)$.
5. $x'' = x, y'' = y$. $T = T^{-1}$. The set is a group.

§6.

1. $x' = x/(2 - x), y' = 2y/(2 - x)$.

§7.

1. (a) $x'y' = 1$, *v.* point on c is $(0, 0)$, none on c';
   (b) $y' = x'^2$, *v.* point on c' is $(0, 0)$, none on c;
   (c) $x'^2 + y'^2 = 1$, *v.* points on c' are $(0, \pm 1)$, none on c;
   (d) $y'^2 - x'^2 = 1$, *v.* points on c are $(0, \pm 1)$, same for c';
   (e) $(abx')^2 - (ay')^2 = b^2$, *v.* points on c are $(0, \pm b)$, none on c';
   (f) $x'^2 + y'^2 = x'$, *v.* point on c' is $(0, 0)$, none on c.
3. (a) The tangent at $(0, 0)$ is $y = 0$. Its image is $y' = 0$;
   (b) The tangent at $(0, 1)$ is $y = 1$. Its image is $x' = y'$.

CHAPTER VII

§2.

1. (a) $r' = 1/r, \theta' = \theta$ when $r \neq 0$; $r' = 0$ when $r = 0$;
   (b) $x' = x/(x^2 + y^2), y' = y/(x^2 + y^2)$ if x, y are not both 0; $x' = y' = 0$ if $x = y = 0$.

§4.

1. No. Yes.
3. (a) 2 simple closed curves with 1 common point;
   (b) 3 simple arcs, two of which do not meet one another, but meet the third, each in 1 point;
   (c) a simple arc and a simple closed curve meeting in 1 point;
   (d) 2 simple closed curves with 3 common points;
   (e) 2 simple arcs with no common point;
   (f) 3 simple arcs, just two of which meet, in 1 point.

§5.

3. (a) S is the plane, excluding the line $x = 0$; same for S'.

§6.

5. Their meeting point would be the image of distinct points of the circle and T would not be 1-1.

§8.

1. (a) Quadrants 1 and 4, axes included; not topological;
(b) Quadrant 1, axes included; not topological;
(c) Quadrants 1 and 4, excluding y-axis; topological;
(d) Interior of region bounded by lines $x = 0$, $y = 0$, $y = 2$; topological.
5. (a) a simple arc not meeting c; more precisely, a semicircumference of the sphere without its endpoints;
(b) a simple arc interior to γ; more precisely, either a diameter of γ or a semiellipse, endpoints excluded in each case.

§9.

1. (a) and (e): $(0, \pm 1)$; (b) and (f): $(\pm 1, 0)$;
(c) and (g): $(\frac{1}{2}\sqrt{2}, \frac{1}{2}\sqrt{2}), (-\frac{1}{2}\sqrt{2}, -\frac{1}{2}\sqrt{2})$;
(d) and (h): $(\frac{1}{5}\sqrt{5}, \frac{2}{5}\sqrt{5}), (-\frac{1}{5}\sqrt{5}, -\frac{2}{5}\sqrt{5})$.
5. (a) a semiellipse and a point on it; (b) two intersecting semiellipses; (c) two nonintersecting semiellipses, one of which is a diameter of γ; (d) two nonintersecting semiellipses, neither a diameter of γ; (e) two non-intersecting semiellipses which, with their excluded endpoints, form a complete ellipse.
7. (a) $x' = e^k$; (b) $x'y' = 1$; (c) $x'y' = e$; (d) $x' = y'$; (e) $x' = ey'$.

§11.

1. (b) It should not completely encompass the torus.

§12.

5. The simple closed curves which result when to each semicircumference of H are added the endpoints which had been excluded from it, and these endpoints are identified.

CHAPTER VIII

§2.

1. (a) ideal; (b) Euclidean; (c) none.

§6.

7. Great circles.

§8.

1. It consists of 3 concurrent lines, their common point, and 3 other points, one on each line.
3. A certain figure consists of 3 nonconcurrent lines and their points of intersection.

§10.

5. Four points A, B, C, D on a projective line are in one of the orders $ABCD$ or $ABDC$ if $(AB, CD) > 0$, and in the order $ACBD$ if $(AB, CD) < 0$.

§11.

1. 3
7. It is the figure consisting of 5 lines, no 3 concurrent, and their 10 points of intersection.

§12.

5. Two vertices of a complete quadrangle are separated harmonically by the diagonal point on their line and the point of intersection of this line and the line through the other diagonal points.

§13.

3. Two pairs of lines through a point form a harmonic set if and only if there exists a complete quadrilateral having one pair, e and f, as diagonals, and the other pair, g and h, as the lines joining point ef to the vertices on the third diagonal.

§16.

3. If the successive sides of a simple hexagon pass alternately through 2 points, the 3 lines joining the pairs of opposite vertices are concurrent.

§17.

5. (a) The points in which two of the bisectors meet the opposite sides (extended) are on a line parallel to the third bisector.
   (b) Each bisector is parallel to the opposite side.

§18.

1. No.

CHAPTER IX

§2.

1. (a) $(\frac{1}{2}, \frac{1}{3}, 1)$, $(3, 2, 6)$; (b) $(1, 0, 1)$, $(2, 0, 2)$; (c) $(0, 0, k)$; (d) $(1, -\frac{2}{3}, 0)$, $(3, -2, 0)$; (e) $(k, 0, 0)$; (f) $(0, k, 0)$.
7. Not necessarily.
9. (a) $(0, 1, 0)$, $(1, 0, 0)$; (b) $(1, \pm 1, 0)$; (c) $(\pm a, b, 0)$; (d) $(\pm b, a, 0)$.
11. $(-B \pm \sqrt{B^2 - 4AC}, 2A, 0)$ if $A \neq 0$; $(1, 0, 0)$, $(0, 1, 0)$ if $A = C = 0$.
13. $(0, 0, 2)$; $(0, 1, 0)$.

§3.

1. (a) $2x_1 + 3x_3 = x_2$; (b) $3x_1 + 2x_2 = 6x_3$; (c) $x_1 = 5x_3$; (d) $x_2 = 3x_3$; (e) $x_1 + x_2 = 0$; (f) $4x_1 - 3x_2 = 0$.
5. $(3, 2, 0)$, $(3, 2, 0)$, $(0, 1, 0)$.
7. (a) $(0, 0, 1)$; (b) $(0, 1, 0)$; (c) $(1, 0, 0)$; (d) $(0, 1, 0)$; (e) $(1, 0, 0)$; (f) $(0, 0, 1)$; (g) $(-2, 3, 7)$; (h) $(-3, 1, 0)$.
17. The plane, excluding the line $x_1 = 0$.

§4.

1. (a) $(6, -7, -2)$; (b) $(6, -11, 2)$.
3. (a) -1; (b) $-\frac{1}{2}$; (c) $-\frac{2}{5}$.

§5.

1. (a) $\frac{1}{2}$; (b) $\frac{2}{3}$; (c) -1; (d) -1.

§6.

1. (a) $(18, -3, 2)$; (b) $(2, -1, 3)$; (c) $x_1' = 0$; (d) $3x_1' + 4x_2' - 13x_3' = 0$; (e) $x_1 - x_2 - x_3 = 0$; (f) $x_2' - 3x_3' = 0$.

§7.

5. (a) $x_1' = x_1 \cos\alpha - x_2 \sin\alpha$, $x_2' = x_1 \sin\alpha + x_2 \cos\alpha$, $x_3' = x_3$; (c) Each is left fixed.

§8.

1. When $B^2 - 4AC < 0$.
3. (a) $x_1^2 = 0$; (b) $x_1x_2 = 0$.
9. $(1, 0, 0)$, $(0, 1, 0)$.

§9.

1. $x - 2y + 5 = 0$.
3. $x_1 - 2x_2 + 6x_3 = 0$ at $(2, 1, 0)$, $3x_1 + 2x_2 - 2x_3 = 0$ at $(-2, 3, 0)$.

§10.

1. (a) $x_2x_3^2 = x_1^3$, $(0, 1, 0)$, simple closed curve;
   (c) $x_1^2x_3 = x_2^3$, $(1, 0, 0)$, simple closed curve;
   (e) $(0, 1, 0)$, simple closed curve;
   (f) $(0, 1, 0)$, $(\pm 1, 1, 0)$, 3 simple closed curves with 1 common point.
3. (b) 3.
5. (a), (b), (d) simple closed curves; (c), (g) a point; (e) 2 lines; (f) no locus.
9. It is.

§11.

1. (a) $[k, -k, 2k]$; (b) $[3k, k, 0]$; (c) $[0, 0, k]$; (d) $[3k, 2k, -6k]$; (e) $[k, 0, 0]$; (f) $[0, 3k, -5k]$.
3. (a) $[0, 1, 0]$; (b) $[0, 0, 1]$; (c) $[2, 1, 0]$.
5. (a) through the origin; (b) the ideal line and the lines associated with the perpendiculars to the x-axis.
7. $[2, 4, 8]$; $[0, 0, 1]$; none.

§12.

1. (a) $2u_1 + u_2 - 3u_3 = 0$; (b) $u_3 = 0$; (c) $u_1 = 0$; (d) $u_1 + 2u_2 = 0$; (e) $u_2 = 0$.
5. (a) $4u_1 + 3u_2 + 12u_3 = 0$; (b) $5u_1 + 3u_2 - u_3 = 0$.
7. $[1, -2, 4]$.

§13.

1. (a) $3x_1 - x_2 = 0$; (b) $x_1 + 3x_2 - 2x_3 = 0$; (c) $4x_1 - 3x_2 + x_3 = 0$; (d) $8x_1 - x_2 - x_3 = 0$.
3. (a) 1; (b) -1; (c) $\frac{1}{2}$.

§14.

1. $x_1 = -5x_1' - 5x_2'$, $x_2 = 3x_1' - 3x_2' + 3x_3'$, $x_3 = -7x_1' + 2x_2' + 3x_3'$.
   $u_1' = -5u_1 + 3u_2 - 7u_3$, $u_2' = -5u_1 - 3u_2 + 2u_3$, $u_3' = 3u_2 + 3u_3$.
   $u_1 = -5u_1' + 3u_2' - 7u_3'$, $u_2 = -5u_1' - 3u_2' + 2u_3'$, $u_3 = 3u_2' + 3u_3'$.
   (a) $(5, 7, 2)$; (b) $[13, 7, 3]$; (c) $11x_1' + 14x_2' + 6x_3' = 0$;
   (d) $5u_1' - 2u_2' + 8u_3' = 0$.

3. (a) $u_1 = 2u_1' + 2u_2'$, $u_2 = 5u_1' - u_2' - 2u_3'$, $u_3 = 3u_1' + u_2' - 2u_3'$;
(b) $9u_1' - u_2' - 2u_3' = 0$.

§16.

1. Lines through O.
3. The ideal line.

§17.

3. (a) $u_1^2 - u_2^2 = u_3^2$; (b) $u_1^2 + 4u_2u_3 = 0$; (c) $u_1^2 + 2u_2^2 = 2u_3^2$;
(d) $4u_1u_2 + u_3^2 = 0$.
5. $u_1^2 + 4u_2^2 - 4u_1u_2 - 4u_2u_3 = 0$.

§18.

5. Three.

§19.

1. [5, −3, −1] and (0, 1, −3).

§20.

1. $$\begin{aligned} &3x_1x_1 + 2x_1x_2 - x_1x_3 \\ &2x_2x_1 + x_2x_2 + 3x_2x_3 \\ -&x_3x_1 + 3x_3x_2 - x_3x_3; \end{aligned}$$
nondegenerate.
3. $2u_1 + u_2 - u_3 = 0$, $u_1 + u_2 - 3u_3 = 0$.
5. $25u_1 + 9u_2 - 4u_3 = 0$.
7. (a) $x_2^2 = 4x_1x_3$; (b) $3x_3^2 = x_1x_2$; (c) $x_1^2 + x_2^2 = 9x_3^2$.

§21.

3. $\begin{pmatrix} 2 \\ -21 \\ 23 \end{pmatrix}$.
5. $\begin{pmatrix} \frac{4}{5} & -\frac{1}{5} & 0 \\ -\frac{3}{5} & \frac{2}{5} & 0 \\ 0 & 0 & 1 \end{pmatrix}$, $\begin{pmatrix} 4 & -1 & 0 \\ -3 & 2 & 0 \\ 0 & 0 & 5 \end{pmatrix}$.
9. (a) affinity; (b) direct similarity; (c) and (d) opposite similarity; (e) direct motion; (f) opposite motion.

§22.

7. (a) point, point; (b) 3 points, 3 points; (c) 3 collinear points, 3 points; (d) 4 points, no 3 collinear, and 4 points; (e) simple closed curve in each case; (f) figure shaped like 8 in each case; **(g)** 2 **nondegenerate** conics meeting in 4 points, 2 simple closed curves meeting in 4 points.

CHAPTER X

§3.

9. (a) 3; (b) 2; (c) 3.

Bibliography

The books listed below are divided into four categories, corresponding to the main geometries considered in the text, and treat or develop those geometries differently than the text does. The books on Euclidean geometry offer rigorous presentations of that system; the others, except Reference 15, develop their respective systems independently of Euclidean geometry, Reference 13 proceeding algebraically, and the others axiomatically. The listing for each category is approximately in order of increasing length and difficulty of treatment.

EUCLIDEAN GEOMETRY

1. Young, J. W., *Fundamental Concepts of Algebra and Geometry*. New York, Macmillan, 1911.
2. Robinson, G. de B., *The Foundations of Geometry*. Toronto, University of Toronto Press, 1940.
3. Keedy, M. L., and Nelson, C. W., *Geometry, a Modern Introduction*. Cambridge, Mass., Addison-Wesley, 1965.
4. Veblen, O., "The Foundations of Geometry," in J. W. A. Young, ed., *Monographs on Topics of Modern Mathematics*. New York, Longmans Green, 1911.
5. Hilbert, D., *The Foundations of Geometry*. Chicago, Open Court, 1902.
6. Forder, H. G., *The Foundations of Euclidean Geometry*. London, Cambridge University Press, 1927.

AFFINE GEOMETRY

7. Levi, H., *Foundations of Geometry and Trigonometry*. Englewood-Cliffs, N. J., Prentice-Hall, 1960.
8. Coxeter, H. S. M., *Introduction to Geometry*. New York, Wiley, 1961.

PROJECTIVE GEOMETRY

9. Forder, H. G., *Geometry*. London, Hutchinson's University Library, 1950.
10. Robinson, G. de B., *The Foundations of Geometry*. Toronto, University of Toronto Press, 1940.

11. Seidenberg, A., *Lectures in Projective Geometry.* Princeton, New Jersey, Van Nostrand, 1962.

12. Coxeter, H. S. M., *The Real Projective Plane.* New York, McGraw-Hill, 1949.

13. Busemann, H., and Kelly, P. J., *Projective Geometry and Projective Metrics.* New York, Academic Press, 1953.

14. Veblen, O., and Young, J. W., *Projective Geometry.* Boston, Ginn, 1910 (vol. 1), 1918 (vol. 2).

TOPOLOGY

15. Courant, R., and Robbins, H., *What Is Mathematics?* New York, Oxford, 1941.

16. Borsuk, K., and Szmielew, W., *Foundations of Geometry.* Amsterdam, North-Holland, 1960.

17. Patterson, E. M., *Topology*, New York, Interscience, 1959.

18. Simmons, G. F., *Introduction to Topology and Modern Analysis.* New York, McGraw-Hill, 1963.

19. Pervin, W. J., *Foundations of General Topology.* New York, Academic Press, 1964.

20. Kuratowski, K., *Introduction to Set Theory and Topology*, New York, Pergamon, 1961.

Index

Adjoint matrix, 350
Affine coordinates, 120
  distance, 116
  equivalence, 107
  figure, 99
  geometry, 101, 113, 307
  group, 84, 304
  incidence relation, 113, 114
  invariant, 10
  plane, 115
  property, 98
  transformations, 83, 304
Angle, 18
  of a vector, 25
Angle-size, 71, 81
Associative, 6, 45, 349
Axiom, on congruence, 116
  of continuity, 115
Axioms for affine geometry, 113
  for betweenness, 113, 114
Axis, of a glide-reflection, 63
  of a perspectivity, 266, 358
  of a range, 359
  of a reflection, 54

Between-relation, 16–18, 113, 114
Bounded model, Euclidean plane, 212
  Euclidean space, 228
Brianchon hexagon, 374
  point, 374
Brianchon's theorem, 185, 373, 374
Brink, R. W., 181

Campbell, H., 43
Center, of a pencil, 359
  of a radial transformation, 73
  of a rotation, 34
Central projection, of a line, 130, 132
  of a plane, 131, 137
Ceva's theorem, 102, 105
Circular model of Euclidean plane, 212, 216
  of projective plane, 226
Class of a curve, 329
Closed curve, 197
Cofactor, 381
Collinearity, 16
  condition for, 104, 290
Collineation, 275, 301, 339
  matrix equation, 348
  in summation notation, 344
  vector equation, 348
Commutative, 6
Commutative group, 9
Comparison of segments, affine, 118
  Euclidean, 116
Complete figures, 252, 253
Component transformations, 6
Components of a vector, 32
Compounding of transformations, 6
Concurrence, condition for, 104, 290
Congruence, affine, 117
  Euclidean, 116
Congruent figures, 19, 20
  transformations, 16, 306
Conic, Euclidean, 69, 93
  projective, 273, 308, 345, 362, 363
Conic section, 69, 93
Connected curve, 196
  surface, 210

Connectivity, 196
Construction, harmonic conjugate, 256
point of contact, 375
tangent line, 367, 368
Continuous function, 190
mapping, 190, 192, 353
transformation of lines, 354
Contraction of the plane, 73
Converse of Desargues's theorem, 188
Convex figure, 101
Correlation, 275, 339, 340, 345
Correspondents, 3
Coxeter, H. S. M., 277
Cramer's Rule, 385
Cross-ratio, Euclidean lines, 150, 157
Euclidean points, 136, 144, 145
extended lines, 245
ideal elements, 243
Cross-ratio and conics, 158, 357
Cross-ratio and cyclic order, 250
Cubic curve, 319
Curve of degree *n*, 94, 319
Cyclic group, 9
Cyclic order, 208
and cross-ratio, 250
on a projective line, 249
and topological transformations, 209

Degenerate conic section, 94
extended conic, 273
line conic, 336
Deleted segment, 202
semicircle, 202
Desargues's theorem, 183, 258
converse, 188
dual, 260
Descartes, R., 197
Destroyed properties, 9
Determinant, definition, 376, 377
expansion by cofactors, 381, 382
expansion by minors, 381
Laplace's development, 383
of a fractional transformation, 171
of a linear transformation, 14
of a matrix, 44
of a projective transformation, 175
of a system of equations, 384, 385
Diagonal points, 253
Diagonals (diagonal lines), 252

Diameter, ellipse, 108
hyperbola, 109
parabola, 109
Difference of two vectors, 32
Direct motions, 53
similarity transformations, 74, 77
transformations, 53
Direction, of a projection, 124
of a translation, 26
of a vector, 25
Disconnected curve, 196
Discriminant of a conic, 345
Displacement, 51, 52
Distance, affine, 116
Euclidean, 16
Distance-preserving mapping, 16
Dual, 242
of Desargues's theorem, 260
Duality, principal of, 242
Dürer, A., 163

Eccentricity, 69
Eisenhart, L. P., 181, 285
Endless curve, 196
surface, 210
Endpoint of a ray, 18
Endpoints of a segment, 16
Equal vectors, 22
Equation(s):
affine transformation, 83, 304
collineation, 344, 348
continuous mapping, 192
correlation, 340, 345
displacement, 51, 52
Euclidean conic, 93
glide-reflection, 61
line conic, 336
motion, 65, 306
point, 323
projective conic, 312
projective line, 288, 290
projective transformation, 174, 301
reflection, 56, 308
rotation, 47, 308
similarity transformation, 77, 306
tangent to a conic, 181, 316
topological mapping, 192
Equiaffine group, 81, 98
transformation, 98
Equiareal group, 81, 98

Equivalent figures, 106
Euclidean geometry, 1
 line, 230
 metric group, 307
 point, 230
Euclid's *Elements*, 1
Euler, L., 304
Expansion, by cofactors, 381, 382
 by minors, 381
 of the plane, 73
Explicit form, 12
Extended conic, 270
 hyperbola, 271
 line, 230, 231
 parabola, 271
 plane, 232
 space, 231
 x- and *y*-axes, 282
Exterior of a triangle, 19

Finite group, 31
 model of the Euclidean plane, 212
 points, 304
Fixed elements, 4
 points, 4, 5
Folium of Descartes, 197
Fractional transformation, 171
Function, 3
Fundamental Theorem of Projective Geometry, 264

Geometries of the projective plane, 307
Glide-reflection, 63
Graustein, W. C., 96, 269, 346
Group, 8

Half-plane, 19
Harmonic conjugate, 148, 155
 division, 148, 155
 progression, 148
 separation, 149
 set, 148, 155
Hemispheric model, Euclidean plane, 212
 projective plane, 228
Hexagrammum mysticum, 367
Hilbert, D., 102, 113, 116
Hjelmslev's theorem, 70
Homeomorph, 191, 200
 of Euclidean plane, 209
 of line and circle, 201
Homeomorphic, 191, 200
Homeomorphism, 190, 199
 of Euclidean plane, 192
 of projective plane, 353, 354
Homogeneous coordinates, of a line, 321
 of a point, 280, 282
Homogeneous equation, 286
 of second degree, 312
Homogeneous strain, 97

Ideal line, 232
 plane, 236
 point, 230, 231, 282
Identification of points, 205
Identity mapping, 5
Image (in a mapping), 3
Implicit form, 12
Incidence relation, 115, 198, 199
Indicator of a conic, 93
Infinite points, 304
Interior point, of an angle, 19
 of a segment, 16
 of a triangle, 18
Into, use of, 3
Invariant properties, 9
Inverse, of a linear transformation, 14
 of a matrix, 45, 350
 of a transformation, 5
Inverting a transformation, 6
Involutory, 36
Isometry, 16
Isomorphic, 30

Klein, F., 2
Kepler, J., 187

Laplace's development, 383
Lemniscate, 197
Length, of a vector, 25, 26
 of a segment, 117, 119

Limit line, 291, 322
  point, 189, 284
Line conic, 336, 371, 372
Line coordinates, homogeneous, 321, 322
  nonhomogeneous, 330
Line curve, 329
  of degree n, 329
Linear combination, 106, 291, 294, 325
  order, 207
  system of equations, 383
  transformation, 13
Lobachevsky, N. I., 2

Mapping, 3, 4
  continuous, 190, 192, 353
  one-to-one, 5
  topological, 190, 199, 353
Matrix, 41
  multiplication, 43, 303
  of a collineation, 348
  of a determinant, 378
  of a projectivity, 303
  of a rotation, 41
  product by a vector, 42
Matrix equation, of a collineation, 348
  of a rotation, 42
Maxwell, E. A., 357
Menelaus's theorem, 102, 104
Metric geometry, Euclidean plane, 81
  projective plane, 307
Metric groups, 81
  invariants, 10
  properties, 1, 81
Minor, 380, 382
Möbius band, 222
Model, of an extended plane, 234
  of the ideal plane, 236
  of a projective plane, 236
  see also Circular model and Spherical model
Modified semiellipse, 227
Motions, of the Euclidean plane, 16, 65
  of the projective plane, 306
Mystic hexagram, 367

n-ic, 319
Noncollinearity, 17, 112, 276
Nondegenerate conic, Euclidean plane, 80
  extended plane, 272, 336
Nonhomogeneous equation, 286
  coordinates, 280, 330
Nonmetric geometry, 81
  invariant, 10
  property, 1
Nonsingular matrix, 44
  affine transformation, 85
Nonzero vector, 26
Normal distribution curve, 109
Notation, between-relation, 16
  cross-ratio, 136
  perspectivity (projection), 239
  projectivity, 264
  segment, 16
Null vector, 26, 33

"On" language, 242
One-dimensional perspectivity, 263
  projectivity, 263
  strain, 91, 96
One-sided surface, 223
One-to-one correspondence, 5
  transformation, 5
Onto, use of, 3
Open curve, 197
  surface, 210
Opposite motions, 55, 59, 60, 61, 306
  similarity transformations, 74, 77, 306
  transformation, 53
Order, cyclic, 208
  linear, 207
  on a circle, 208
  on a Euclidean line, 207
  on a projective line, 249
Order relation, 115
Ordinary lines, 167
  points, 167
Originals, 3
Orthogonal projection, 11
  of a line, 123, 124
Osgood, W. F., 96
Outside of a triangle, 19

Pappus's theorem, 183, 262, 265, 266
Parallel figures, 24

line and plane, 126
planes, 126, 233
projection of a line, 124
projection of a plane, 127
Pascal hexagon, 367
line, 367
Pascal's theorem, 185, 274, 365, 367
Pencil of lines, 169, 242
Perspective correspondence, 359
Perspective from a line, 184, 261
from a point, 184, 261
Perspectivity, 263, 266, 358
Plane family of parallel lines, 126
Plane motions, 16
Plücker line, 367
Point, equation of, 323
Euclidean, 230
of contact, 336, 375
ordinary, 167
projective, 237
Point conic, 336, 357, 362, 363
coordinates, 322
curve, 329
transformation, 4
Pointwise fixed, 55
Polar of a point, 343
Polarity, 343
Pole of a line, 343
Preserved property, 9
Primitive transformation, 87
Principle of Duality, 242
Product, of matrices, 43, 303
of a matrix and a vector, 42
of a scalar and a vector, 293
Projections, applications, 160
in Euclidean space, 123–143
in projective space, 238–240
notation, 239
Projective conic, 273, 312
correspondence, 358
equivalence, 181, 278
geometry of the Euclidean plane, 182
of the projective plane, 275
group, 177, 276, 342
invariants, 10
line, 233, 288
pencils, 359
plane, 225, 236
point, 237
property, 144, 241, 354
ranges, 359
space, 237
Projective transformations, 10, 98, 165
Euclidean plane into itself, 166, 174, 177
projective plane into itself, 275, 301, 327
one-dimensional, 263
two-dimensional, 165
in line coordinates, 327
Projectivity, in a projective plane, 275
one-dimensional, 263, 266
two-dimensional, 275
Punctured curves, 196
sphere, 216

Quadrangle, complete, 252
Quadric cone, 143
Quadrilateral, complete, 252
simple, 252
Quartic curve, 319
Quintic curve, 319

Radial transformation, 73
Range of points, 359
Rank of a matrix, 378
Ratio, of directed distances, 89
of distances, 89
of division, lines, 150
points, 21 89, 90
of a radial transformation, 73
of a similarity, 71
Ray, 18
Reflection, of a circle, 4
of the Euclidean plane, 54, 56, 61
of the projective plane, 308
Resultants, 6
of affine transformations, 84
of projective transformations, 176
of rotations and translations, 49
Riemann, B., 2
Rigid motion, 51
Rotation, of a circle, 4
of the Euclidean plane, 34, 38, 47
of the projective plane, 308

Salmon, G., 159, 316, 339
Scalar, 31

Scalar equations of a translation, 32
Segment, 16, 115
  length of, 117, 119
Self-corresponding element, 360
Self-dual, 252
Semicircumference, 213
Sense, of an angle, 24
  on a circle, 9
  on a line, 88
  of a triangle, 27
Separation, 147
  harmonic, 149
  on a projective line, 249–251
Similar figures, 78
Similarity group, 72, 307
  transformations, 71, 77, 306
Simple arc, 198
  closed curve, 198
  polygon, 252
Single-valued transformation, 5
Singular matrix, 44
  affine transformation, 85
Skew, 126
Solution of a linear system, 383
Space family of lines, 126
Spherical model, Euclidean plane, 216
  Euclidean space, 229
  projective space, 238
Square matrix, 46
Steiner, J., 357
Steiner point, 367
Subgroup, 8
Sum of two vectors, 29, 32
Summation notation, 343
Symmetry, to an axis, 58
  to a point, 36, 40
System of linear equations, 385

Tangent to a conic, 181, 274, 316, 367
Theorem of Brianchon, 185, 274, 373, 374
  Ceva, 102, 105
  Desargues, 183, 258
  Hjelmslev, 70
  Menelaus, 102, 104
  Pappus, 183, 265, 266
  Pascal, 185, 274, 365, 367
Three-dimensional model, 229, 238
  vector, 31
Topological equivalence, 191, 200
  geometry, 1, 191
  group, 191
  image, 191, 200
  invariant, 191
  property, 191
  transformation, 190, 199, 353, 354
Topology, 1, 191
Torus, 221
Transformation, 3
  identity, 5
  inverse, 5
  one-to-one, 5
  point, 4
  resultant, 6
  see also Mapping, Translation, Affine, etc.
Translation, 25, 308
Triangle (projective plane), 252
Two-dimensional affine coordinates, 120
  distance system, 119
  model, 216, 226, 234, 236
  projective transformation, 165
  vector, 31
Two-sided surface, 223

Undefined term, 115
Unit matrix, 44, 350
  segment, 116, 119

Vanishing lines, 138, 167
  points, 133, 167
Variable line, 291, 322
  point, 190
Veblen, O., 102, 110, 276, 277
Vector, 22
  equation, 32, 348
  of a translation, 25
  product by a matrix, 42
    by a scalar, 293
  sum and difference, 29, 32
  with two components, 31, 32
    three components, 31, 348

Whitehead, J. H. C., 110

Young, J. W., 276, 277

Zero vector, 26, 33